Groundwater Pollution: Sources, Mechanisms, and Prevention

Groundwater Pollution: Sources, Mechanisms, and Prevention

Guest Editors

Pantelis Sidiropoulos
Augustina Clara Alexander

MDPI

Basel • Beijing • Wuhan • Barcelona • Belgrade • Novi Sad • Cluj • Manchester

Guest Editors
Pantelis Sidiropoulos
Rural and Surveying Engineering
Aristotle University of Thessaloniki
Thessaloniki
Greece

Augustina Clara Alexander
Water Resources Engineering
University of Dar es Salaam
Dar es Salaam
Tanzania

Editorial Office
MDPI AG
Grosspeteranlage 5
4052 Basel, Switzerland

This is a reprint of the Special Issue, published open access by the journal *Hydrology* (ISSN 2306-5338), freely accessible at: https://www.mdpi.com/journal/hydrology/special_issues/TQC3GXVQ0L.

For citation purposes, cite each article independently as indicated on the article page online and as indicated below:

Lastname, A.A.; Lastname, B.B. Article Title. *Journal Name* **Year**, *Volume Number*, Page Range.

ISBN 978-3-7258-3305-4 (Hbk)
ISBN 978-3-7258-3306-1 (PDF)
https://doi.org/10.3390/books978-3-7258-3306-1

Contents

About the Editors

Pantelis Sidiropoulos

Dr. Pantelis Sidiropoulos is an Assistant Professor of Simulation and Management of Groundwater Resources at the Lab of Hydraulic Works and Environmental Management of the Rural and Surveying Engineering Department of Aristotle University of Thessaloniki (2023–present). He was a Lecturer of the Civil Engineering Department of University of Thessaly (2015–2023) and of the Civil Engineering Department of the University of Applied Science of Thessaly (2015–2018). He is a registered professional civil engineer in Greece and is a member of the Technical Chamber of Greece, the Association of Civil Engineers of Greece, the International Association of Hydrological Sciences, the Greek Committee for Water Resources Management, the Hellenic Hydrotechnical Association and the Hellenic Ecological Society. He has participated in many National and European research projects and an abundance of hydrological and hydraulic studies and water resource management projects. He is the author and co-author of more than one hundred referred international journal publications, conference proceedings, books and technical reports. His research interests include the following: (a) the deterministic and stochastic analysis, modeling, management and forecasting of groundwater resources; (b) the spatiotemporal analysis of groundwater quantity and quality variables; (c) water resource management; (d) applications of GIS; (e) the analysis, modeling, and forecasting of floods; (f) the hydraulic simulation of water distribution systems (DISs); and (g) the design and supervision of hydrotechnical projects. His teaching activities include the following: undergraduate programs: Water Resources Management, Groundwater Hydraulics, River Hydraulics and River Training, and Water Supply and Sewerage Networks; and graduate program: Groundwater Resources.

Augustina Clara Alexander

Dr. Augustina Clara Alexander is a Senior Lecturer in the Department of Water Resources Engineering and the Director of the Directorate of Internationalization, Convocation, and Advancement (2022–present) at the University of Dar es Salaam. She previously served as Head of the Department of Water Resources Engineering (2019–2022) and was a Teaching Assistant at the Tshwane University of Technology, South Africa (2015). As a registered professional engineer in Tanzania, Dr. Alexander is an active member of several professional bodies, including the Institute of Engineers Tanzania, the Africa Groundwater Network, and the Intergovernmental Hydrological Programme (IHP). She is also the Deputy Regional Coordinator of the FRIEND network and a board member of the Engineers Registration Board (ERB), where she chairs the Professional Conduct and Ethics Committee. Her research focuses on groundwater pollution monitoring and remediation, groundwater–surface water interactions, climate change impacts on water resources, watershed management, and sustainable sanitation. She has led and contributed to multiple research projects aimed at enhancing water resource resilience, integrating surface and groundwater management, and advancing pollution monitoring and remediation. Dr. Alexander has authored and co-authored numerous peer-reviewed journal articles, conference papers, and technical reports. Her contributions to water research and engineering have earned her several prestigious accolades, including the 2024 OWSD–Elsevier Foundation Award for Early Career Women Scientists in Water, Sanitation, and Hygiene.

Preface

The Special Issue entitled "Groundwater Pollution: Sources, Mechanisms, and Prevention" aims to present new research contributions in groundwater contamination worldwide. It focuses on the sources, effects, and exposure of natural and artificial groundwater pollutants, hydrological processes, and hydrochemical properties of groundwater, human health risk assessment, climate change impact, monitoring and remote sensing techniques, simulation and optimization modeling, and contemporary trends in management and pollution mitigation, prevention and remediation strategies. The published papers refer to study areas from all over the world as they are located in Africa, America, Asia, and Europe. Almost fifty authors contributed to the writing of the included manuscripts from seventeen universities.

This Special Issue is addressed to scientists, researchers, engineers, stakeholders and decision makers occupied with groundwater resources.

The motivation for launching this Special Issue is the dissemination of knowledge, methods and techniques that are implemented worldwide to gain a better understanding of groundwater pollution sources and mechanisms, which will lead to its prevention.

Pantelis Sidiropoulos and Augustina Clara Alexander
Guest Editors

hydrology MDPI

Editorial

Groundwater Pollution: Sources, Mechanisms, and Prevention

Pantelis Sidiropoulos

School of Rural and Surveying Engineering, Faculty of Engineering, Aristotle University of Thessaloniki, 54124 Thessaloniki, Greece; pasidirop@topo.auth.gr

Groundwater resources are vital for ecosystems and human health and prosperity. Groundwater pollution continues to increase, further limiting the potential of groundwater resources for use. The sources of contaminants can be natural (e.g., salinity or arsenic) or anthropogenic (excess fertilizers, pesticides, industrial chemicals, sewage effluent). Groundwater pollution may emerge from point sources (which can be easily identified spatially) and nonpoint sources, which are more difficult to identify, measure, and control than the former. To deal with this phenomenon, the sources of pollution and the mechanisms of the fate and transport of pollutants in groundwater must be identified. This process is important because it forms the basis for formulating appropriate prevention and mitigation measures. The latter is preferable for any remediation effort, as adverse effects on the environment and the health of living organisms will be prevented. However, since groundwater contamination is less obvious than surface water pollution, it frequently goes unnoticed. In this case, a remediation strategy must be utilized.

For groundwater protection and restoration, scientists and engineers worldwide are continuously studying related phenomena, proposing many solutions for understanding and managing groundwater. The evolution of technology has boosted this effort in recent years with the rise of many contemporary tools such as simulation models, remote sensing, monitoring instruments, etc.

This Special Issue, entitled "Groundwater Pollution: Sources, Mechanisms, and Prevention", contains publications from around the world that present new research contributions in the area of groundwater contamination, focusing on (i) the sources, effects, and exposure of natural and artificial groundwater pollutants, hydrological processes, and the hydrochemical properties of groundwater; (ii) the intrinsic and specific vulnerability of groundwater to pollution; (iii) human health risk assessments; (iv) the recent trends in pollution management, mitigation, prevention, and remediation strategies.

The eleven manuscripts published in this Special Issue (two focusing on sites in Africa, two in America, two in Asia, four in Europe, and one theoretical) cover a variety of modern applications for different problems to be mitigated and different scales of analysis/intervention. Specifically, five papers deal with monitoring [1–5], five papers with simulation and/or optimization [6–10] and one paper [11] deals with remote sensing.

An overview of the articles in this Special Issue is presented (References).

Breall et al. [1] analyzed the quality and the potential noncarcinogenic health risks related to nitrate in groundwater in the El Milia plain, Kebir Rhumel Basin, Algeria. Moran's I and the ordinary kriging (OK) interpolation technique were used to examine the spatial distribution pattern of the hydrochemical parameters in the groundwater. The hydrochemical parameters Ca, Cl, and HCO_3 showed a strong spatial autocorrelation in the El Milia plain, indicating the spatial dependence and clustering of these parameters in the groundwater. The groundwater quality was evaluated using the entropy water quality index (EWQI). The results showed that approximately 86% of the total groundwater samples in the study area fell within the moderate groundwater quality category. The potential noncarcinogenic health risks for adults and children associated with nitrate contamination of groundwater through the drinking water pathway were assessed using the hazard

Citation: Sidiropoulos, P. Groundwater Pollution: Sources, Mechanisms, and Prevention. *Hydrology* **2024**, *11*, 98. https://doi.org/10.3390/hydrology11070098

Received: 28 June 2024
Revised: 3 July 2024
Accepted: 4 July 2024
Published: 5 July 2024

quotient (HQ). The results revealed that approximately 5.7% of the total groundwater samples exceeded the HQ limit for adults, indicating potential health risks. Moreover, a higher percentage, 14.28%, of the total groundwater samples exceeded the HQ limit for children, highlighting their increased vulnerability to noncarcinogenic health hazards associated with nitrate contamination in the study area.

Ismail et al. [2] evaluated the groundwater potentiality and its suitability for various uses in a region of Middle Egypt using hydrochemistry and electrical resistivity studies. The groundwater potential in the study area was evaluated based on 24 VESs (vertical electrical soundings), and its quality was determined based on the analyses of 57 groundwater samples. The EC (salinity index), Na% (salt hazard), SAR (ratio of sodium adsorption), chloride risks, SSP (soluble sodium percentage), MH (magnesium hazard), and other indicators were used to determine whether the collected water samples were suitable for irrigation. Four layers in the study area were mentioned in the geoelectrical cross-sections that have been constructed. Most of the groundwater samples that were collected were safe for drinking; however, none of them were fit for home usage because of their extreme hardness. According to the SAR and US diagram, RSC, KR, and PI, most groundwater samples from the Pleistocene and Eocene aquifers were fit for irrigation.

In Zohud et al.'s paper [3], the quality of the groundwater resources in the West Bank region in Palestine was evaluated using the water quality index (WQI) and human health risk (HHR) assessment. In the West Bank, groundwater is particularly valuable because of its scarcity and inaccessibility, and due to the nature of the area's aquifers, is currently regarded as being at a high risk of pollution. The water quality in this area is also of wide concern, with its effects being directly linked to human health. Certain parts of the West Bank groundwater suffer from high concentrations of nitrate and potassium. In total, 38.8% of the nitrate and 10% of the potassium concentrations in the well samples exceeded the permissible limit set by the WHO and PSI, and, therefore, health problems may arise as a limiting factor for life quality and welfare in this region. Moreover, 87.7% of the samples were classified as having very hard water. The WQI values showed that 78% of the well samples were of good quality. The health risk assessments were evaluated for fluoride and nitrate in drinking water for adults, children, and infants. The main values of the estimated Total Hazard Index (THI) obtained from the analyzed data on the health risk assessments revealed a diverse effect on the local population based on age. The ranges of the THI in all the sampling locations varied considerably and extended from 0.093 to 3.01 for adults, 0.29 to 3.08 for children, and 0.302 to 3.21 for infants. These results largely indicate that infants are more exposed to health risks.

Kovač et al. [4] proposed an analysis of the distance between the measured and assumed location of a point source of pollution in groundwater as a function of the variance of the estimation error. The research took place in the north of the Republic of Croatia and included the area of the Varaždin wellfield, which was closed due to an excessively high groundwater nitrate concentration. Seven different interpolation methods were used to create spatial distribution models. Each method provided a different model, a different variance of the estimation error, and estimates of the location of the pollution source. The results showed a nonlinear and monotonic relationship between the distance and the variance of the estimation error, so logarithmic and rational quadratic models were fitted to the scatter point data. The models were linearized, a t-test was performed, and the results showed that the models can be considered reliable, which was confirmed by the values of the coefficients of determination of the linearized models. The obtained results can be used in planning additional research to determine the measured location of the pollution source. The research methodology used is universal and can be applied to other locations where high concentrations of certain contaminants have been detected in groundwater in alluvial aquifers.

Correa-González et al. [5] proposed a methodology to assess nitrate in groundwater from diffuse sources considering the spatiotemporal patterns of hydrological systems using a coupled SWAT/MODFLOW/MT3DMS model. The application of the model was carried out using a simplified simulation scheme of hydrological and agricultural systems because of the limited spatial and temporal data. The study area included the Cuitzeo Lake basin in the superficial flow form and the Morelia–Querendaro aquifer in the groundwater flow form. The results within the methodology included the surface runoff, groundwater levels, and nitrate concentrations and indicated that the historical and simulated nitrate concentrations were obtained within acceptable values of the statistical parameters and, therefore, were considered adequate.

Siarkos et al. [6] presented an integrated framework to assess the environmental and economic impact of fertilizer restrictions in a nitrate-contaminated aquifer located in the southwestern part of the Chalkidiki peninsula, Northern Greece. For this task, various scenarios involving reductions (10%, 20%, 30%, 40%, and 50%) in fertilizer application were investigated, evaluated, and ranked to determine the most suitable option. The environmental assessment considered the occurrences of nitrates in the groundwater, with a specific emphasis on the nitrate concentrations in the water supply wells, as obtained by a nitrate fate and transport model, while the economic analysis focused on the losses experienced by farmers due to reduced fertilizer usage. The results showed that a 30% reduction in fertilization was the most appropriate option for the area being studied, highlighting the importance of adopting such an approach when confronted with conflicting outcomes among alternatives.

The study by Lyra et al. [7] examined the impacts of climate change and sea level rise on coastal aquifers, focusing on the influence of the components of the water cycle on seawater intrusion, and the evolution of this phenomenon in the future. The simulation of coastal water resources was performed using an integrated modeling system (IMS), designed for agricultural coastal watersheds, which consisted of inter-connected models of surface hydrology (UTHBAL), groundwater hydrology (MODFLOW), and seawater intrusion (SEAWAT). Climatic models for the adverse impact scenario (RCP8.5) and the medium impact scenario (RCP4.5) of climate change were used. The transient boundary head conditions were set to the coastal boundary to dynamically represent the rise in the sea level due to climate change. The response of the groundwater in the coastal Almyros Basin, located in central Greece, was simulated from 1991 to 2100. The findings indicated that seawater intrusion will be advanced in the future, in both climate change scenarios. The models show varying patterns of groundwater recharge, with varying uncertainty projected for the future, and sensitivity to time in the fluctuation of the components of the water cycle.

Perdikaki et al. [8] investigated the mechanism of seawater intrusion and the performance of free and open-source codes for the simulation of variable density flow problems in coastal aquifers. For this purpose, the research focused on the Marathon Watershed, located in the northeastern tip of Attica, Greece. For the simulation of the groundwater system, MODFLOW, MT3DMS, and SEAWAT codes were implemented, while sensitivity analysis and calibration processes were carried out with UCODE. Hydraulic head calibration was performed on the MODFLOW model, and the TDS concentration is validated in the SEAWAT model. The calibrated parameters of the MODFLOW model were obtained for the variable density flow simulation with SEAWAT. The MODFLOW and SEAWAT hydraulic head outputs were analyzed and compared to one another. The outcome of this analysis showed that SEAWAT produces slightly better results in terms of the hydraulic heads, and the authors concluded that parameter transferability can take place between the two models. For the seawater intrusion assessment, the use of the SEAWAT code revealed that the aquifer is subjected to passive and passive–active seawater intrusion during the wet and dry seasons, respectively. Finally, an irregularly shaped saltwater wedge was developed at a specific area, associated with the hydraulic parameters of the aquifer.

Kontos [9] investigated the optimal remediation process in an aquifer using Modflow 6 software and genetic algorithms. A theoretical confined aquifer was polluted over a long period of time by unnoticed leakage from a pipeline conveying leachate from an adjacent landfill to a wastewater treatment plant. When the extended leakage and groundwater pollution were discovered, the optimal planning of the remediation strategy was investigated using the pump-and-treat method and/or hydrodynamic control of the pollution. The practical goal was to find the optimal locations and flow rates of two additional pumping wells, which would pump the polluted water and/or control pollution, protecting an existing drinking water pumping well, securing its full operation even during the remediation process with the minimum possible cost, simply represented by the pumped water volume of the additional wells. The remediation process was considered complete when the maximum concentration in the aquifer dropped below a certain limit. The Modflow software (handled by the Flopy Python package: Version 4) simulated the flow field and advective–dispersive mass transport, and a genetic algorithm was used as the optimization tool. The coupled simulation–optimization model, Modflow-GA, complemented by a sophisticated post-processing results analysis, provided optimal and alternate sub-optimal remediation strategies for the decision-makers to select from.

The study by Ibn Salam et al. [10] aimed to elucidate an aquifer's contaminant transport mechanisms by determining longitudinal and transverse dispersivities through inverse modeling. The aquifer being studies was in the municipal wellfield in Collierville, Tennessee, and it was contaminated with trichloroethylene (TCE) and hexavalent chromium (Cr (VI)) due to industrial operations dating back to the 1970s and 1980s. Utilizing MT3DMS for contaminant transport simulation, based on a well-calibrated groundwater flow model, and leveraging Python's multiprocessing library for efficiency, the study employed a trial-and-error methodology. The key findings revealed that the longitudinal dispersivity values ranged from 5.5 m near the source to 20.5 m further away, with horizontal and vertical transverse dispersivities between 0.28 m and 3.88 m and 0.03 m and 0.08 m, respectively. These insights into the aquifer's dispersivity coefficients, which reflect the scale-dependent nature of the longitudinal dispersivity, will be crucial for optimizing remediation strategies and achieving cleanup goals. This study underscored the importance of accurate parameter estimation in contaminant transport modeling and contributed to a better understanding of contaminant dynamics in the Collierville wellfield.

EL-Bana et al. [11] proposed an integrated remote sensing and GIS-based technique for mapping groundwater recharge zones. To identify groundwater potential zones, the study was conducted in Central Saudi Arabia, southwest of Riyadh. The analysis employed a multi-criteria approach that relied on remote sensing and geographic information systems. The variables employed in this technique included geology, rainfall, elevation, slope, aspect, hillshade, drainage density, lineaments density, and land use/land cover (LULC). The Analytical Hierarchical Process (AHP) was used for assigning weights to the parameters, and the corresponding significance of each parameter's several classes for groundwater potentiality. Different groundwater potential zones were identified by the study: very high (16.8%), high (30%), medium (26.7%), low (18.6%), and very low (7.9%). Only two of the observation wells were located in the "medium" potential zone, but the other ten wells were observed in the "very high" and "high" potential zones, according to the validation survey. Consequently, the results may demonstrate that the approach used, which combined improved conceptualization with AHP to define and map groundwater potential zones, has a greater chance of producing accurate results and can be used to reduce the threat of drought in broader arid regions.A plethora of focused and contemporary knowledge on groundwater pollution is presented in this Special Issue. The papers published herein contribute significantly to the advancement of groundwater pollution research at three levels: sources, mechanisms, and prevention. They promote the combination of widely used methodologies, such as remote sensing, and the use of indicators to optimally assess the status of an aquifer and find the sources of pollution. The creation of groundwater simulations with appropriate software and new methodologies in the context of integrated management will help scientists better understand the mechanisms of groundwater pollution. Finally, through the proposal of innovative strategies and contemporary practices such as genetic algorithms, not only is optimal prevention aimed for but also the remediation of groundwater pollution. One of the most important reasons why this Special Issue was created was so that the research within it gained as much visibility as possible. We hope that all the proposed methodologies and research on groundwater pollution presented in this Special Issue will be reflected in operational strategies and projects by decision-makers and stakeholders, ensuring groundwater protection and restoration.

Conflicts of Interest: The author declare no conflict of interest.

References

1. Brella, D.; Belkhiri, L.; Tiri, A.; Salhi, H.; Lakouas, F.E.; Nouibet, R.; Amrane, A.; Merdoud, R.; Mouni, L. Identification of the Groundwater Quality and Potential Noncarcinogenic Health Risk Assessment of Nitrate in the Groundwater of El Milia Plain, Kebir Rhumel Basin, Algeria. *Hydrology* **2023**, *10*, 171. [CrossRef]
2. Ismail, E.; Alexakis, D.E.; Heleika, M.A.; Hashem, M.; Ahmed, M.S.; Hamdy, D.; Ali, A. Applying Geophysical and Hydrogeochemical Methods to Evaluate Groundwater Potential and Quality in Middle Egypt. *Hydrology* **2023**, *10*, 173. [CrossRef]
3. Zohud, A.; Alam, L.; Goh, C.T. Evaluation of Groundwater Quality Using the Water Quality Index (WQI) and Human Health Risk (HHR) Assessment in West Bank, Palestine. *Hydrology* **2023**, *10*, 198. [CrossRef]
4. Kovač, I.; Šrajbek, M.; Klišanin, N.; Gilja, G. Analysis of the Distance between the Measured and Assumed Location of a Point Source of Pollution in Groundwater as a Function of the Variance of the Estimation Error. *Hydrology* **2023**, *10*, 199. [CrossRef]
5. Correa-González, A.; Hernández-Bedolla, J.; Martínez-Cinco, M.A.; Sánchez-Quispe, S.T.; Hernández-Hernández, M.A. Assessment of Nitrate in Groundwater from Diffuse Sources Considering Spatiotemporal Patterns of Hydrological Systems Using a Coupled SWAT/MODFLOW/MT3DMS Model. *Hydrology* **2023**, *10*, 209. [CrossRef]
6. Siarkos, I.; Mallios, Z.; Latinopoulos, P. An Integrated Framework to Assess the Environmental and Economic Impact of Fertilizer Restrictions in a Nitrate-Contaminated Aquifer. *Hydrology* **2024**, *11*, 8. [CrossRef]
7. Lyra, A.; Loukas, A.; Sidiropoulos, P.; Mylopoulos, N. Climatic Modeling of Seawater Intrusion in Coastal Aquifers: Understanding the Climate Change Impacts. *Hydrology* **2024**, *11*, 49. [CrossRef]
8. Perdikaki, M.; Chrysanthopoulos, E.; Markantonis, K.; Kallioras, A. Groundwater Flow Model Calibration Using Variable Density Modeling for Coastal Aquifer Management. *Hydrology* **2024**, *11*, 59. [CrossRef]
9. Kontos, Y.N. A Simulation–Optimization Model for Optimal Aquifer Remediation, Using Genetic Algorithms and MODFLOW. *Hydrology* **2024**, *11*, 60. [CrossRef]
10. Ibn Salam, M.; Waldron, B.; Schoefernacker, S.; Jazaei, F. Determination of Contaminant Transport Parameters for a Local Aquifer by Numerical Modeling of Two Plumes: Trichloroethylene and Hexavalent Chromium. *Hydrology* **2024**, *11*, 88. [CrossRef]
11. EL-Bana, E.M.M.; Alogayell, H.M.; Sheta, M.H.; Abdelfattah, M. An Integrated Remote Sensing and GIS-Based Technique for Mapping Groundwater Recharge Zones: A Case Study of SW Riyadh, Central Saudi Arabia. *Hydrology* **2024**, *11*, 38. [CrossRef]

hydrology

MDPI

Article

Identification of the Groundwater Quality and Potential Noncarcinogenic Health Risk Assessment of Nitrate in the Groundwater of El Milia Plain, Kebir Rhumel Basin, Algeria

Djouhaina Brella [1], Lazhar Belkhiri [1], Ammar Tiri [1], Hichem Salhi [1], Fatma Elhadj Lakouas [1], Razki Nouibet [1], Adeltif Amrane [2], Ryma Merdoud [3] and Lotfi Mouni [3,*]

[1] Laboratory of Applied Research in Hydraulics, University of Mustapha Ben Boulaid Batna 2, Batna 5000, Algeria; d.brella@univ-batna2.dz (D.B.); belkhiri.la@gmail.com or l.belkhiri@univ-batna2.dz (L.B.); a.tiri@univ-batna2.dz (A.T.); h.salhi@univ-batna2.dz (H.S.); f.elhadjlakous@univ-batna2.dz (F.E.L.); r.nouibet@univ-batna2.dz (R.N.)

[2] Ecole Nationale Supérieure de Chimie de Rennes, Université de Rennes, CNRS, ISCR—UMR 6226, 35000 Rennes, France; abdeltif.amrane@univ-rennes1.fr

[3] Laboratoire de Gestion et Valorisation des Ressources Naturelles et Assurance Qualité, Faculté SNVST, Université Akli Mohand Oulhadj, Bouira 10000, Algeria; r.merdoud@univ-bouira.dz

* Correspondence: l.mouni@univ-bouira.dz

Abstract: In this study, we analyzed the quality and the potential noncarcinogenic health risk of nitrate in groundwater in the El Milia plain, Kebir Rhumel Basin, Algeria. Moran's *I* and the ordinary kriging (OK) interpolation technique were used to examine the spatial distribution pattern of the hydrochemical parameters in the groundwater. It was found that the hydrochemical parameters Ca, Cl, and HCO_3 showed strong spatial autocorrelation in the El Milia plain, indicating a spatial dependence and clustering of these parameters in the groundwater. The groundwater quality was evaluated using the entropy water quality index (EWQI). The results showed that approximately 86% of the total groundwater samples in the study area fall within the moderate groundwater quality category. The spatial map of the EWQI values indicated an increasing trend from the south-west to the northeast, following the direction of groundwater flow. The highest EWQI values were observed near El Milia city in the center of the plain. This spatial pattern suggests variations in groundwater quality across the study area, with potentially higher risks near the city center. The potential noncarcinogenic health risks associated with nitrate contamination in groundwater for adults and children through the drinking water pathway were assessed using the hazard quotient (HQ). The results revealed that approximately 5.7% of the total groundwater samples exceeded the HQ limit for adults, indicating potential health risks. Moreover, a higher percentage, 14.28%, of the total groundwater samples exceeded the HQ limit for children, highlighting their increased vulnerability to noncarcinogenic health hazards associated with nitrate contamination in the study area. Taking timely action and ensuring strict compliance with regulations in groundwater management are crucial for protecting public health, preserving the environment, addressing water scarcity, and achieving sustainable development goals.

Keywords: groundwater quality; health risk assessment; Moran's *I*; ordinary kriging interpolation; El Milia plain; Kebir Rhumel Basin

Citation: Brella, D.; Belkhiri, L.; Tiri, A.; Salhi, H.; Lakouas, F.E.; Nouibet, R.; Amrane, A.; Merdoud, R.; Mouni, L. Identification of the Groundwater Quality and Potential Noncarcinogenic Health Risk Assessment of Nitrate in the Groundwater of El Milia Plain, Kebir Rhumel Basin, Algeria. *Hydrology* **2023**, *10*, 171. https://doi.org/10.3390/hydrology10080171

Academic Editors: Pantelis Sidiropoulos and Augustina Clara Alexander

Received: 16 July 2023
Revised: 9 August 2023
Accepted: 10 August 2023
Published: 14 August 2023

1. Introduction

Groundwater plays a vital role in water supply systems worldwide, serving various purposes such as drinking water, irrigation, and industrial use [1–11]. The demand for groundwater has increased significantly due to population growth, urbanization, and industrial activities. Excessive pumping of groundwater beyond its natural recharge rate has resulted in the depletion of aquifers in many regions, especially in arid and semi-arid areas where surface water resources and precipitation are scarce [7,12,13]. This overexploitation can lead to a drop in water tables, land subsidence, and the drying up of wells. As groundwater gets depleted, there is an increased risk of contamination. Several human activities and geological factors have led to challenges in groundwater utilization, particularly in the presence of harmful substances such as nitrate (NO_3). To address these challenges and ensure the sustainable use of groundwater resources, it is necessary to enhance groundwater quality monitoring and assessment efforts in these regions.

Because groundwater quality is closely related to human health hazards, developing an appropriate tool to assess groundwater quality for drinking purposes has become a critical study in recent years [14,15]. The entropy weighted water quality index (EWQI) is a widely used mathematical tool for the evaluation of groundwater quality [16–18]. This method can measure the weights of groundwater hydrochemical attributes while ignoring artificial weight distribution, allowing for clear demarcation of groundwater quality classes.

Groundwater contaminated by nitrate has a direct negative effect on human health and agricultural activities. Nitrate can enter groundwater through multiple pathways, with the most significant sources being point and non-point sources, and natural processes can also contribute to their presence. The most common point sources of nitrate contamination in groundwater are septic tanks, dairy lagoons, wastewater effluent percolation, and livestock waste [19,20]. Elevated nitrate levels in groundwater can result from non-point sources such as fertilizers, pesticides, and manure application, as well as natural processes through rock–water interactions [13,21–23]. Also, nitrate contamination in groundwater can result from nitrogen fixation by legume plants and microbes. In addition, the extensive use of nitrogen-based fertilizers in agricultural areas is a major contributor to nitrate contamination in groundwater. When plants grow, they primarily take up nitrogen in the form of nitrate, an oxidized form of dissolved nitrogen. However, with intensive farming and continuous use of fertilizers, the soil may lose its natural ability to retain nitrates effectively. Consequently, excess nitrates can be washed into the groundwater by rainfall or irrigation, leading to elevated nitrate levels. Nitrates in groundwater pose health risks when they enter the food chain via contaminated groundwater and surface water. Drinking water with elevated nitrate levels can be harmful, particularly for vulnerable populations like infants and pregnant women. The World Health Organization [24] limit for nitrate in drinking water that is safe for drinking is <45 mg/L.

The health risk assessment method has been employed in numerous studies on pollutants in groundwater and soils worldwide [5,7,12,25]. In recent years, some studies have underscored the human health risk assessment of nitrate in groundwater [26,27]. This method provides a systematic and robust approach to evaluate the potential health risks associated with exposure to organic and inorganic pollutants in the environment, including those present in contaminated groundwater [3,5,28–30]. It considers factors such as the concentration and toxicity of contaminants, exposure pathways (e.g., drinking water consumption, dermal contact), exposure duration, and population characteristics. The results of the human health risk assessment method inform decision-making and risk management strategies. By quantifying the potential health risks associated with groundwater contamination, stakeholders can make informed decisions regarding contamination prevention, remediation measures, water treatment, and land-use planning. Risk assessment outcomes help prioritize actions to minimize exposure and protect public health. The estimation of health risks caused by concentrations of nitrate in drinking water has never been conducted in this study area. Therefore, in this study we aimed to evaluate the groundwater quality and assess the potential noncarcinogenic health risk of nitrate in the groundwater of the El Milia plain, Kebir Rhumel Basin.

2. Study Area

The El Milia plain is situated in the Kebir Rhumel Basin, which is located in the northeastern part of the province of Jijel, Algeria (Figure 1). The study area is located between 36°40′–36°47′ N latitude and 6°10′–6°20′ E longitude. The Kebir Rhumel Basin is a geological basin known for its agricultural significance, and it encompasses several plains, including the El Milia plain. The El Milia region experiences a Mediterranean climate with an average annual temperature and precipitation of 17 °C and 930 mm, respectively [4,7,31]. The El Milia plain is characterized by its fertile soils, which make it suitable for agricultural activities. The region is known for the cultivation of various crops, including cereals, vegetables, fruits, and olives. Agriculture plays a vital role in the economy of the area, and the El Milia plain is known for its agricultural productivity.

Figure 1. (**a**) location of the study area; (**b**) groundwater samples; (**c**) piezometric map; and (**d**) geological map of the study area.

The study area is located in the alluvial plain of the Mio-Plio-Quaternary (Figure 1). This designation indicates that the aquifer consists of sedimentary deposits from the Miocene, Pliocene, and Quaternary periods. Alluvial aquifers are typically composed of porous materials such as sand, gravel, and silt, which can store and transmit groundwater [7,31]. The potentiometric study conducted in the El Milia plain provides important insights into the groundwater flow patterns and hydrogeological characteristics of the area [7,31]. The potentiometric study reveals that the main axes of groundwater flow converge towards the center of the El Milia plain. This indicates that the groundwater in the area moves towards a central point, likely due to the topography and geological features of the region. Convergence of groundwater flow can have implications for the distribution and availability of groundwater resources in the plain (Figure 1).

3. Materials and Methods

3.1. Data Collection and Analysis Methods

Thirty-three groundwater samples were collected from wells in the El Milia plain during April 2015. The selected wells are used for domestic, agricultural, and domestic/agricultural purposes and are uniformly distributed over the study area. The hydrochemical parameter analyzed was described by Belkhiri et al. [4]. Figure 1 shows the locations of the groundwater samples collected. In situ measurements of pH and electrical conductivity (EC) were conducted at each sampling well. Major ions (Ca, Mg, Na, K, Cl, SO_4, HCO_3, and NO_3) in the groundwater samples were measured in the laboratory.

3.2. Moran's I (Index) Statistic

Moran's I statistic is a correlation coefficient that examines the spatial autocorrelation of the datasets [31–34]. It is defined as

$$\mu(x) = \frac{N}{W}\frac{\sum_{i=1}^{N}\sum_{j=1}^{N} w_{ij}(x_i - \overline{x})(x_j - \overline{x})}{\sum_{i=1}^{N}(x_i - \overline{x})^2} \quad (1)$$

where N is the number of spatial units indexed by i and j, x is the variable of interest, $\overline{x}$ is the mean of x_i, w_{ij} are the elements of a matrix of spatial weights with zeroes on the diagonal (i.e., $w_{ii} = 0$), and W is the sum of all w_{ij}

$$W = \sum_{i=1}^{N}\sum_{j=1}^{N} w_{ij} \quad (2)$$

Resampling and randomization null hypotheses have been tested following the discussion of Goodchild [35]. To calculate the global Moran's I statistic, the function Moran's I in R was used with three arguments such as the coordinates of the observations, the number of nearest neighbors, and the variable for which the statistic is calculated.

3.3. Entropy Water Quality Index (EWQI)

The entropy water quality index (EWQI) is a robust technique that is widely used to provide precise and comprehensive information about the overall groundwater quality for human consumption and drinking purposes [16,18]. The EWQI offers a comprehensive assessment of groundwater quality by considering multiple water quality parameters. By combining and weighting these parameters, the EWQI provides an overall evaluation of groundwater quality. The weighting of each hydrochemical parameter is determined on the basis of its entropy value, which reflects the amount of variation or uncertainty in the data [36–38]. The process to calculate the EWQI is described below.

In this method, one must first assign an entropy weight to each hydrochemical parameter. For *n* number of groundwater samples, every groundwater sample has m number of hydrochemical parameters, generating an eigenvalue matrix (*X*).

$$X = \begin{bmatrix} x_{11} & x_{12} & \cdots & x_{1m} \\ x_{21} & x_{22} & \cdots & x_{2m} \\ \vdots & \vdots & \vdots & \vdots \\ x_{n1} & x_{n2} & \cdots & x_{nm} \end{bmatrix} (i = 1, 2, \ldots, n \text{ and } j = 1, 2, \ldots, m) \quad (3)$$

The standardized value y_{ij} can be expressed following Equation (4), and the standard matrix is expressed as $Y = (y_{ij})$ $(n \times m)$ as shown in Equation (5).

$$y_{ij} = \frac{x_{ij} - (x_{ij})_{min}}{(x_{ij})_{max} - (x_{ij})_{min}} \quad (4)$$

$$Y = \begin{bmatrix} y_{11} & y_{12} & \cdots & y_{1m} \\ y_{21} & y_{22} & \cdots & y_{2m} \\ \vdots & \vdots & \vdots & \vdots \\ y_{n1} & y_{n2} & \cdots & y_{nm} \end{bmatrix} \quad (5)$$

where $(x_{ij})_{min}$ and $(x_{ij})_{max}$ are the minimum and maximum value of the hydrochemical parameter of the groundwater samples, respectively.

Then the ratio of the index value of *i* groundwater sample and *j* index can be expressed as Equation (6).

$$P_{ij} = \frac{1 + y_{ij}}{\sum_{i=1}^{n} (1 + y_{ij})} \quad (6)$$

Information entropy (e_j) can be expressed as Equation (7).

$$e_j = -\frac{1}{\ln(n)} \sum_{i=1}^{n} P_{ij} \ln (P_{ij}) \quad (7)$$

As the value of e_j becomes smaller, the effect of index *j* becomes bigger. The entropy weight (w_j) is given in Equation (8).

$$w_j = \frac{1 - e_j}{\sum_{j=1}^{m} (1 - e_j)} \quad (8)$$

For the next step, assign a quality rating scale for every hydrochemical parameter (q_j).

$$q_j = \frac{C_j}{S_j} \times 100 \quad (9)$$

$$q_{pH} = \begin{cases} \frac{C_{pH} - 7}{8.5 - 7} & C_{pH} > 7 \\ \frac{7 - C_{pH}}{8.5 - 7} & C_{pH} < 7 \end{cases} \quad (10)$$

where C_j is the concentration of each hydrochemical parameter (mg/L), C_{pH} is the measured pH value, and S_j is the permissible limit of each hydrochemical parameter given by the World Health Organization [24].

After calculating all the parameters of equations from 2 to 8, the last EWQI can be expressed as

$$EWQI = \sum_{j=1}^{m} w_j q_j \quad (11)$$

According to the EWQI value, groundwater quality can be classified as: excellent (EWQI ≤ 25), good (25 < EWQI ≤ 50), moderate (50 < EWQI ≤ 100), poor (100 < EWQI ≤ 150), and extremely poor (EWQI > 150).

3.4. Health Risk Assessment

Nitrate in groundwater is generally considered to be a noncarcinogenic substance. However, it can still pose risks to human health through different exposure pathways, primarily via drinking water (e.g., oral pathway) and skin contact (e.g., bathing) [12,21,39]. Nitrate can be absorbed by the human body through the consumption of contaminated groundwater. In the digestive system, nitrate can be converted to nitrite, which can further react with other compounds in the body to form nitrosamines. High levels of nitrosamines have been associated with various health concerns. Moreover, excessive levels of nitrate in drinking water can lead to methemoglobinemia, also known as "blue baby syndrome," which affects the ability of blood to carry oxygen, particularly in infants. In addition, the nitrate can be absorbed through the skin and enter the bloodstream. However, the extent of absorption through the skin is generally lower compared to ingestion. Because human health risks from drinking are much higher than that through skin contact, only the health risk of the drinking water pathway is appraised [40,41]. Therefore, the current study considered only the effect of the drinking water pathway on children and adults. In this study, the hazard quotient (HQ) was used to measure the noncancer risk of the nitrate parameter in groundwater. The HQ was calculated as follows [42–44]:

$$HQ = \frac{C_{NO_3} \times IR \times EF \times ED}{BW \times AT \times RfD} \quad (12)$$

where C_{NO_3} is the observed concentration of NO_3 (mg/L), IR is the ingestion rate (L/Day; for child: 1.5; adult: 2.5), EF is the exposure frequency (Days/Year; for both child and adult: 365), ED is the exposure duration (Years, for child: 12, and adult: 30), BW is the body weight (kg, for child: 36, and adult: 70), and AT is the average time (days/year, for both child and adult: ED × 365). The RfD (reference dose) value for NO_3 is 1.6 mg/kg/day [12,45,46].

4. Results and Discussion

4.1. General Characteristics of Hydrochemical Parameters

The statistical analysis of the hydrochemical parameters in groundwater is highly valuable for understanding the enrichment and variation in hydrochemical components in groundwater. It helps identify patterns, characterize variations, assess trends, evaluate relationships, develop spatial maps, and support data-driven decision making in groundwater management and protection efforts. A statistical summary of hydrochemical parameters for groundwater in the El Milia plain was discussed by Belkhiri et al. [7]. The results presented in Table 1, based on the pH values, indicate that the groundwater ranges from slightly acidic to slightly alkaline in nature. The majority of the groundwater samples exhibited high values of electrical conductivity (EC), above the WHO permissible limit [24], indicating that these groundwater samples are classified as saline water quality. Eighty percent of the total groundwater samples had Ca concentrations above the limits set by the World Health Organization [24]. Additionally, 14% of the samples had Mg concentrations above the WHO limits. All the groundwater samples for both Na and K had concentrations lower than the World Health Organization standard levels. The mean concentrations of anions in the groundwater samples were 150.08, 125.97, 232.02, and 23.47 mg/L for Cl, HCO_3, SO_4, and NO_3, respectively. From the results, it can be observed that only one groundwater sample for chloride and bicarbonate, and two groundwater samples for sulfate and nitrate, were lower than the World Health Organization [24] standard levels. On the other hand, the remaining groundwater samples were within the WHO standard for drinking water.

Table 1. Statistical summary of hydrochemical parameters and EWQI.

	Min	Max	Mean	SD	Cv	S_j	e_j	w_j
pH	6.00	7.50	6.77	0.26	0.04	8.5-7	0.9982	0.0530
EC	228	1411	821	279	0.34	500	0.9966	0.0997
Ca	16.19	198.18	95.56	32.50	0.34	75	0.9979	0.0606
Mg	11.02	61.32	37.39	12.01	0.32	50	0.9966	0.0993
Na	12.47	38.14	24.57	4.99	0.20	200	0.9976	0.0698
K	1.78	5.45	3.51	0.71	0.20	12	0.9976	0.0698
Cl	63.90	255.60	150.08	53.12	0.35	250	0.9951	0.1441
SO_4	61.39	270.00	125.97	49.93	0.40	250	0.9957	0.1252
HCO_3	97.60	524.60	232.02	90.22	0.39	500	0.9966	0.0989
NO_3	0.02	47.54	23.47	14.77	0.63	45	0.9938	0.1795
EWQI	41.54	94.42	63.35	12.51	0.20			

The unit of each hydrochemical element is mg/L, except for pH and EC, μS/cm for EC. S_j: the WHO permissible limit. e_j: Information entropy. w_j: entropy weight. Min: Minimum; Max: Maximum; SD: Standard deviation; Cv: Coefficient of variation.

4.2. Spatial Distribution Pattern of Hydrochemical Parameters

In this study, the Moran's I was used to analyze the spatial distribution pattern of hydrochemical parameters in groundwater. Moran's I is a statistical measure that assesses the degree of spatial autocorrelation and helps identify any spatial patterns in the hydrochemical parameters. The values of Moran's I typically range from −1 to +1. Positive values indicate a higher degree of spatial autocorrelation, suggesting a pattern of spatial clustering. This means that neighboring spatial units tend to have similar values for the hydrochemical parameters. This indicates that there is spatial dependence in the data. On the other hand, negative values of Moran's I indicate a lower degree of spatial autocorrelation. This suggests that the data are more dispersed or randomly distributed in space, with neighboring spatial units having dissimilar values. A lower negative value indicates a stronger dispersal or randomness in the distribution of the hydrochemical parameters. A Moran's I value of zero in spatial autocorrelation analysis indicates that the results are completely randomly distributed in space. This means that there is no spatial autocorrelation or pattern observed in the data [33,47].

The results of the spatial autocorrelation analysis based on Moran's I values for each hydrochemical parameter in the study area are presented in Table 2. From the results, it can be observed that the Moran's I values range from −0.0950 (Mg) to 0.3837 (HCO_3). Three hydrochemical parameters, namely Ca, Cl, and HCO_3, exhibit strong spatial autocorrelation in the El Milia plain. Each of these hydrochemical parameters has a Moran's I value more than or close to 0.3. Furthermore, it is clear that all three of these hydrochemical parameters have Z-score values larger than 2.58 and p-values less than 0.01. This indicates that the high and/or low values of each hydrochemical parameter are geographically clustered in the dataset.

Table 2. Moran's I values for each hydrochemical parameter.

	Moran' I	Expected I	Z Resampling	p-Value Resampling	Z Randomization	p-Value Randomization
pH	0.1556	−0.0294	2.2510	2.44×10^{-2}	2.3744	0.0176
EC	0.0999	−0.0294	1.5736	1.16×10^{-1}	1.5647	0.1176
Ca	0.2695	−0.0294	3.6360	2.77×10^{-4}	3.7808	0.0002
Mg	−0.0950	−0.0294	−0.7974	4.25×10^{-1}	−0.7940	0.4272
Na	−0.0806	−0.0294	−0.6223	5.34×10^{-1}	−0.6352	0.5253
K	−0.0805	−0.0294	−0.6216	5.34×10^{-1}	−0.6345	0.5258
Cl	0.3360	−0.0294	4.4453	8.78×10^{-6}	4.3863	0.0000
SO_4	0.0810	−0.0294	1.3435	1.79×10^{-1}	1.3828	0.1667
HCO_3	0.3837	−0.0294	5.0257	5.02×10^{-7}	5.1597	0.0000
NO_3	−0.0738	−0.0294	−0.5403	5.89×10^{-1}	−0.5336	0.5936

4.3. Spatial Interpolation of the Hydrochemical Parameters

In the present study, the spatial distribution of the three hydrochemical parameters, namely Ca, Cl, and HCO_3, was analyzed using the ordinary kriging (OK) interpolation technique, and the results are displayed in Figure 2. As shown in Figure 3, the spatial distribution of calcium in the El Milia plain indicates that the highest Ca values, exceeding 100 mg/L, are primarily concentrated in the central region of the plain. Furthermore, these high Ca values tend to follow the direction of groundwater flow in the area. This suggests that the movement of groundwater is influencing the distribution of calcium concentrations. The observed spatial distribution of chloride and bicarbonate concentrations in the El Milia plain shows an increase in values from the western to the eastern part of the plain. The concentrations of Cl reach up to 200 mg/L, while the concentrations of HCO_3 reach up to 300 mg/L in the same direction. The spatial trend of increasing Cl and HCO_3 concentrations from west to east may be indicative of the direction of groundwater flow in the El Milia plain. As groundwater moves through the subsurface, it can interact with different geological formations, which can influence the solute composition, including Ca, Cl, and HCO_3, in the groundwater. Geological factors such as the composition and mineralogy of the aquifer materials can play an important role in the release and transport of Ca, Cl, and HCO_3. Another factor to consider is the impact of human activities on the hydrochemical composition of groundwater. Industrial processes, agricultural practices, and domestic wastewater can introduce Ca, Cl, and HCO_3 into the groundwater system. These anthropogenic inputs can contribute to the spatial patterns and variations observed in the hydrochemical distribution.

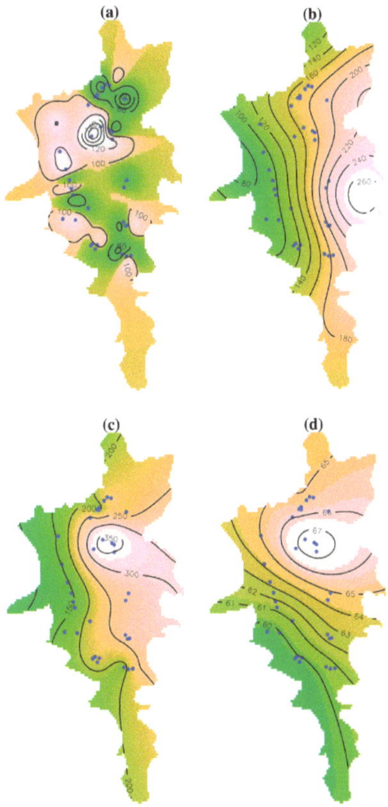

Figure 2. Spatial distribution map of (**a**) Ca; (**b**) Cl; and (**c**) HCO_3 parameters (mg/L); and (**d**) EWQI using OK method.

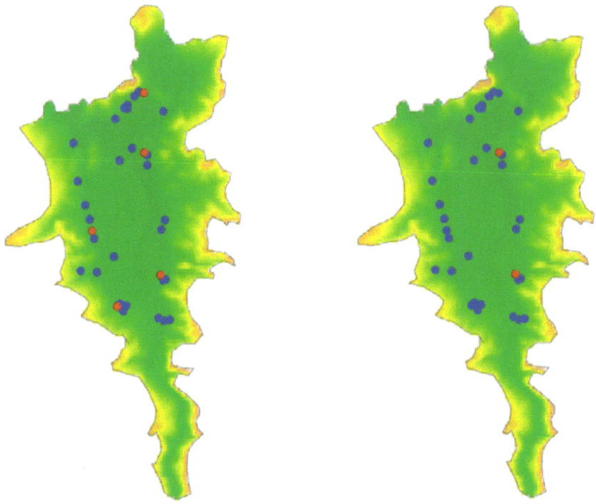

Figure 3. Spatial distribution of HQ for children and adults. The red and blue dots represent medium risk ($1 < HQ \leq 4$) and no risk ($HQ \leq 1$), respectively.

4.4. Comprehensive Assessment of Groundwater Quality

The application of the entropy water quality index (EWQI) method can provide a comprehensive understanding of the overall groundwater quality in the El Milia plan. The EWQI method is a robust technique used to assess and summarize the overall quality of water resources, including groundwater. In this method, the hydrochemical parameters, including pH, EC, Ca, Mg, Na, K, Cl, SO_4, HCO_3, and NO_3, were involved in the calculation of the EWQI. Table 2 summarized the general statistics of EWQI in the study area. The percentage of the total groundwater samples for each category of EWQI is presented in Table 3. The spatial distribution map of EWQI using the ordinary kriging method is displayed in Figure 2.

Table 3. The EWQI range and percentage of groundwater samples in the study area.

Range	Water Type	No. of Samples	% of Samples
$EWQI \leq 25$	Excellent	0	0
$25 < EWQI \leq 50$	Good	5	14
$50 < EWQI \leq 100$	Moderate	30	86
$100 < EWQI \leq 150$	Poor	0	0
$EWQI > 150$	Extremely Poor	0	0

From the results, it was observed that the values of EWQI in the groundwater ranged from 41.54 to 94.42, with a mean value of 63.35. These values provide an overview of the overall groundwater quality in the study area. The distribution of EWQI values indicates that a majority, approximately 86% of the groundwater samples, fall within the moderate groundwater quality category. This suggests that most of the groundwater in the study area is classified as having a moderate environmental groundwater water quality. This implies that there might be some level of contamination or pollutants present in the groundwater, albeit not to an extent that classifies it as poor or very poor in terms of quality. On the other hand, only a smaller portion, around 14% of the total groundwater samples, fall within the good groundwater quality category. This indicates that a relatively small proportion of the groundwater samples meet the criteria for good environmental groundwater quality. These samples likely exhibit lower levels of contamination or pollutants compared to the majority of the groundwater samples. This implies that these specific groundwater samples meet the criteria for good environmental groundwater quality.

The spatial distribution of the EWQI in the study area provides valuable insights into the variability of groundwater quality across different locations. The spatial distribution of EWQI indicates a trend of increasing values from the southwest to the northeast part of the study area. This pattern follows the direction of groundwater flow, suggesting a potential influence of groundwater movement on groundwater quality. In other words, as the groundwater flows from the southwest to the northeast, the EWQI values tend to increase. The highest values of EWQI are observed in the center of the plain, particularly near El Milia city. This suggests that the groundwater quality in that particular area is relatively poor compared to other parts of the study area. The factors contributing to the high EWQI values, such as industrial discharges, agricultural runoff, and other human activities, indicate potential pollution sources in the study area. Excessive release of nutrients like nitrogen from industries can lead to nutrient pollution in water bodies, causing harmful algal blooms and oxygen depletion. The runoff from agricultural fields can carry excess fertilizers containing nitrogen, leading to nutrient pollution in water bodies and causing eutrophication. Agricultural activities involve the use of pesticides and herbicides, which can be washed off of fields and enter water bodies, affecting aquatic life and human health. Improperly treated or untreated sewage and wastewater from residential and commercial sources can introduce pathogens, organic matter, and nutrients into water bodies. Furthermore, these sources could introduce pollutants or contaminants into the groundwater, affecting its quality. Identifying and understanding these pollution sources is crucial for effective groundwater management and protection.

4.5. Noncarcinogenic Health Risk Assessment of Nitrate

Nitrate is commonly used as a good indicator of groundwater quality and vulnerability, especially in agricultural regions where it can be associated with fertilizer use and other agricultural activities [12,21,39]. The presence of high levels of nitrate in groundwater can pose potential health risks, particularly through the drinking water pathway [40,41]. In the present study, the hazard quotient (HQ) was used to assess the noncancer health risk assessment with the nitrate parameter in groundwater, specifically for children and adults through the drinking water pathway. The HQ is a commonly used method in health risk assessment to estimate the potential adverse effects of exposure to a particular chemical or contaminant.

Table 4 presents a statistical summary of the HQ values of nitrate for children and adults. As presented in Table 4, the HQ values for children ranged between 0.00052 and 1.23802, with a mean value of 0.61127. The HQ values for adults varied from 0.00045 to 1.06116 with a mean of 0.52395. This indicates that the HQ values for children are higher than those for adults, indicating in turn that the hazards to human health by nitrate are higher among children than among adults.

Table 4. Statistical summary of HQ for children and adults.

HQ	Min	Max	Mean	SD	CV
$HQ_{Children}$	0.00052	1.23802	0.61127	0.38451	0.62903
HQ_{Adults}	0.00045	1.06116	0.52395	0.32958	0.62903

To quantify the degree of noncarcinogenic health risk, Xiao et al. [41] classified the HQ values into three categories: low risk (HQ $\leq$ 1), medium risk (1 < HQ values $\leq$ 4), and high risk (HQ > 4). Table 5 presents the classification of groundwater quality based on HQ for children and adults. The spatial map of the HQ values for children and adults in the study area is shown in Figure 3. It can be observed that the health risk results revealed that 14% and 6% of the groundwater samples exceeded the acceptable limit for noncarcinogenic risk of 1.0 for children and adults, respectively. Thus, the evaluation of the HQ reveals that children were more vulnerable to noncarcinogenic health hazards than adults in the study area.

Table 5. The HQ range and percentage of groundwater samples in the study area.

Range	Risk	Child		Adult	
		No. of Samples	% of Samples	No. of Samples	% of Samples
$HQ \leq 1$	No risk	30	86	33	94
$1 < HQ \leq 4$	Medium	5	14	2	6
$HQ > 4$	High	0	0	0	0

These results indicate that nitrate contamination in the groundwater of the study area is a potential risk to human health, especially for children. It suggests that long-term consumption of contaminated groundwater as drinking water can be harmful. To mitigate this health risk, the treatment of discharged waste water and the management of fertilizer applications are recommended.

5. Conclusions

In this study, we analyzed the quality and the potential noncarcinogenic health risk of nitrate in the groundwater in the El Milia plain, Kebir Rhumel Basin, Algeria. In order to understand the spatial distribution pattern of the hydrochemical parameters in the groundwater, the Moran's *I* and the irdinary kriging (OK) interpolation technique were applied. It was found that the hydrochemical parameters Ca, Cl, and HCO_3 exhibited strong spatial autocorrelation in the El Milia plain, indicating a spatial dependence and clustering of these hydrochemical parameters in the groundwater. The entropy water quality index (EWQI) method was employed to evaluate the groundwater quality in the study area. The results indicated that approximately 86% of the total groundwater samples fell under the moderate water category based on the EWQI classification. This suggests that the majority of the groundwater samples in the El Milia plain have moderate environmental groundwater quality. The spatial map of the EWQI values revealed a gradual increase from the southwest to the northeast, following the direction of groundwater flow. The highest values were observed near El Milia city in the center of the plain. This spatial pattern suggests variations in groundwater quality across the study area, with potential implications for water resource management and protection.

The hazard quotient (HQ) was used to assess the potential noncarcinogenic health risks associated with nitrate contamination in groundwater for both adults and children through the drinking water pathway. From the results, it was observed that approximately 6% of the groundwater samples exceeded the limit for adults, while a higher percentage, 14%, exceeded the limit for children. This highlights the increased vulnerability of children to noncarcinogenic health hazards associated with the contaminated groundwater in the El Milia plain. Therefore, the findings suggest that the nitrate contamination of groundwater in the El Milia plain poses a human health risk, particularly through long-term consumption of contaminated groundwater as drinking water. In order to mitigate this health risk, we recommend implementing management approaches that focus on limiting nitrate contamination. This may include improving the treatment of discharged wastewater and implementing better management practices for fertilizer application.

Author Contributions: Methodology, L.B., A.T. and H.S.; software, D.B. and R.M.; validation, L.M.; formal analysis, D.B. and R.N.; investigation, F.E.L. and R.N.; resources, R.M.; data curation, A.T. and H.S.; writing—original draft preparation, L.B., H.S., F.E.L. and L.M.; writing—review and editing, A.A.; supervision, L.B., A.A. and L.M.; project administration, A.T. and R.N. All authors have read and agreed to the published version of the manuscript.

Funding: This study received no external funding.

Data Availability Statement: Not applicable.

Acknowledgments: The authors wish to thank all who assisted in conducting this study.

Conflicts of Interest: The authors declare no conflict of interest.

References

1. Belkhiri, L.; Boudoukha, A.; Mouni, L.; Baouz, T. Application of multivariate statistical methods and inverse geochemical modeling for characterization of groundwater—A case study: Ain Azel plain (Algeria). *Geoderma* **2010**, *159*, 390–398.
2. Belkhiri, L.; Narany, T.S. Using multivariate statistical analysis, geostatistical techniques and structural equation modeling to identify spatial variability of groundwater quality. *Water Resour. Manag.* **2015**, *29*, 2073–2089. [CrossRef]
3. Belkhiri, L.; Mouni, L.; Narany, T.S.; Tiri, A. Evaluation of potential health risk of heavy metals in groundwater using the integration of indicator kriging and multivariate statistical methods. *Groundw. Sustain. Dev.* **2017**, *4*, 12–22. [CrossRef]
4. Belkhiri, L.; Mouni, L.; Tiri, A.; Narany, T.S.; Nouibet, R. Spatial analysis of groundwater quality using self-organizing maps. *Groundw. Sustain. Dev.* **2018**, *7*, 121–132. [CrossRef]
5. Tiri, A.; Belkhiri, L.; Mouni, L. Evaluation of surface water quality for drinking purposes using fuzzy inference system. *Groundw. Sustain. Dev.* **2018**, *6*, 235–244. [CrossRef]
6. Wang, W.; Zhang, Z.; Duan, L.; Wang, Z.; Zhao, Y.; Zhang, Q.; Dai, M.; Liu, H.; Zheng, X. Response of the groundwater system in the Guanzhong Basin (central China) to climate change and human activities. *Hydrogeol. J.* **2018**, *26*, 1429–1441.
7. Belkhiri, L.; Tiri, A.; Mouni, L. Spatial distribution of the groundwater quality using kriging and Co-kriging interpolations. *Groundw. Sustain. Dev.* **2020**, *11*, 100473. [CrossRef]
8. Wu, J.; Wang, L.; Wang, S.; Tian, R.; Xue, C.; Feng, W.; Li, Y. Spatiotemporal variation of groundwater quality in an arid area experiencing long-term paper wastewater irrigation, northwest China. *Environ. Earth Sci.* **2017**, *76*, 460. [CrossRef]
9. Li, P.; He, X.; Li, Y.; Xiang, G. Occurrence and health implication of fluoride in groundwater of loess aquifer in the Chinese Loess Plateau: A case study of tongchuan, northwest China. *Expo. Health* **2019**, *11*, 95–107. [CrossRef]
10. Karunanidhi, D.; Aravinthasamy, P.; Deepali, M.; Subramani, T.; Bellows, B.C.; Li, P. Groundwater quality evolution based on geochemical modeling and aptness testing for ingestion using entropy water quality and total hazard indexes in an urban-industrial area (Tiruppur) of Southern India. *Environ. Sci. Pollut. Res.* **2021**, *28*, 18523–18538. [CrossRef]
11. Tiri, A.; Belkhiri, L.; Asma, M.; Mouni, L. Suitability and assessment of surface water for irrigation purpose. In *Water chemistry*; IntechOpen: London, UK, 2020; pp. 1–18.
12. Liu, J.; Peng, Y.; Li, C.; Gao, Z.; Chen, S. Characterization of the hydrochemistry of water resources of the Weibei Plain, Northern China, as well as an assessment of the risk of high groundwater nitrate levels to human health. *Environ. Pollut.* **2021**, *268*, 115947. [CrossRef]
13. Shukla, S.; Saxena, A. Appraisal of groundwater quality with human health risk assessment in parts of indo-gangetic alluvial plain, north India. *Arch. Environ. Contam. Toxicol.* **2021**, *80*, 55–73. [CrossRef]
14. Singaraja, C. Relevance of water quality index for groundwater quality evaluation: Thoothukudi District, Tamil Nadu, India. *Appl. Water Sci.* **2017**, *7*, 2157–2173. [CrossRef]
15. Belkhiri, L.; Mouni, L.; Tiri, A.; Narany, T.S. Evaluation of groundwater quality and its suitability for drinking and agricultural purposes using self-organizing maps. *Int. J. Geol. Environ. Eng.* **2017**, *11*, 705–709.
16. Shannon, C.E. A mathematical theory of communication. *Bell. Syst. Tech. J.* **1948**, *27*, 623–656. [CrossRef]
17. Adimalla, N. Controlling factors and mechanism of groundwater quality variation in semiarid region of South India: An approach of water quality index (WQI) and health risk assessment (HRA). *Environ. Geochem. Health* **2020**, *42*, 1725–1752. [CrossRef]
18. Gao, Y.Y.; Qian, H.; Ren, W.H.; Wang, H.K.; Liu, F.X.; Yang, F.X. Hydrogeochemical characterization and quality assessment of groundwater based on integrated-weight water quality index in a concentrated urban area. *J. Clean. Prod.* **2020**, *260*, 121006. [CrossRef]
19. Kalaivanan, K.; Gurugnanam, B.; Pourghasemi, H.R.; Suresh, M.; Kumaravel, S. Spatial assessment of groundwater quality using water quality index and hydrochemical indices in the Kodavanar sub-basin, Tamil Nadu, India. *Sustain. Water Resour. Manag.* **2018**, *4*, 627–641. [CrossRef]
20. Nawale, V.P.; Malpe, D.B.; Marghade, D.; Yenkie, R. Non-carcinogenic health risk assessment with source identification of nitrate and fluoride polluted groundwater of Wardha sub-basin, central India. *Ecotoxicol. Environ. Saf.* **2021**, *208*, 111548. [CrossRef]
21. Zhai, Y.; Lei, Y.; Wu, J.; Teng, Y.; Wang, J.; Zhao, X.; Pan, X. Does the groundwater nitrate pollution in China pose a risk to human health? A critical review of published data. *Environ. Sci. Pollut. Res.* **2017**, *24*, 3640–3653. [CrossRef]
22. Kom, K.P.; Gurugnanam, B.; Bairavi, S. Non-carcinogenic health risk assessment of nitrate and fluoride contamination in the groundwater of Noyyal basin, India. *Geod. Geodyn.* **2022**, *13*, 619–631. [CrossRef]
23. Noori, R.; Farahani, F.; Jun, C.; Aradpour, S.; Bateni, S.M.; Ghazban, F.; Hosseinzadeh, M.; Maghrebi, M.; Naseh, M.R.V.; Abolfathi, S. A non-threshold model to estimate carcinogenic risk of nitrate-nitrite in drinking water. *J. Clean. Prod.* **2022**, *363*, 132432. [CrossRef]
24. WHO (World Health Organization). *Guidelines for Drinking-Water Quality*, 4th ed.; WHO: Geneva, Switzerland, 2017.
25. Shil, S.; Singh, U.K. Health risk assessment and spatial variations of dissolved heavy metals and metalloids in a tropical river basin system. *Ecol. Indic.* **2019**, *106*, 105455. [CrossRef]
26. Zhang, Y.; Wu, J.; Xu, B. Human health risk assessment of groundwater nitrogen pollution in Jinghui canal irrigation area of the loess region, northwest China. *Environ. Earth Sci.* **2018**, *77*, 273. [CrossRef]
27. Adimalla, N. Spatial distribution, exposure, and potential health risk assessment from nitrate in drinking water from semi-arid region of South India. *Hum. Ecol. Risk. Assess.* **2019**, *26*, 310–334. [CrossRef]

28. Zhang, Q.; Qian, H.; Xu, P.; Hou, K.; Yang, F. Groundwater quality assessment using a new integrated-weight water quality index (IWQI) and driver analysis in the Jiaokou Irrigation District, China. *Ecotoxicol. Environ. Saf.* **2021**, *212*, 111992. [CrossRef]
29. Liu, J.; Gao, Z.; Feng, J.; Wang, M. Identification of the hydrochemical features, genesis, water quality and potential health hazards of groundwater in Dawen River Basin, North China. *Ecol. Indic.* **2023**, *149*, 110175. [CrossRef]
30. Singh, G.; Rishi, M.S.; Herojeet, R.; Kaur, L.; Sharma, K. Evaluation of groundwater quality and human health risks from fluoride and nitrate in semi-arid region of northern India. Environ. *Geochem. Health* **2020**, *42*, 1833–1862. [CrossRef]
31. Moran, P.A.P. The interpretation of statistical maps. *J. R. Stat. Society* **1948**, *10*, 243–251. [CrossRef]
32. Moran, P.A.P. Notes on continuous stochastic phenomena. *Biometrika* **1950**, *37*, 17–23. [CrossRef]
33. Cliff, A.D.; Ord, J.K. *Spatial Autocorrelation*; Pion: London, UK, 1973.
34. Cliff, A.D.; Ord, J.K. *Spatial Processes: Models and Applications*; Pion: London, UK, 1981.
35. Goodchild, M.F. *Spatial Autocorrelation*; Catmog 47; Geo Books: Kattappana, India, 1986.
36. Adimalla, N. Application of the entropy weighted water quality index (EWQI) and the pollution index of groundwater (PIG) to assess groundwater quality for drinking purposes: A case study in a rural area of telangana state, India. *Arch. Environ. Contam. Toxicol.* **2021**, *80*, 31–40. [CrossRef]
37. Li, X.; Huang, X.; Zhang, Y.H. Spatio-temporal Analysis of Groundwater Chemistry, Quality and Potential Human Health Risks in the Pinggu Basin of North China Plain: Evidence from High-Resolution Monitoring Dataset of 2015–2017. *Sci. Total Environ.* **2021**, *800*, 149568. [CrossRef]
38. Wu, C.; Wu, X.; Qian, C.; Zhu, G. Hydrogeochemistry and groundwater quality assessment of high fluoride levels in the Yanchi endorheic region, northwest China. *Appl. Geochem.* **2018**, *98*, 404–417. [CrossRef]
39. Sheng, D.; Meng, X.; Wen, X.; Wu, J.; Yu, H.; Wu, M.; Zhou, T. Hydrchemical characteristics, quality and health risk assessment of nitrate enriched coastal groundwater in northern China. *J. Clean. Prod.* **2023**, *403*, 136872. [CrossRef]
40. Chen, J.; Wu, H.; Qian, H. Groundwater nitrate contamination and associated health risk for the rural communities in an agricultural area of ningxia, northwest China. *Expo. Health* **2016**, *8*, 349–359. [CrossRef]
41. Liu, J.; Gao, Z.; Zhang, Y.; Sun, Z.; Sun, T.; Fan, H.; Wu, B.; Li, M.; Qian, L. Hydrochemical evaluation of groundwater quality and human health risk assessment of nitrate in the largest peninsula of China based on high-density sampling: A case study of Weifang. *J. Clean. Prod.* **2021**, *322*, 12916. [CrossRef]
42. USEPA (U.S. Environmental Protection Agency). *Risk Assessment Guidance for Superfund Volume I: Human Health Evaluation Manual (Part A)*; Office of Emergency and Remedial Response: Washington, DC, USA, 1989.
43. USEPA (U.S. Environmental Protection Agency). *Exposure Factors Handbook*; General Factors; Office of Research and Development: Washington, DC, USA, 1997; Volume 1.
44. USEPA (U.S. Environmental Protection Agency). *Supplemental Guidance for Developing Soil Screening Levels for Superfund Sites*; Office of Emergency and Remedial Response: Washington, DC, USA, 2002.
45. Tran, D.A.; Tsujimura, M.; Loc, H.H.; Dang, D.H.; Le Vo, P.; Ha, D.T.; Nguyen, T.V. Groundwater quality evaluation and health risk assessment in coastal lowland areas of the Mekong Delta, Vietnam. *Groundw. Sustain. Dev.* **2021**, *15*, 100679. [CrossRef]
46. Xiao, Y.; Liu, K.; Hao, Q.; Xiao, D.; Zhu, Y.; Yin, S.; Zhang, Y. Hydrogeochemical insights into the signatures, genesis and sustainable perspective of nitrate enriched groundwater in the piedmont of Hutuo watershed, China. *Catena* **2022**, *212*, 106020. [CrossRef]
47. Wang, Z.; Liu, Y.; Zhang, Y.; Liu, Y.; Wang, B.; Zhang, G. Spatially varying relationships between land subsidence and urbanization: A case study in Wuhan, China. *Remote Sens.* **2022**, *14*, 291. [CrossRef]

hydrology MDPI

Article

Applying Geophysical and Hydrogeochemical Methods to Evaluate Groundwater Potential and Quality in Middle Egypt

Esam Ismail [1], Dimitrios E. Alexakis [2,*], Mohamed Abou Heleika [1], Mohamed Hashem [1], Mohamed S. Ahmed [3], Doha Hamdy [1] and Ahmed Ali [1]

[1] Geology Department, Faculty of Science, Minia University, Minia 61519, Egypt; essam.ismail@mu.edu.eg (E.I.); heleika@mu.edu.eg (M.A.H.); esamismail79@yahoo.com (D.H.); ahmad.ali@mu.edu.eg (A.A.)
[2] Laboratory of Geoenvironmental Science and Environmental Quality Assurance, Department of Civil Engineering, School of Engineering, University of West Attica, 250 Thivon & P. Ralli Str., GR 12241 Athens, Greece
[3] Geology and Geophysics Department, College of Science, King Saud University, Riyadh 11451, Saudi Arabia; mohahmed@ksu.edu.sa
* Correspondence: d.alexakis@uniwa.gr

Abstract: The El-Minia district is a location of interest for future urban development. Using hydrochemistry and electrical resistivity studies, this work aimed to evaluate the groundwater potentiality and it's suitable for various uses. The groundwater potential in the study area was evaluated based on 24 VESs (vertical electrical soundings), and its quality was determined based on the analyses of 57 groundwater samples. EC (salinity index), Na% (salt hazard), SAR (ratio of sodium adsorption), chloride risks, SSP (soluble sodium percentage), MH (magnesium hazard), and other indicators were used to determine whether the collected water samples were suitable for irrigation. Four layers in the study area are mentioned in the geoelectrical cross-sections that have been constructed. The first is made up of silt and clay from the Nile River, while the second is made up of sandy clay, which has a resistivity range of 15 to 32 Ohm.m and a range thickness of 2 to 68 m. Dry limestone makes up the third layer; its resistivity ranges from 1222 to 3000 Ohm.m and its thickness varies between 75 and 95 m. The Eocene aquifer in the research area is represented by the final layer, which has a thickness of more than 250 m and resistivity values that range from 602 to 860 Ohm.m. Most groundwater samples that were collected are safe for drinking; however, none of them are fit for home usage because of their extreme hardness. According to the SAR and US diagram, RSC, KR, and PI, most groundwater samples from the Pleistocene and Eocene aquifers are fit for irrigation.

Keywords: hydro-geochemistry; electrical resistivity; groundwater quality; irrigation uses; Minia; Egypt

Citation: Ismail, E.; Alexakis, D.E.; Heleika, M.A.; Hashem, M.; Ahmed, M.S.; Hamdy, D.; Ali, A. Applying Geophysical and Hydrogeochemical Methods to Evaluate Groundwater Potential and Quality in Middle Egypt. *Hydrology* **2023**, *10*, 173. https://doi.org/10.3390/hydrology10080173

Academic Editors: Pantelis Sidiropoulos and Augustina Clara Alexander

Received: 11 July 2023
Revised: 7 August 2023
Accepted: 16 August 2023
Published: 18 August 2023

1. Introduction

Water quality degradation has become a global issue due to its fundamental ability to cause significant changes in the hydrological cycle [1–4]. Water quality concerns that are complex and diverse require immediate attention. There has been widespread interest and response. Intensive anthropogenic activities and rapid economic growth have put additional strain on the environment and ecosystems, resulting in soil and water resource degradation and severely limiting sustainability [2–6]. Pollution in both groundwater and surface water is caused by the presence of chemical substances [3–9]. These chemical components considerably impact whether or not water is appropriate for human consumption as well as industrial and agricultural purposes. Due to population growth, rising demands for home water use, and increased industrial and agricultural activity. Pollution has decreased the amount of water that is available, which has led to over-pumping. In the last few decades, the Eastern Desert has attracted the attention of different investments, especially in petroleum, tourism, mining, and agricultural activities. It occupies a portion of the arid to the semi-arid belt of Egypt. Water management presents unique issues in arid and semi-arid areas. These areas present significant demands to produce and manage freshwater resources since they are, by definition, locations where water is most scarce.

Egypt's over-population and the significant rise in water demand for urban, industrial, and agricultural growth are two of the country's many water resource challenges. With the expansion of urban and agricultural development, groundwater quality and quantity are regularly impacted by increased groundwater abstraction and new discharge sources. Groundwater exploration greatly benefits from geophysical research, especially geoelectrical techniques. These techniques are widely used to establish the groundwater carrier, layer thickness, and depth of the groundwater level. The fundamental hydrogeochemical activities that govern the groundwater environment and its connection to its migration are clearly understood thanks to groundwater geochemical investigations.

The management of groundwater resources offers a thorough grasp of the aquifer's hydraulic properties and changes in chemical composition.

The area under investigation was thoroughly investigated for hydrogeology and hydro-geochemistry, as well as to assess groundwater quality and its suitability for various applications, using various methods. In the research area, overuse seriously threatens groundwater, particularly for irrigation. Groundwater pollution and abstraction have considerably increased due to industrial and domestic use, which has greatly impacted the hydrologic system. Due to human activities such as wastewater leaching and extensive fertilizer use in agriculture, groundwater pollution is also increasing. In order to safeguard the water resources in the research area, geochemical and geophysical technologies are effective tools for establishing integrated and sustainable water management strategies. The evaluation of groundwater for different purposes has been performed by many researchers, including [10–16]. In Egypt, the evaluation of groundwater has been by many scholars [17–29].

The area under investigation is located along the Eastern portion of the Nile Valley facing El-Minia and the northern part of Assiut districts. Wadi El-Sheikh bounds it at the north and Wadi El-Assuit at the south. The area under research is restricted between latitudes 27°10′ N–28°48′ N and longitudes 30°30′ E–31°30′ E (Figure 1).

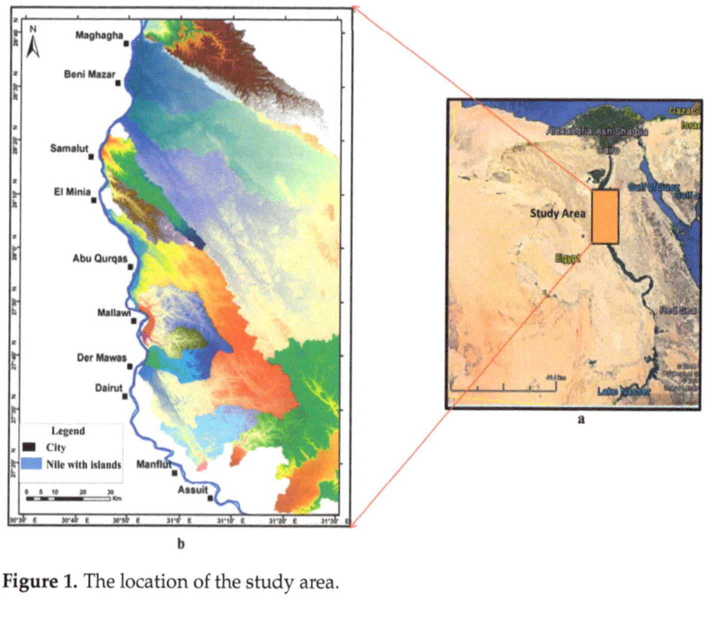

Figure 1. The location of the study area.

In order to identify and characterize the Pleistocene and Eocene aquifers, hydrogeochemical and geoelectrical measurements were taken in this study. The second goal was to examine groundwater potential for agricultural and household use.

2. Geological Settings

The study of the area's surface and subsurface geology is essential for understanding the water potentiality, water migration through various aquifers, and hydrogeologic configuration in terms of aquifer characteristics and leaching processes. Generally, a sedimentary succession dating from the Eocene to the Holocene dominates the studied area. The geological succession in the study area is mainly built of the following rock units from the bottom to the top: (a) Eocene rocks: the Eocene deposits are shales, limestone, chalky limestone, and argillaceous limestone. These rocks are the oldest units cropping out in the study area [30]; (b) Oligocene rocks: the Oligocene rocks are represented by different basaltic flows, which are mainly restricted to the Alawy El-Zurq and Lessan El-Baqara [31]; and (c) Pliocene deposits comprise an upper fluviatile sequence from the late Pliocene and a lower marine sequence from the early Pliocene. The lower marine sequence may cause salinization for the wells tapping it, while the fluviatile sequence is a freshwater-bearing formation [32]. The Pliocene deposits were reported from several deep wells [33].

The Pleistocene deposits are a thick succession that unconformably overlies the Nile Valley's Pliocene or older deposits and the surrounding deserts and underlies the Holocene Nile silts. The Pleistocene sediments are generally friable, highly porous, and permeable. The produced water of the Pleistocene aquifer is extracted from the Qena Formation (which acts as the Nile Valley's primary groundwater aquifer).

A variety of unconsolidated sediments that formed under various environmental conditions make up the Holocene deposits. Dunes and Nile silts serve as representations of these deposits. The top layer of the old cultivated area in the floodplain comprises deposits of silt and clay called Nile silts. Their thickness ranges from location to location and gets thinner near the Nile Valley's margins, where these deposits are laid out unevenly on top of the eroded surface of Pleistocene sediments. Dunes are composed of well-sorted loose quartz sand grains mixed with heavy minerals. In addition, these deposits cover the marginal areas adjacent to the western portion of the cultivated land.

The groundwater of the Pleistocene aquifer occurs under free conditions in major parts of the study area, except in some localities near the Nile River he groundwater is present under unconfined to semi-confined conditions, where it is overlain by Holocene silt clay. The possible recharge of the Pleistocene aquifer in the area under study occurs from upward leakage of underlying Eocene aquifer and surface runoff due to flash floods, while the discharge occurs through pumping wells used mainly for irrigation. The water of the Eocene aquifer occurs under unconfined conditions where the permeable Quaternary sediments overlie it. The possible recharge of the Eocene aquifer may occur from the direct recharge by downward seepage through percolation of the atmospheric precipitation and the occasional flood flashes and direct recharge from the percolation of irrigation water and the lateral seepage of the overlying younger aquifers. The discharge of this aquifer mainly occurs through the pumping wells for irrigation purposes, as well as the seepage towards the Nile River through the fractures.

3. Materials and Methods

In the area studied, 57 groundwater samples (22 from the Eocene aquifer and 35 from the Pleistocene aquifer) were collected in 2020 (Figure 2). The coordinates and the ground elevation of the studied water points were noted using the GPS (Global Positioning System). After an hour of pumping, water samples were collected from the pumps and put into plastic bottles that had already been cleaned. With the assistance of the well owners, the depth to the water table, the total depth of drilling wells, and the groundwater level were all recorded for each point where samples were taken. Water table and water level contour maps were constructed by using ARC-GIS software program version 10.3 to interpret the water flow. After sampling, the Ultrameter SM101 equipment was used to test temperature, electrical conductivity (EC), total dissolved solids (TDS), pH, and TDS. At the Ministry of Agriculture Laboratories in El-Minia Governorate, Egypt, all of the chemical analyses of the collected samples were carried out using the methods recommended by the American Public Health Association. Volumetric titration was used to test magnesium (Mg^{2+}), calcium (Ca^{2+}), and bicarbonate (HCO_3^-); a flame photometer was used to detect chloride (Cl^-), potassium (K^+), and sodium (Na^+); and a UV spectrophotometer was used to assess sulphate (SO_4^{2-}). Several indices were applied to evaluate the suitability of groundwater in the study area for drinking and irrigation uses. The outcomes were contrasted with drinking water quality guidelines set by the WHO (World Health Organization). The equations in Table 1 were used to determine EC, Na%, SAR, RSC, MH, KR, and PI from the correlation of analytical data by projecting several graphical representations [34–37]. The classification of the groundwater quality for different uses was established. Twenty-four vertical electrical sounds were measured. The Schlumberger setup method and a Syscal-Pro instrument were used to obtain these VES, with the measuring process' current spacing varying between 300 and 1000 m. For each VES profile, a distance between the potential electrodes of 0.5 m to 50 m was gradually increased. Three inversion programs (ATO by Zohdy [38], RESIST by Velpen and Resist [39] and RESIXP by Davis et al. [40]) were used to interpret the electrical data. The lithological information constrained the starting models for these programs from the drilled wells, including the thickness and resistivity of each lithologic layer in the study area.

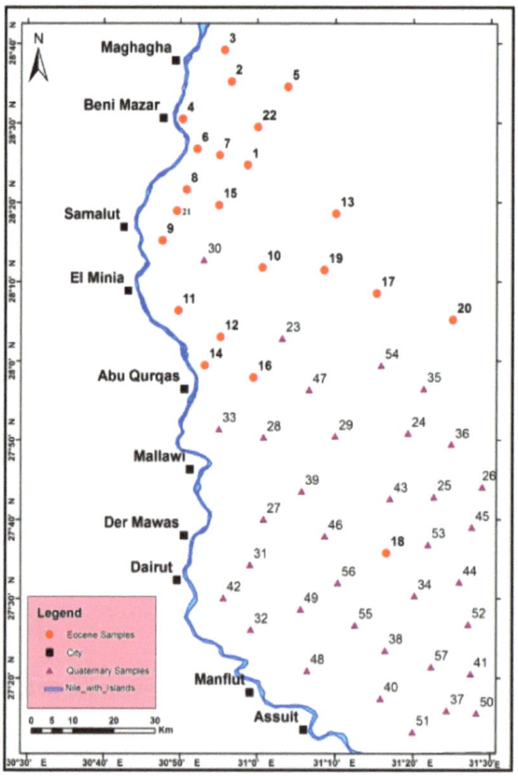

Figure 2. Map showing the location of wells in the study area.

Table 1. The mathematical formulas used for calculating the irrigation quality parameters.

Item	Equation	References
TH	TH = 2.497 Ca^{2+} + 4.115 Mg^{2+} ions in mg/L	[41]
SAR	$SAR = Na/\sqrt{(Ca+Mg)/2}$ all ions in meq/L	[35]
Na (%)	$Na\% = \frac{(Na+K)}{(Ca+Mg+Na+K)} \times 100$ all ions in meq/L	[37]
RSC	$RSC = \left(HCO_3^- + CO_3^{2-}\right) - \left(Ca^{2+} + Mg^{2+}\right)$ all ions in meq/L	[35]
MH	$MH = \frac{Mg}{(Ca+Mg)} \times 100$ all ions in meq/L	[42]
SSP	SSP = ((Na^+ + K^+)/(K^+ + Na^+ + Ca^{2+} + Mg^{2+})) × 100 all ions in meq/L	[43]
PS	$PS = Cl + \sqrt{SO4}$ all ions in meq/L	[44]
KR	$KR = \frac{Na}{(Ca+Mg)}$ all ions in meq/L	[45]

4. Results and Discussion

4.1. Electrical Resistivity Studies

Using the interpreted resistivity data, four geoelectrical cross-sections were created (Figure 3). These sections have been constructed to reflect the underlying arrangement in terms of each layer's type, thickness, and depth.

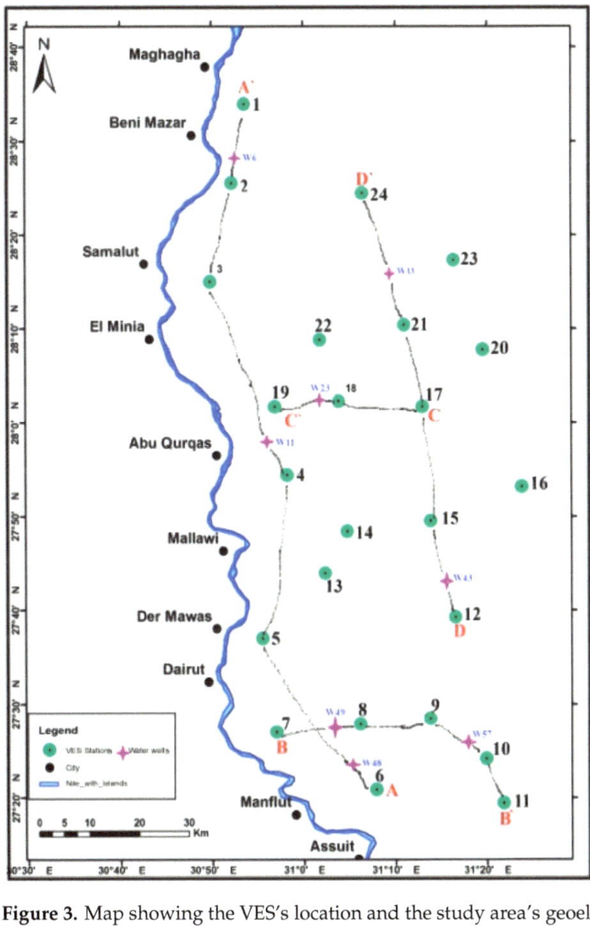

Figure 3. Map showing the VES's location and the study area's geoelectrical cross-sections.

4.1.1. Geoelectrical Cross-Section A-A′

This section runs in a south–north direction (Figure 4), parallel to the Nile River from the eastern side, and consists mainly of four geoelectrical layers. The first geoelectrical layer extends to a depth of 10 m below the ground surface and displays resistivity ranges ranging from 4229 to 5171 Ohm.m. The second layer exhibits resistivity values varying between 52 and 113 Ohm.m. By comparing wells in the study area, such as well No. 6, 11, and 48, it can be seen that this layer's thickness ranges from 2 to 68 m and is formed from sandy clay deposits. This layer is not recorded at VESs No. 3 and 4 as a result of the weathering process. The third layer of this geoelectrical cross-section is represented by dry limestone. It exhibits resistivity values varying from 1312 to 2617 Ohm.m. The thickness of this layer ranges from 22 to 170 m. The last geoelectrical layer of this section is characterized by resistivity values ranging between 613 and 889 Ohm.m and a thickness that reaches over 200 m. This layer is considered a water-bearing formation and is composed of limestone sediments of the Samault formation; it seems to be affected by fault action.

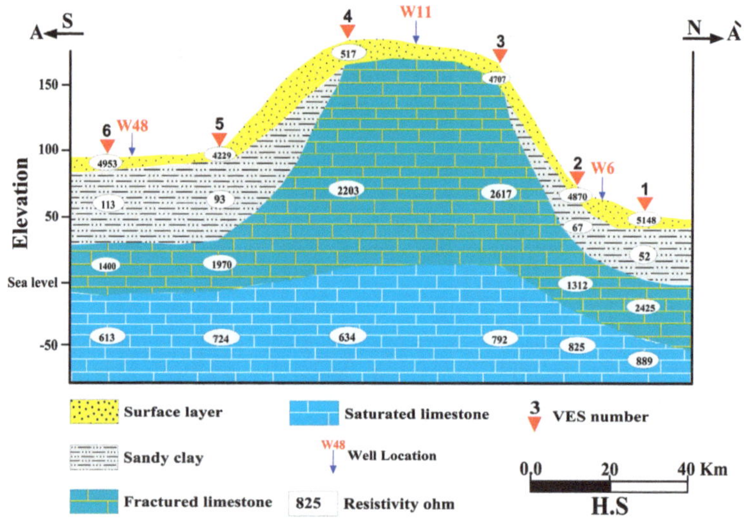

Figure 4. Geoelectric cross-section A-A′.

4.1.2. Geoelectrical Cross-Section B-B′

This section runs perpendicular to the mouth of Wadi El-Omrani, Wadi El-Ibrahimi, and Wadi El-Assuiti. Five geoelectrical layers were recorded in this section (Figure 5). Nile silt and clay make up the top layer, which ranges in thickness from 2 to 26 m and resistivity from 450 to 6000 Ohm.m. The thickness of the second layer is between 10 and 152 m, and the resistivity ranges from 15 to 32 Ohm.m. This layer was removed entirely by the effect of weathering in the eastern part of this section under VES stations No. 10 and 11. Sand and clay deposits make up the third layer in the cross-section, and the estimated resistivity values range from 50 to 82 Ohm.m. This layer ranges in thickness from 25 to 33 m. This layer is also, entirely removed by the effect of weathering in the eastern part of this section under VES stations No. 10 and 11. The resistivity of the fourth layer ranges from 1222 to 3000 Ohm.m, and the thickness ranges between 75 and 95 m. It is composed of dry limestone. The Eocene aquifer is represented by the last layer, which has a predicted thickness of more than 200 m and resistivity ranges from 613 to 820 Ohm.m.

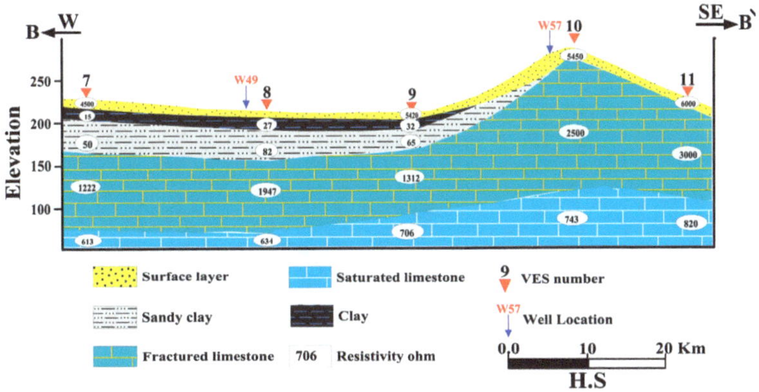

Figure 5. Geoelectrical cross-section B-B′.

4.1.3. Geoelectrical Cross-Section C-C′

This section runs in the E–W direction and the rock succession of this section consists of three layers (Figure 6). The surface layer consists of Nile silt and clay, ranges in resistivity from 4563 to 6000 Ohm.m, and has thickness ranges between 17 and 35 m. Dry limestone is represented in the second layer, with resistivity ranging from 1632 to 2430 Ohm.m. The thickness of this layer is relatively large; it varies between 103 and 157 m. The Eocene aquifer is represented in the last layer; its calculated thickness exceeds 200 m and resistivity ranges from 682 to 812 Ohm.

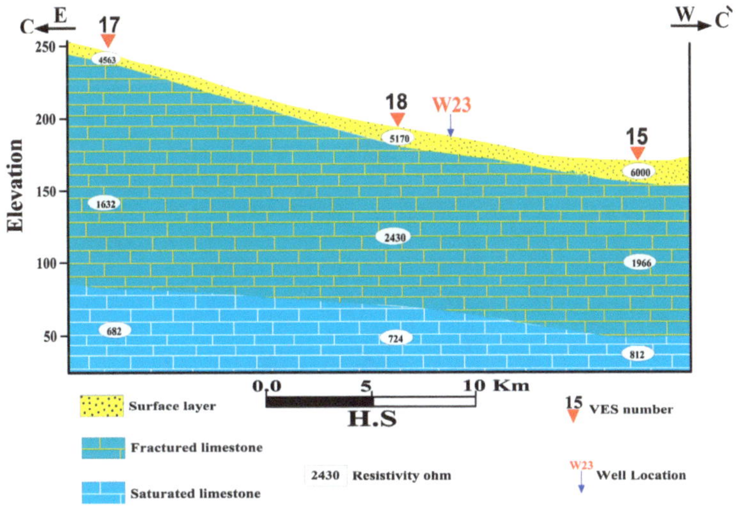

Figure 6. Geoelectrical cross-section C-C′.

4.1.4. Geoelectrical Cross-Section D-D′

This cross-section runs in the S-N direction. It includes VES stations No. 12, 15, 17, 21, 24, and wells No. 13 and 43 (Figure 7). Three geoelectrical layers can be recognized in this section, the first one ranging in thickness from 5 to 21 m and resistivity from 4520 to 5877 Ohm.m. The second one has resistivity varying from 1440 to 2903 Ohm.m and thickness varying between 115 and 157 m, and is composed of dry limestone. The Eocene aquifer is represented by the last layer, which is formed of marly limestone. The thickness of marly limestone exceeds 250 m and its resistivity ranges from 602 to 860 Ohm.m.

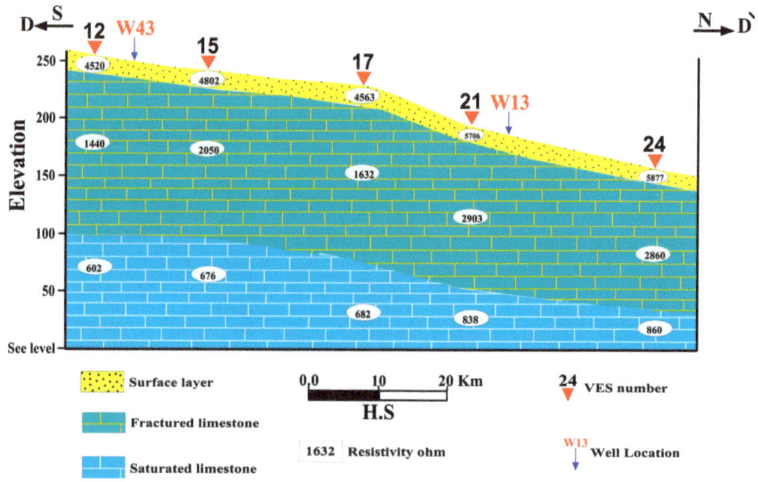

Figure 7. Geoelectrical cross-section D-D′.

4.2. Groundwater Level

The depth of water of the Pleistocene and Eocene aquifer in the study area ranges from 18.5 to 93 m and 15 and 91 m, respectively. The water level varies between 8.6 and 74 m for the Pleistocene aquifer and 18.5 and 175 m for the Eocene aquifer. The study area's groundwater flow mainly occurs in topographically low directions, such as west and north towards the Nile River (Figure 8a,b).

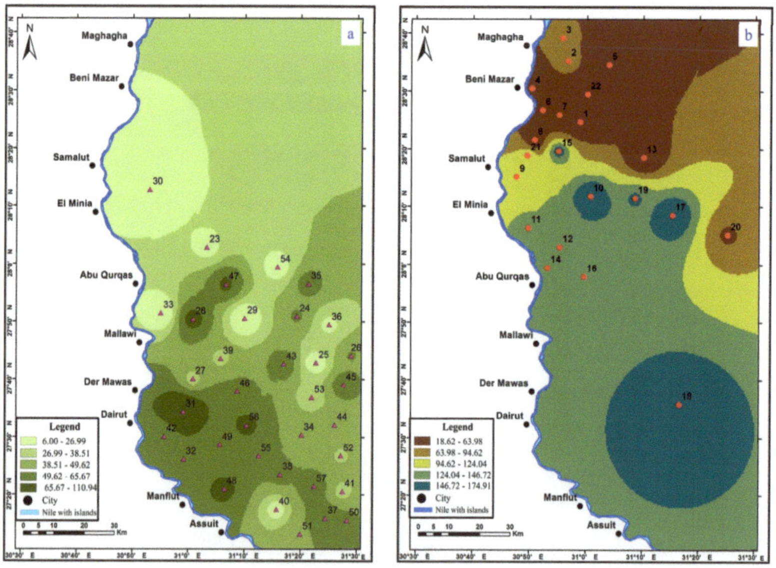

Figure 8. (**a**) Water level zonation map of the studied Pleistocene water samples (**b**) Water level zonation map of the studied Eocene water samples.

4.3. Hydrogeochemical Properties of Groundwater

The measured pH values of the water of the Pleistocene aquifer range from 7.74 and 8.60, with an average of 8.18, while in the Eocene aquifer, pH in water samples vary between 7.84 and 8.20, with an average of 7.98. All the pH values of the water samples of the two aquifers indicate a slightly alkaline media (natural groundwater). The collected Pleistocene water samples presented EC values that vary between 1230 and 9659 micro-mhos, averaging 3135 micro-mhos. The EC values of the Eocene water samples vary between 600 and 2200 micro-mhos, with an average of 1253 micro-mhos. In the study area, the collected samples of Pleistocene water have salinities varying between 787 and 6182 mg/L, with an average of 2006 mg/L. Leaching processes, the effect of direct surface evaporation through irrigation water before penetration to the aquifer, and the heavy pumping of groundwater from the aquifer are the driving forces of the high water salinity. In the collected water samples of the Eocene wells, TDS values ranging from 384 to 1408 mg/L, with an average of 802 mg/L, were measured. The high TDS values of some Eocene wells can be attributed to the dissolution of salts such as halite and gypsum. According to Hem [46], 6% of the Pleistocene wells and 82% of the Eocene wells show TDS values less than 1000 mg/L, reflecting the fresh category. A total of 83% of the Pleistocene wells and 18% of the Eocene wells are slightly saline. The rest of the Pleistocene samples (11%) are moderately saline.

Figure 9 shows the salinity zonation maps of the collected Pleistocene and Eocene samples, these maps indicate increased values in east, north, south, and central areas. Generally, the zonation maps show increases in salinity from west to east, reflecting the impact of limestone deposits being leached and dissolved by water and the fact that the eastern region is far from the Nile River (source of recharge).

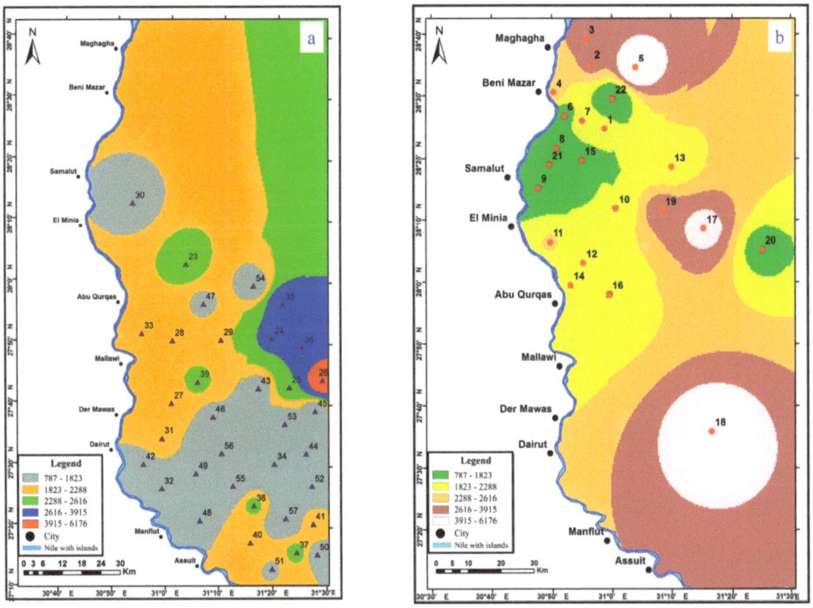

Figure 9. (**a**) TDS zonation map of the studied Pleistocene water samples; (**b**) TDS zonation map of the studied Eocene water samples.

The total hardness ranges from 260 to 2537 mg/L for the collected Pleistocene water samples, with an average of 770 mg/L, and varies between 101 and 716 mg/L for the collected Eocene water samples, with an average of 280 mg/L. According to Hem [46], the Pleistocene water samples and 95% of the Eocene water samples are very hard water, and the rest of the Eocene water samples (5%) are moderately hard water. The leaching and dissolution processes of Ca- and Mg-bearing deposits (limestone) are responsible for the high values of total hardness.

Sodium is the dominant cation and sulphate is the dominant anion in the water collected from Pleistocene aquifer. Na_2SO_4 and NaCl are the two main chemical water types identified in the Pleistocene wells. The presence of the Na_2SO_4 water type is attributed to the intensive application of fertilizers in the study area. In the Eocene water, sodium is the dominant cation, while the dominant anion is chloride, and the dominant chemical water type is NaCl. The sodium chloride water type reflects the ultimate phase of metasomatism and the impact of the dissolution of the marine Eocene rocks.

The Piper trilinear chart [34] was created to analyze a water's origin and the source of its dissolved salts and to explain various processes that impact the characteristics of groundwater. Figure 10 illustrates how the Pleistocene and Eocene samples in the study area fall into two categories based on the chemical data of the obtained groundwater samples. Alkalies are more abundant than alkaline earth in the first category, and strong acids are more abundant than weak acids in the second category.

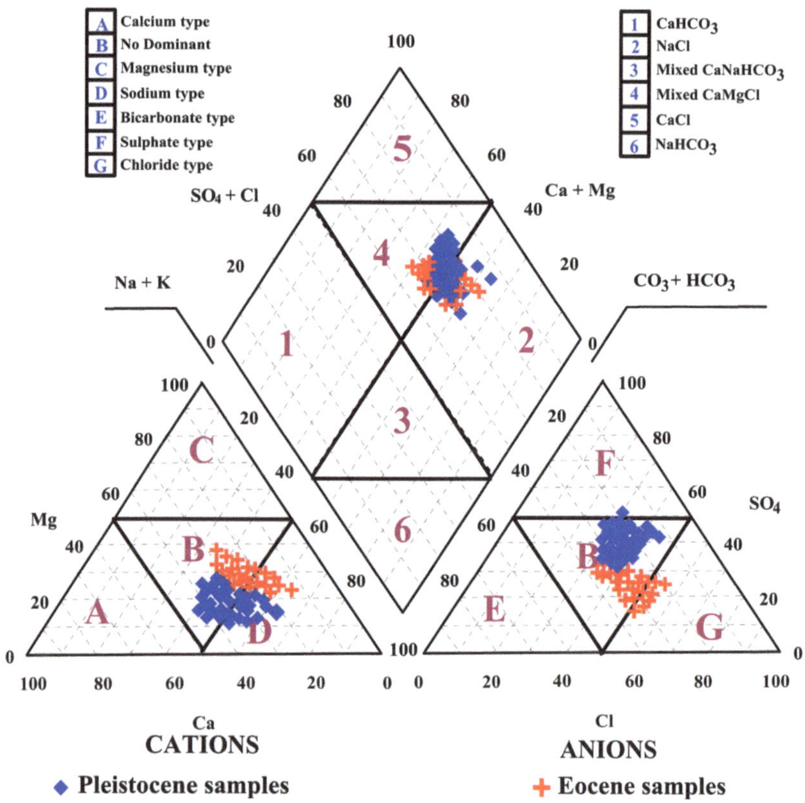

Figure 10. Piper's diagram of the collected groundwater samples.

An evaluation of the functional sources of dissolved ions is performed by plotting samples in relation to the variation in the ratios of Na/(Na + Ca) and Cl/(Cl + HCO_3) as functions of TDS. The collected groundwater wells were plotted on a Gibbs diagram (Figure 11). The Gibbs diagram shows that the groundwater quality is affected by the chemical weathering of rock-forming minerals through the dissolution of the aquifer materials.

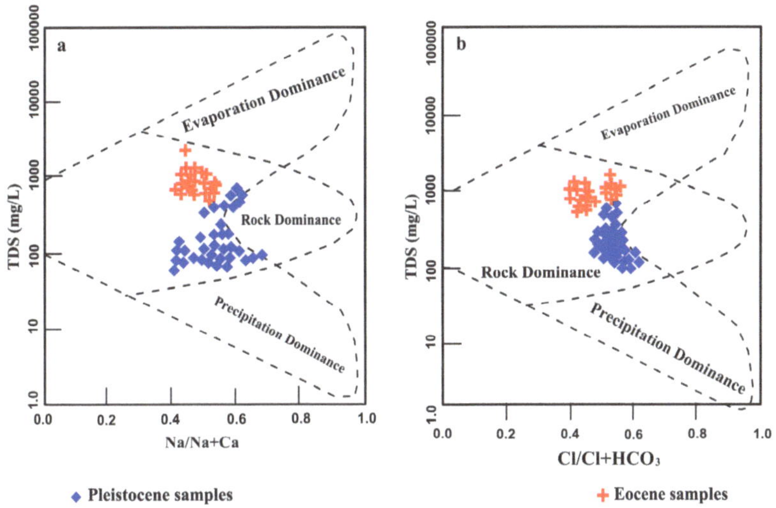

Figure 11. Gibbs's diagram of the collected groundwater samples.

4.4. Assessment of Water Quality for Domestic and Drinking Uses

The majority of the tested groundwater was tasteless, odorless, and colorless. All of the Eocene wells, with the exception of 5%, are suitable for drinking (TDS 1200 mg/L), in comparison with the guidelines of the WHO and the Egyptian maximum permissible limit, while 80 % of the Pleistocene are unsafe for drinking (TDS > 1200 mg/L), and the remaining 20 % are suitable. According to Sawyer and McCarthy [47], the collected Pleistocene water samples have total hardness values ranging from 260 to 770 mg/L, which are hard to very hard water, while the Eocene water samples have total hardness values ranging from 101 to 280 mg/L, which is moderate to hard water. This indicates that all the collected groundwater is unfit for domestic use. The increase in total hardness is attributed to the abundance of Ca^{+2} and Mg^{+2} within the aquifer material.

4.5. Evaluation of Water Quality for Irrigation

The evaluation of water for agriculture depends on the water's quality, the soil's composition, and the farming methods used. EC, Na%, SAR, RSC, KR, and PI parameters were used to determine the suitability of water for irrigation. Regarding the TDS, 5% of the Eocene water can be used for irrigation, while 95% of the Eocene and 54% of the Pleistocene wells can be used for irrigation with more problems. A total of 46% of the Pleistocene wells are not acceptable for irrigation.

According to Bauder et al. [48], 95% of the Eocene and 23% of the Pleistocene wells are acceptable for irrigation, while 77% of the Pleistocene and 5% of the Eocene are suitable for irrigation use (Table 2).

Table 2. Classification of the water samples based on electrical conductivity (EC).

Class. No	EC (μS/cm)	Water Class	Pleistocene Water Samples (%)	Eocene Water Samples (%)
1	<250	Excellent	-	-
2	251–750	Good	-	5
3	751–2000	Permissible	23	90
4	2001–3000	Doubtful	31	5
5	>3000	Unsuitable	46	-

Based on Richards' [35] classification, all the studied water samples show SAR values lower than 10 and fall into the "Excellent" water category for agriculture. The U.S. salinity laboratory staff used SAR and salinity hazards to assess irrigation water in 1954. Hence, the Eocene water is in class C3-S1, except for one sample located in class C2-S1 (Figure 12). Generally, this water is acceptable for irrigation. In the Pleistocene, 23% are in class C3-S1, 31% are in class C3-S2, which is acceptable for irrigation, and 46% are in class C4-S2, which is unsuitable for agriculture.

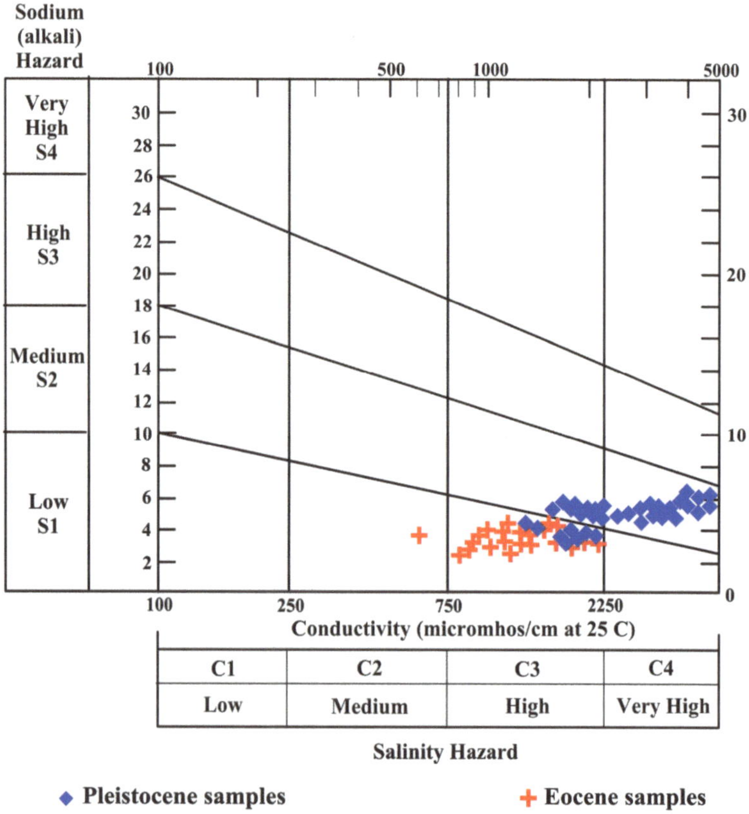

Figure 12. US Salinity Laboratory (USSL) diagram of the collected groundwater samples.

4.5.1. Sodium Hazard (Na%)

Figure 13 illustrates that 5% of the Eocene wells are categorized as excellent to good; 90% of the Eocene and 14% of the Pleistocene are categorized as good to permissible; 34% of the Pleistocene are categorized as permissible to doubtful; 5% of the Eocene and 29% of the Pleistocene are categorized as doubtful to unsuitable, while 23% of the Pleistocene are categorized as unacceptable for irrigation purposes.

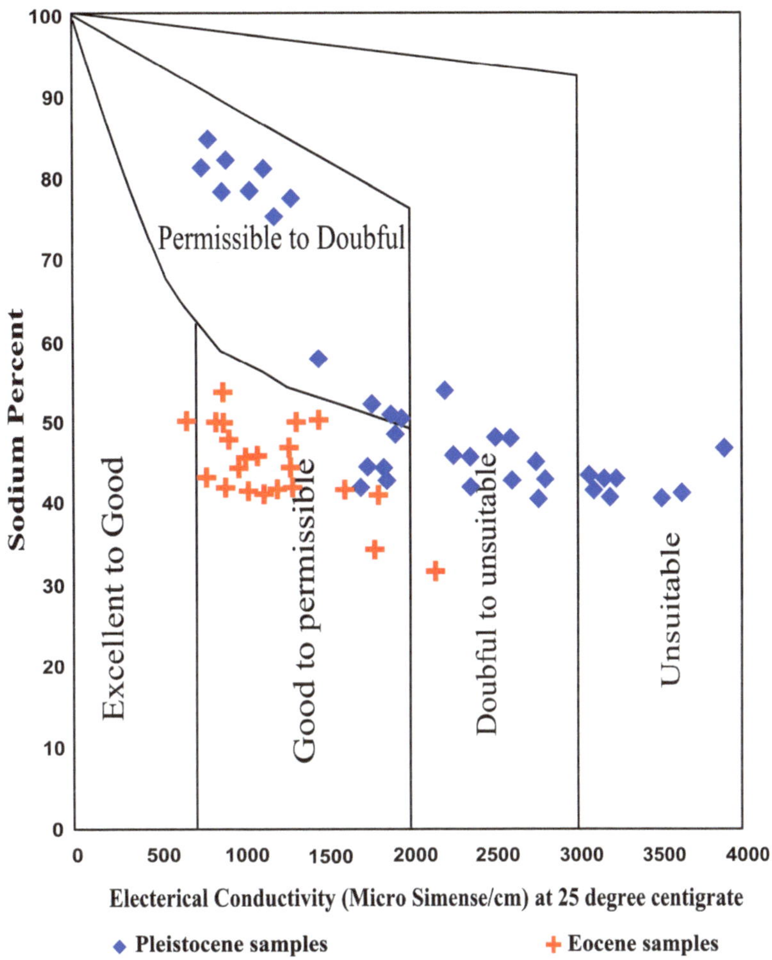

Figure 13. Wilcox diagram for the studied wells.

4.5.2. Residual Sodium Carbonate (RSC)

RSC can be estimated by the formula introduced by Eaton [49] and Ragunath [50], as follows:

$$RSC = (CO_3 + HCO_3) - (Ca + Mg) \quad (epm)$$

Water can be categorized as safe (1.25), slightly appropriate (1.25–2.5), or unsuitable (>2.5) based on the RSC values. A total of 3% of the collected Pleistocene water wells and 64% of the Eocene wells are acceptable for irrigation according to the data gathered (Table 3), whereas 97% of the Pleistocene and 36% of the Eocene wells are not.

Table 3. Classification of collected water samples based on RSC values.

RSC	Water Class	Pleistocene Water Samples (%)	Eocene Water Samples (%)
<1.25	Safe	-	5
1.25–2.5	suitable	3	59
>2.5	unsuitable	97	36

4.5.3. Permeability Index (PI)

Based on the PI values of Doneen's chart [51] (Figure 14), the groundwater can be classified as classes 1 (Excellent > 75%), 2 (Good 25–75%), and 3 (<25% Unsuitable). All of the collected Pleistocene water samples and 82% of the Eocene water samples fall in class 1, while the rest (18%) of the Eocene water samples fall in class 2 (Table 4).

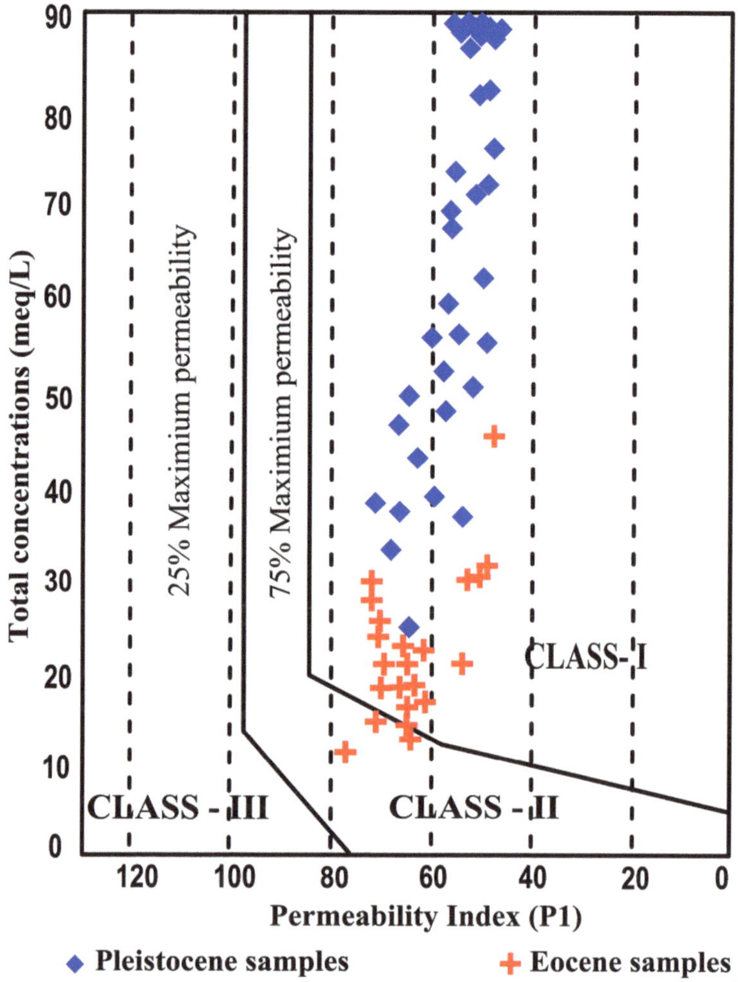

Figure 14. Classification of collected water based on permeability index (PI).

Table 4. Classification of collected samples based on the permeability index (PI).

Class. No	PI (%)	Water Class	Pleistocene Water Samples (%)	Eocene Water Samples (%)
1	>75	Excellent	100	82
2	75–25	Good	-	18
3	<25	Unsuitable	-	-

4.5.4. Kelly's Ratio (KR)

According to Kelly [52] and Paliwal [53], who used sodium measured against Ca^{2+} and Mg^{2+} to calculate this parameter, 55% of Eocene and 69% of Pleistocene wells are suitable for irrigation, while 45% of Eocene and 31% of Pleistocene wells are unsuitable for irrigation.

4.5.5. Soluble Sodium Percentage (SSP)

According to Joshi et al. [43], SSP is the ratio of sodium to all cations multiplied by 100. The Soluble salt Percentage (SSP) is a crucial metric for analyzing the salt hazard in irrigation water quality assessments. SSP results (Table 5) show that 14% of the collected Eocene samples fall into the category of "Good", while all Pleistocene samples and 86% of Eocene samples fall into the "Fair" category.

Table 5. Classification of collected samples based on soluble sodium percentage (SSP).

SSP (epm)	Water Quality	Pleistocene Water Samples (%)	Eocene Water Samples (%)
<20	Excellent quality	-	-
20–40	Good quality	-	14
40–80	Fair	100	86
>80	Poor	-	-

4.5.6. Potential Salinity (PS)

Doneen [54,55] described PS as the sum of the chloride and sulphate concentrations. Groundwater samples' potential salinity was divided into: Excellent to Good (5), Good to Injurious (5–10), and Injurious to Unsatisfactory (>10). According to this classification (Table 6), 9% of the Eocene wells fall into the "Excellent to Good" category, 86% of the Eocene and 23% of the Pleistocene wells fall into the "Good to Injurious" category, and 5% of the Eocene and 77% of the Pleistocene wells fall into the "Injurious to Unsatisfactory" category.

Table 6. Classification of collected water samples based on Potential salinity (PS).

Class. No	PS (epm)	Water Class	Pleistocene Water Samples (%)	Eocene Water Samples (%)
1	<5	Excellent to Good	100	82
2	5–10	Good to Injurious	-	81
3	>10	Injurious to Unsatisfactory	-	-

Wilcox [37] categorized irrigation groundwater by integrating the electrical conductivity and sodium content. Based on this classification, 18% of the collected Eocene water is of the good type, while all of the Pleistocene and 82% of the Eocene samples are permissible.

5. Conclusions

Four geological formations of the study area are mentioned in the geoelectrical cross-sections that have been constructed. The first is made up of silt and clay from the Nile River, while the second is made up of sandy clay, which has a range of resistivity of 15 to 32 Ohm.m and a range of thickness of 2 to 68 m. Dry limestone makes up the third layer; its resistivity ranges from 1222 to 3000 Ohm.m and its thickness varies between 75 and 95 m. The Eocene aquifer in the area studied is represented by the final layer, which has a thickness of more than 250 m and resistivity values ranging from 602 to 860 Ohm.m. While most Pleistocene wells are undesirable due to their high salt content, most Eocene wells are suitable for drinking. Due to their high hardness values, the collected water is unsuitable for home use. Based on the RSC, KR, SAR, and PI, most of the studied wells (Pleistocene and Eocene) are acceptable for irrigation.

Author Contributions: Conceptualization, E.I., M.A.H. and D.H.; methodology, E.I., A.A., M.S.A., M.H. and D.H.; validation, E.I., M.A.H., M.S.A. and A.A.; data curation, E.I. and D.H.; writing—review and editing, E.I., A.A., M.A.H., M.S.A., D.E.A. and M.H.; visualization, E.I., A.A., M.A.H. and D.E.A.; supervision, E.I. and M.A.H. All authors have read and agreed to the published version of the manuscript.

Funding: This work is funded by the Researchers Supporting Project number (RSP2023R455), King Saud University, Riyadh, Saudi Arabia.

Data Availability Statement: An Appendix includes the chemical analyses and the hydrochemical parameters of the collected groundwater samples is associated.

Acknowledgments: The editor and the anonymous reviewers are thanked for their constructive comments which enhanced the manuscript. This work is funded by Researchers Supporting Project number (RSP2023R455), King Saud University, Riyadh, Saudi Arabia.

Conflicts of Interest: The authors declare no conflict of interest.

References

1. Sidiropoulos, P.; Dalezios, N.R.; Loukas, A.; Mylopoulos, N.; Spiliotopoulos, M.; Faraslis, I.N.; Alpanakis, N.; Sakellariou, S. Quantitative Classification of Desertification Severity for Degraded Aquifer Based on Remotely Sensed Drought Assessment. *Hydrology* **2021**, *8*, 47. [CrossRef]
2. Sidiropoulos, P.; Mylopoulos, N.; Lyra, A.; Tziatzios, G.A.; Loukas, A. Risk analysis framework for the optimum remediation of a contaminated aquifer under uncertainty: Application in Lake Karla aquifer, Thessaly, Greece. *Stoch. Environ. Res. Risk Assess* **2023**, *37*, 1281–1302. [CrossRef]
3. Cedrick, M.M.; Alexander, A.; Nobert, J.; Mbudi, C.N.U.D. Modeling groundwater flow under chaotic urbanization constraints in Kinshasa Capital Region (D.R. Congo). *Phys. Chem. Earth* **2021**, *124*, 102985. [CrossRef]
4. Alexakis, D.E. Applying Factor Analysis and the CCME Water Quality Index for Assessing Groundwater Quality of an Aegean Island (Rhodes, Greece). *Geosciences* **2022**, *12*, 384. [CrossRef]
5. Alexakis, D.; Kiskira, K.; Gamvroula, D.; Emmanouil, C.; Psomopoulos, C. Evaluating toxic element contamination sources in groundwater bodies of two Mediterranean sites. *Environ. Sci. Pollut. Res.* **2021**, *28*, 34400–34409. [CrossRef]
6. Gamvroula, D.E.; Alexakis, D.E. Evaluating the Performance of Water Quality Indices: Application in Surface Water of Lake Union, Washington State-USA. *Hydrology* **2022**, *9*, 116. [CrossRef]
7. Tsitsis, C.; Alexakis, D.E.; Moustris, K.; Gamvroula, D.E. Combining Artificial Neural Network and Driver–Pressure–State–Impact–Response Approach for Evaluating a Mediterranean Lake. *Water* **2023**, *15*, 266. [CrossRef]
8. Alexakis, D.E. Meta-Evaluation of water quality indices. Application into groundwater resources. *Water* **2020**, *12*, 1890. [CrossRef]
9. Alexakis, D.E. Linking DPSIR model and water quality indices to achieve sustainable development goals in groundwater resources. *Hydrology* **2021**, *8*, 90. [CrossRef]
10. Liu, J.; Feng, J.; Gao, Z.; Wang, M.; Li, G.; Shi, M.; Zhanget, H. Hydrochemical characteristics and quality assessment of groundwater for drinking and irrigation purposes in the Futuan River Basin, China. *Arab. J. Geosci.* **2019**, *12*, 560. [CrossRef]
11. Rasul, A.K. Hydrochemistry and quality assessment of Derbendikhan Reservoir, Kurdistan Region, Northeastern Iraq. *Arab. J. Geosci.* **2019**, *12*, 312. [CrossRef]
12. Berhe, B.A. Evaluation of groundwater and surface water quality suitability for drinking and agricultural purposes in Kombolcha town area, eastern Amhara region, Ethiopia. *Appl. Water Sci.* **2020**, *10*, 127. [CrossRef]
13. Al Maliki, A.A.; Abbass, Z.D.; Hussain, H.M.; Al-Ansari, N. Assessment of the groundwater suitability for irrigation near Al Kufa City and preparing the final water quality maps using spatial distribution tools. *Environ. Earth Sci.* **2020**, *79*, 330. [CrossRef]
14. Nagarajan, M.; Gauns, A.; Lalitha, R.; Baskar, M. GIS-based assessment of groundwater quality for drinking and irrigation by water quality index. *Int. J. Curr. Microbiol. App. Sci.* **2020**, *9*, 2361–2370. [CrossRef]
15. Asadi, E.; Isazadeh, M.; Samadianfard, S.; Ramli, M.F.; Mosavi, A.; Nabipou, N.; Shamshirband, S.; Hajnal, E.; Chau, K.W. Groundwater quality assessment for sustainable drinking and irrigation. *Sustainability* **2020**, *12*, 177. [CrossRef]
16. Ismail, E.; El-Sayed, E.; Sakr, S.; Youssef, E. Characteristic of groundwater potentialities in West Nile Valley South, Minia Governorate, Egypt. *Arab. J. Geosci.* **2017**, *10*, 521. [CrossRef]
17. Ismail, E.; Zaki, R.; Rapantova, N.; Licbinska, M.; Sharawi, H. Hydrogeochemical characteristics of the groundwater in the quaternary aquifer of western fringes of El-Minia Governorate, Egypt using an integration of geochemical modeling and geo-statistical techniques. *Desalin. Water Treat.* **2020**, *189*, 134–151. [CrossRef]
18. Ismail, E.; El-Rawy, M. Assessment of groundwater quality in West Sohag, Egypt. *Desalin. Water Treat.* **2018**, *123*, 101–108. [CrossRef]
19. Abd-Elaty, I.; Abd-Elhamid, H.F.; Nezhad, M.M. Numerical analysis of physical barriers systems efficiency in controlling saltwater intrusion in coastal aquifers. *Environ. Sci. Pollut. Res.* **2019**, *26*, 35882–35899. [CrossRef]

20. Snousy, M.G.; Ismail, E.; Zaki, R. Trace element occurrence and distribution problems in the irrigation water at El-Minia district, north Upper Egypt. *Arab. J. Geosci.* **2019**, *12*, 582. [CrossRef]
21. Geriesh, M.H.; Mansour, B.M.H.; Farouk, H. Assessment of drinking water quality along Port Said Canal treatment plants, Suez Canal corridor, Egypt. *Arab. J. Geosci.* **2019**, *12*, 738. [CrossRef]
22. Abdelhalim, A.; Sefelnasr, A.; Ismail, E. Numerical modeling technique for groundwater management in Samalut city, Minia Governorate, Egypt. *Arab. J. Geosci.* **2019**, *12*, 124. [CrossRef]
23. Abdelhalim, A.; Sefelnasr, A.; Ismail, E. Response of the interaction between surface water and groundwater to climate change and proposed megastructure. *J. Afr. Earth Sci.* **2020**, *162*, 103723. [CrossRef]
24. Abdalazem, A.; Gamee, M.; Hamdan, A.; Awad, A.; Mohamed, A. Groundwater quality assessment for irrigation in West Edfu region, Aswan Egypt. *Assiut J. Agri. Sci.* **2020**, *51*, 125–149.
25. Sharaky, A.M.; Abdoun, S.H. Assessment of groundwater quality in Bahariya Oasis, Western Desert, Egypt. Environ. *Earth Sci.* **2020**, *79*, 145. [CrossRef]
26. Abotalib, Z.; Heggy, E.; El Bastawesy, M.; Ismail, E.; Gad, A.; Attwa, M. Groundwater mounding: A diagnostic feature for mapping aquifer connectivity in hyper-arid deserts. *Sci. Total Environ.* **2021**, *801*, 149760. [CrossRef]
27. Snousy, M.G.; Jianhua, W.; Fengmei, S.; Abdelhalim, A.; Ismail, E. Groundwater Quality and Its Regulating Geochemical Processes in Assiut Province, Egypt. *Expo. Health* **2022**, *14*, 305–323. [CrossRef]
28. Abotalib, Z.; Abdelhady, A.A.; Heggy, E.; Salem, S.G.; Ismail, E.; Ali, A.; Khalil, M.M. Irreversible and large-scale heavy metal pollution arising from increased damming and untreated water reuse in the Nile Delta. *Earth's Future* **2023**, *11*, e2022EF002987. [CrossRef]
29. Ismail, E.; Snousy, M.G.; Alexakis, D.E.; Abdelhalim, A.; Ahmed, M.S.; Elsayed, E. Diagnosis of Groundwater Quality in North Assiut Province, Egypt, for Drinking and Irrigation Uses by Applying Multivariate Statistics and Hydrochemical Methods. *Water* **2023**, *15*, 2812. [CrossRef]
30. Said, R. *The Geology of Egypt*; Balkema: Rotterdam, The Netherlands, 1990.
31. Zaki, R. Geochemical Characteristics and Element Associations in Different Rock Types of El Minia District, North Upper Egypt. *Sedimentol. Egypt* **2011**, *13*, 349–370.
32. Gamil, R. Hydro-Geochemical Study for the Water Resources in the Desert Fringes of Assiut Governorate, Egypt. Ph.D. Thesis, Faculty of Science, Assuit University, Asyut, Egypt, 2013.
33. Tamer, M.A. Hydrogeology of R. E. A. and Some Changes Occurring as a Result of the Construction of Hydrotechnologic Projects. Ph.D. Thesis, Kiev University, Kyiv, Ukraine, 1974.
34. Piper, A.M. *A Graphic Procedure in the Geochemical Interpretation and Analysis of Water Samples*; USGS, Water Supply 1953, Paper 1454; US Geological Survey: Reston, VA, USA, 1953.
35. Richards, L.A. Diagnosis and improvement of saline and alkaline soils. In *U.S. Department of Agricultural Handbook*; No. 60; US Government Printing Office: Washington, DC, USA, 1954; 160p.
36. Gibbs, R.J. Mechanisms controlling World's water chemistry. *Science* **1970**, *170*, 1088–1090. [CrossRef]
37. Wilcox, V. *Classification and Use of Irrigation Waters*; US Department of Agriculture: Washington, DC, USA, 1955; 19p.
38. Zohdy, A.A.R. A new method for the automatic interpretation of Schlumberger and Wenner sounding curve. *Geophysics* **1989**, *54*, 245–253. [CrossRef]
39. Velpen, V.D.B.; Resist, A.A. *Computer Program for the Interpretation of Resistivity Sounding Curves, An ITC.M. Sc., Research Project*; ITC: Delft, The Netherlands, 1988.
40. Davis, P.A.; Greenhalgh, S.A.; Merrick, N.P. Resistivity Sounding Computations with Any Array using a Single Digital Filter. *Explor. Geophys.* **1980**, *11*, 54–62. [CrossRef]
41. Todd, D.K. *Groundwater Hydrology*, 2nd ed.; Wiley: New York, NY, USA, 1982; 552p.
42. Paliwal, K.V. *Irrigation with Saline Water*; Monogram No. 2 (New Series); IARI: New Delhi, India, 1972; 198p.
43. Joshi, D.M.; Kumar, A.; Agarwal, N. Assessment of the irrigation water quality of River Ganga in Haridwar District India. *J. Chem.* **2009**, *2*, 285–292.
44. Doneen, L.D. *Notes on Water Quality in Agriculture*; Water Science and Engineering, University of California: Davis, CA, USA, 1964; pp. 101–105.
45. Kelly, W.P. Use of Saline Irrigation Water. *Soil Sci.* **1963**, *95*, 355–391. [CrossRef]
46. Hem, J.D. *Study and Interpretation of the Chemical Characteristics of Natural Water*; Water Supply, Paper 2254; USGS: Reston, VA, USA, 1985; pp. 23–25.
47. Sawyer, G.N.; McCarthy, D.L. *Chemistry of Sanitary Engineers*, 2nd ed.; Mc Graw Hill: New York, NY, USA, 1987; 518p.
48. Bauder, T.A.; Waskom, R.M.; Davis, J.G. *Irrigation Water Quality Criteria*; Extension Fact Sheet No. 506; Colorado State University: Fort Collins, CO, USA, 2007. Available online: https://extension.colostate.edu/docs/pubs/crops/00506.pdf (accessed on 17 January 2023).
49. Eaton, F.M. Significance of carbonate in Irrigation Water. *Soil Sci.* **1950**, *69*, 123–133. [CrossRef]
50. Ragunath, H.M. *Groundwater*; Wiley Eastern: New Delhi, India, 1987; 563p.
51. Domenico, P.A.; Schwartz, F.W. *Physical and Chemical Hydrogeology*; Wiley: New York, NY, USA, 1990; pp. 410–420.
52. Kelly, W.P. Permissible composition and concentration of irrigated waters. *Proceeding ASCF* **1940**, *66*, 607.
53. Paliwal, K.V. Effect of gypsum application on the quality of irrigation waters. *Madras. Agric. J.* **1967**, *59*, 646–647.

54. Doneen, L.D. Salination of Soil by Salts in the Irrigation Water. *Am. Geophys. Union Trans.* **1954**, *35*, 943–950. [CrossRef]
55. Doneen, L.D. The influence of crop and soil on percolating waters. In Proceedings of the 1961 Biennial Conference on Groundwater Recharge, Dusseldorf, Germany; 1962; pp. 156–163.

hydrology

MDPI

Article

Evaluation of Groundwater Quality Using the Water Quality Index (WQI) and Human Health Risk (HHR) Assessment in West Bank, Palestine

Ashraf Zohud [1,*], Lubna Alam [2,*] and Choo Ta Goh [1]

1 Institute for Environment and Development (LESTARI), Universiti Kebangsaan Malaysia (The National University of Malaysia), Bangi 43600, Malaysia; gohchoota@ukm.edu.my
2 Institute for the Oceans and Fisheries, University of British Columbia, AERL 2202 Main Mall, Vancouver, BC V6T 1Z4, Canada
* Correspondence: p110470@siswa.ukm.edu.my (A.Z.); lubna.alam@ubc.ca (L.A.)

Abstract: Access to clean and safe water is extremely important, not only in Palestine but also worldwide. In the West Bank, groundwater is particularly valuable because of its scarcity and inaccessibility, and, due to the nature of the area's aquifers, is currently regarded as being at high risk of pollution. Moreover, the water quality in this area is also of wide concern, with its effects being directly linked to human health. Certain parts of the West Bank groundwater suffer from high concentrations of nitrate and potassium. In total, 38.8% of nitrate and 10% of potassium concentrations in well samples exceed the permissible limit set by the WHO and PSI, and, therefore, health problems arise as a limiting factor for life quality and welfare in this region. Moreover, 87.7% of samples are classified as having very hard water. To evaluate the well water in the study area, an assessment was conducted based on the WQI and HHR. Therefore, 49 samples were taken from a group of wells distributed across the study area during the year 2021. The physico-chemical parameters of each sample were analysed. The WQI values showed that 78% of the well samples were of good quality. Moreover, in the classification of the water based on a Piper diagram, 65% of the groundwater was determined to be calcium–magnesium–bicarbonate-type water. Likewise, health risk assessments were evaluated for fluoride and nitrate in drinking water for adults, children, and infants. The main values of the estimated total hazard index (THI) obtained from the analysed data on the health risk assessments revealed a diverse effect on the local population based on age category. The ranges of THI in all sampling locations varied considerably and extended from 0.093 to 3.01 for adults, 0.29 to 3.08 for children, and 0.302 to 3.21 for infants. These results widely indicate that infants are more exposed to health risks.

Keywords: groundwater contamination; physico-chemical parameters; water quality index (WQI); total hazard index (THI); Piper diagram; West Bank

Citation: Zohud, A.; Alam, L.; Goh, C.T. Evaluation of Groundwater Quality Using the Water Quality Index (WQI) and Human Health Risk (HHR) Assessment in West Bank, Palestine. *Hydrology* **2023**, *10*, 198. https://doi.org/10.3390/hydrology10100198

Academic Editors: Pantelis Sidiropoulos and Augustina Clara Alexander

Received: 18 August 2023
Revised: 26 September 2023
Accepted: 3 October 2023
Published: 7 October 2023

1. Introduction

Groundwater remains a major and extremely important source of drinking water around the world. In fact, groundwater supplies around 33% of the world's freshwater needs [1]. Groundwater is becoming increasingly vital in low-precipitation regions that already suffer from water scarcity and uneven distribution [2,3]. In the West Bank, where the majority of the water supply is provided by groundwater, access to safe drinking water is crucial [4]. The Mountain Aquifer is an important water source, the importance of which is widely attributed to both the quantity and high quality of its freshwater [5]. It is mainly composed of karstic and permeable limestone and dolomite formations of the Cenomanian and Turonian ages [6]. Karsts are the fractures and joints in rocks that form conduits and caves. These formations can store groundwater. Moreover, the flow can be rapid through fractures that have been enlarged by dissolution [7]. However, the biggest threat to the water quality in the Mountain Aquifer is pollution due to wastewater and agricultural activity [8]. Various groups of well-known factors related to groundwater pollution in different regions of the world have been studied and detailed, including rapid population growth, untreated sewage, leaking septic tanks and domestic waste, increased industrialization and the overuse of fertilizers in agricultural activities, the dissolution of minerals and the interactions between water and rocks, the evaporation of groundwater, and changes in land cover and changes that occur in the natural environment [9–11].

Groundwater pollutants can be attributed to either inorganic metals, including total dissolved salts (TDSs), heavy metals, cations, and anions, or pathogens, such as bacteria, viruses, and parasites [12,13]. Over the past few decades, groundwater quality has become a serious dispute worldwide, leading to extensive research on quality and health risk assessments carried out by countries such as India, China, and Tunisia [14–16]. Furthermore, 80% of the health risks are connected to the consumption of contaminated water [17]. Zohud [18] reported that limited studies have been conducted on the extent of the health risks associated with the groundwater in the West Bank. Elevated nitrate (NO_3^-) levels surpassing 50 mg/L can result in potential health risks [19]. Rising nitrate levels in groundwater are hazardous to human health and can impact haemoglobin, potentially causing methemoglobinemia disorder, which mainly affects the health of infants and children [20]. The fluoride (F^-) in groundwater usually helps to promote dental health by reducing dental caries when consumed in small doses, while large doses may result in fluoride toxicities [9,21]. The F^- concentration has been found to be within the permissible level in the West Bank, but not in the Gaza Strip [22,23]. The chloride (Cl^-) in groundwater may originate from natural sources or anthropogenic sources, such as wastewaters, cesspits, and fertilizers [24,25]. When the Cl^- concentration rises in groundwater, it develops a saline flavour, which may cause diarrhoea to allergic individuals [26]. Hejaz [27] reported that the Cl^- concentration in the groundwater of the northern West Bank is below the permissible level of 250 mg/L, but that 18% of the NO_3^- concentration is above the permissible level of 50 mg/L. However, Daghara [28] reported that 24% of Cl^- and 21% of NO_3^- concentrations are above the permissible limit in the springs of the West Bank. A high level of sulphate (SO_4^{2-}) in groundwater causes a bitter flavour and may cause diarrhoea, which can lead to dehydration and put infants at severe risk [29]. The minerals in water are essential for the development of the human body, but only when found within the permissible limits [30,31]. Mahmoud [22] reported that there are three main hydrochemical facies in the West Bank: freshwater, freshwater mixed with another water type, and an extreme water type. Using a variety of water quality parameters, the water quality index (WQI) is a useful tool for evaluating water quality [32]. These parameters have been assigned weights according to their importance and health effects [33]. However, we are aware that, in our region, the groundwater quality typically deteriorates during the dry season and improves during the rainy season. Therefore, we deliberately selected the dry season for our study.

This study aimed to (1) analyse the hydrogeochemical properties and hydrochemical facies of the groundwater, as well as their formation mechanisms, and to assess the groundwater quality in the northern West Bank; (2) use the WQI to assess the drinking groundwater quality; and (3) to determine the extent to which the fluoride and nitrate concentrations pose a non-cancerous health risk to adults, children, and infants. Finally, there is a lack of studies evaluating groundwater quality using the WQI and HHR in the West Bank, which makes this study extremely significant. Nonetheless, the outputs of this study are important for stakeholders, including researchers, decision makers, and policymakers, and it will also facilitate the management of water resources and enhance the protection and sustainability of groundwater in the West Bank, Palestine. This is anticipated to be necessary to confront the increasing water scarcity in the coming years.

2. Materials and Methods

2.1. Study Area

The West Bank is a Palestinian territory located in the Mediterranean Region, west of Jordan. It is surrounded by the Dead Sea and the Jordan River on the east and by Israel on three sides. It has a total area of 5860 km^2 and a population of approximately 3,100,000 registered in 2020. The West Bank rocks contain a complex sequence of limestone, dolomite, chalk, and marl [7]. The climate of the West Bank is Mediterranean, which fluctuates from 250 mm of rain in the east and southeast to more than 500 mm in the west (Figure 1). The climate is characterised by cold and rainy winters, and it gradually becomes hot and dry during the summer. The average annual rainfall in the study area is between 450 and 650 mm [34], with a gradual distribution that increases from the east and west towards the mountain areas, where a large amount of this rainfall leaks out to feed the groundwater, and the rest runs into the nearest valleys. The study area is formed by the governorates of the northern West Bank; these include Nablus, Qalqilya, Jenin, Tulkarm, Tubas, and Salfit, which had total populations of about 400,000, 120,000, 335,000, 200,000, 65,000, and 80,000 in the year 2020, respectively [35].

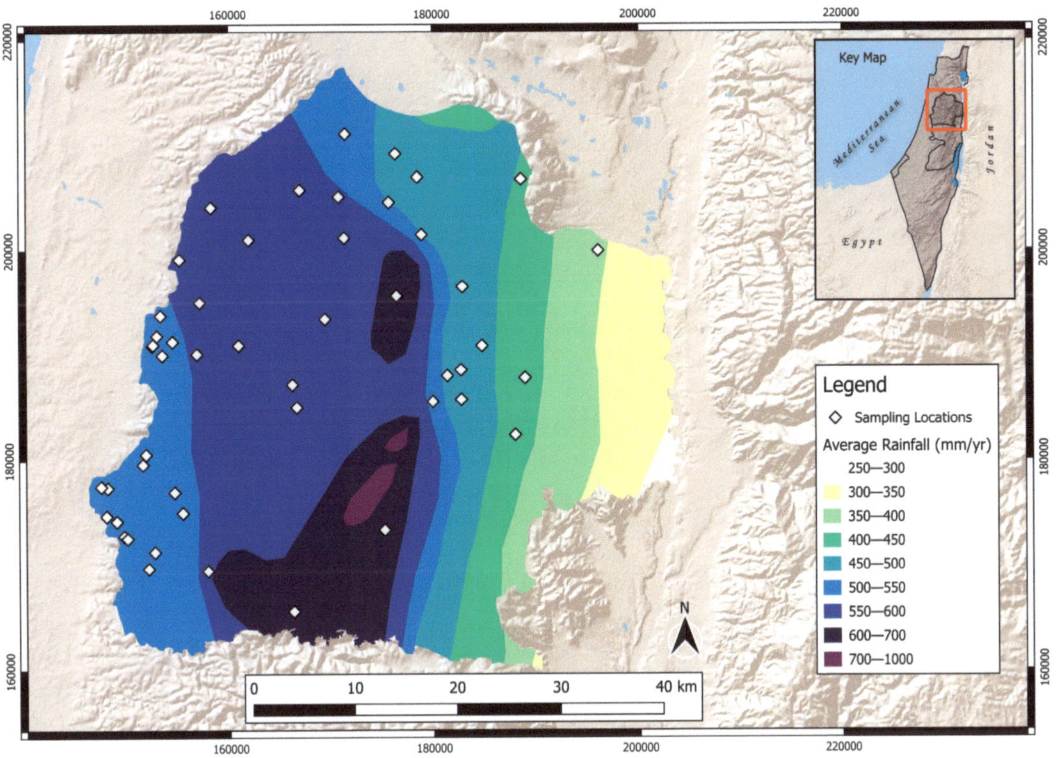

Figure 1. Annual rainfall distribution for the study area.

The Mountain Aquifer is split into three sections: the Eastern, Northeastern, and Western Aquifers (Figure 2). While portions of the Northeastern Aquifer and the Eastern Aquifer flow eastward towards the Jordan River, portions of the Northeastern Aquifer and the Western Aquifer flow westerly towards the Mediterranean Sea [36,37]. The study area is divided into two main aquifers, the Western and Northeastern Aquifers. The governorates of Tulkarm, Qalqilya, Salfit, and the northern part of Nablus are located in the western basin, while Jenin, Tubas, and the southern part of Nablus are located in the northeastern basin. Chalk, dolomite, and limestone from the Upper Cretaceous to Tertiary era make up the majority of the Mountain Aquifer [25]. Due to extensive fracturing and karst channels, the limestone and dolomite rock formations possess relatively high hydraulic conductivities, while marl rocks, characterised by limited fracturing, exhibit significantly lower hydraulic conductivity [38].

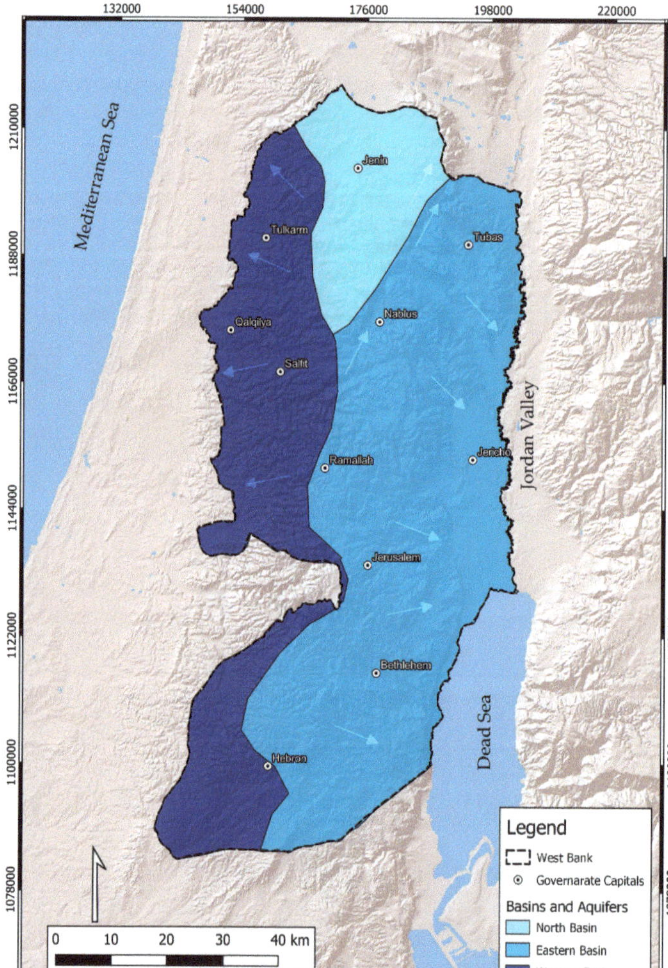

Figure 2. Extent and location of the main aquifer system in the West Bank.

2.2. Groundwater Sampling and Analysis

A total of 49 groundwater samples were collected from the northern part of the West Bank (Figure 3) during the year 2021. Following the WHO guidelines [39], the following parameters were tested: pH, electrical conductivity (EC), total dissolved solids (TDSs), total hardness (TH), chloride, sulphate (SO_4^{2-}), nitrate (NO_3^-), fluoride (F^-), bicarbonate (HCO_3^-), chloride (Cl^-), calcium (Ca^{2+}), magnesium (Mg^{2+}), sodium (Na^+), potassium (K^+), zinc, barium, lead, arsenic, copper, cadmium, aluminium, iron, manganese, chromium, and nickel. The samples were analysed for different physico-chemical parameters following standard methods [40,41].

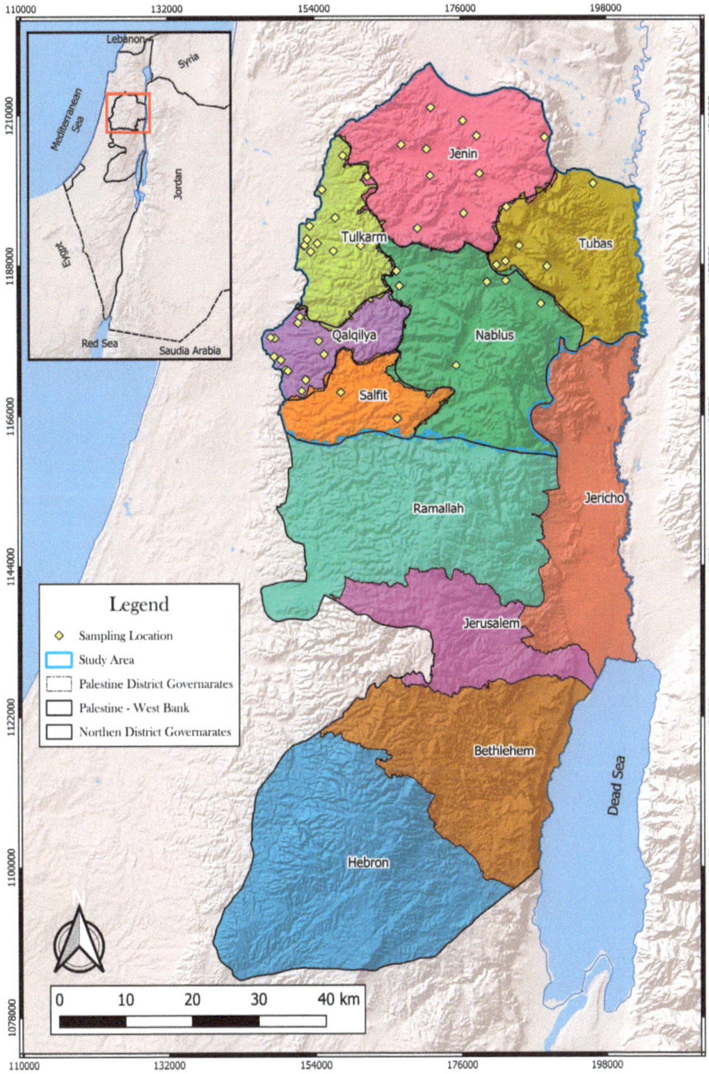

Figure 3. Location map of groundwater samples from different governorates of the study area.

EC, salinity, and pH were measured using an HQ40 portable meter [42]. TDSs were determined by filtering a measured volume of the sample through a standard glass fibre filter [43]. The concentrations of nitrate, sulphate, and fluoride in the water were determined using a UV spectrophotometer [42]. The concentrations of physico-chemical parameters and heavy metals in the water were quantified using an optical emission spectrometer (optima 7300 dv) [42]. The TH of the water was determined via titration with a standard solution of ethylene diamine tetra acetic acid (EDTA), which is a complexing agent [42]. Cl^- was estimated via $AgNO_3$ titration [42]. HCO_3^- was estimated using the volumetric technique, employing hydrosulphuric acid (H_2SO_4), along with phenolphthalein and methyl orange indicators [44].

2.3. Methods

2.3.1. Water Quality Index (WQI)

As it serves as an effective tool for evaluating groundwater quality, the water quality index (WQI) is commonly used worldwide to assess groundwater suitability for drinking [3,45]. In the process of determining the water quality index (WQI), the initial stage involved computing the parameter weights by using Equation (1), as shown in Table 1. Each of these parameters was attributed a weight (wi) ranging from 1 to 5, reflecting its significance and impact on health [33].

$$\mathrm{Wi} = \mathrm{wi}/\sum_{i=1}^{n} \mathrm{wi} \tag{1}$$

$$\mathrm{Qi} = (\mathrm{Ci}/\mathrm{Si}) \tag{2}$$

$$\mathrm{SIi} = \mathrm{Wi} \times \mathrm{Qi} \tag{3}$$

$$\mathrm{WQI} = \sum_{i=1}^{n} \mathrm{SIi} \tag{4}$$

where Wi = the relative weight;

wi = the weight assigned to each specific parameter;

n = the count of parameters.

Moving to the third phase, the calculation of the quality rating scale (Qi) was carried out [46] using Equation (2).

where Qi = the quality rating;

Ci = the concentration of each chemical parameter in each water sample in mg/L;

Si = taken from the PSI guidelines for each chemical parameter.

Table 1. Relative weights of chemical parameters in groundwater wells in study area.

Chemical Parameter	Weight (wi)	Wi (wi/∑wi)	PSI
pH	3	0.083	6.5–7.5
TDS	4	0.111	500
TH	2	0.056	500
Ca^{2+}	3	0.083	75
Mg^{2+}	2	0.056	50
Na^{+}	3	0.083	200
K^{+}	2	0.056	10
Cl^{-}	3	0.083	250
SO_4^{2-}	4	0.111	250
NO_3^{-}	5	0.139	50
F^{-}	5	0.139	1.5
Total	36	1	

Eventually, the water quality sub-index (SIi) of each chemical parameter was computed using Equation (3), and the whole WQI was determined using Equation (4), where SIi denotes the sub-index associated with the ith parameter, while Qi represents the rating determined by the concentration of that specific parameter. The variable 'n' corresponds to the overall count of parameters. As a result, the calculated WQI values were sorted into five different categories [9,47].

2.3.2. Human Health Risk (HHR) Assessment

It is good to carry out a health risk assessment to estimate HHR on the basis of ingestion [48]. A health risk assessment method proposed by the USEPA, which has proven to be an effective tool to evaluate health risk [1], was used and adapted through this work. Nevertheless, this study primarily focused on the oral consumption of drinking water as the key pathway of exposure. The health risk assessment centred on the concentrations of fluoride (F^{-}) and nitrate (NO_3^{-}) as the chosen parameters. The USEPA considered these two parameters as non-carcinogenic risks for human health [48,49]. The average daily dosage (CDI) of F^- and NO_3^- ingested from groundwater in the study area was calculated as part of an exposure assessment using Equation (5) [48]:

$$CDI = \frac{CW \times IR \times ED \times EF}{BW \times AT} \quad (5)$$

where

CDI = the average daily dose of ingestion of F and NO_3 (mg/kg/d);
CW = the F^- and NO_3^- concentrations in the water (mg/L);
IR = the ingestion rate (L/day);
EF = the exposure frequency (365 days/year);
ED = the exposure duration (years);
BW = the average body weight (kg);
AT = the average exposure time in years (body weight × 365).
The symbol values are shown in Table 2.

Table 2. Symbols and their values for adults, children, and infants [48,50,51].

Symbol	Value for Adults	Value for Children	Value for Infants	References
IR	2.5 L/day	0.78 L/day	0.3 L/day	USEPA 2014 [50]
ED	64 years	12 years	<1 year	WHO 2013 [40]
EF	365 days/year	365 days/year	365 days/year	USEPA 2014 [50]
BW	57.5 Kg	18.5 Kg	6.9 Kg	USEPA 2014 [50]
AT	23,360 days	4380 days	365 days	USEPA 2014 [50]

The reference dosage served as a quantification of chronic non-carcinogenic hazards [52]. Adverse toxic effects are more likely to manifest when the exposure dose of the specific contaminant surpasses this reference dosage. Typically, this comparison is expressed through the calculation of a hazard quotient (HQ), as outlined in Equation (6):

$$HQ = \frac{CDI}{RfD} \quad (6)$$

where

RfD = the reference dose for a non-carcinogenic pollutant through the oral intake pathway, which is 0.04 mg/kg/d for F and 1.6 mg/kg/d for NO_3 [52]. Based on Equation (6), the total hazard index (THI) of a non-carcinogenic pollutant is computed using Equation (7):

$$THI = \sum_{i=1}^{n} HQi \quad (7)$$

3. Results

The statistical data of the samples from the 49 wells are shown in Table 3. The pH values of the sampled groundwater wells were found to vary from 7.1 to 8.3, showing properties that are almost neutral to slightly alkaline, and this slightly reflects the water–rock interactions. These pH values fall under the permitted limits of drinking water regulations [26,46], and they are in line with those in two different studies in the same study area [22,27]. When the pH is > 8, the water is usually unsuitable for effective chlorine disinfection, but if the value goes below 6.5, pipe erosion starts to increase [27]. In the West Bank, the predominant rock formation is a carbonaceous one, which is reflected in the water's alkaline quality [6]. The EC values of the wells ranged from 504 to 1261 μS/cm, with a mean of 829.6 μS/cm. All groundwater samples that were collected had an electrical conductivity value within the permissible limit. Moreover, the TDSs of the groundwater ranged from 292 to 731 mg/L, with a mean of 477.4 mg/L; all samples had values within the normal range for drinking water. The high level of TDSs in the well samples is widely attributed to the leaching of salts from soil and cesspit tanks, which could potentially infiltrate the groundwater [53–56]. The anion concentrations displayed the following trend: $HCO_3^- > Cl^- > NO_3^- > SO_4^{2-} > F^-$. However, the cation concentrations displayed the following order: $Ca^{2+} > Na^+ > Mg^{2+} > K^+$. Testing for total alkalinity is crucial for assessing alterations in water alkalinity resulting from human activities [57]. The alkalinity of the tested groundwater wells in the study area was classified as most desirable according to the WHO [17]. The HCO_3^- concentrations of the groundwater samples ranged from 165 to 420 mg/L, with a mean of 312.5. This reflects the karst and carbonate nature of the rocks in the study area. The total hardness (TH) values of the sampled groundwater wells ranged from 260 to 540 mg/L, with a mean of 358. It is clear that 2% of the groundwater exceeded the permissible limit set by the WHO and PSI. According to the TH categorization of groundwater, 87.7% of the samples were classified as having very hard water (Table 3). These values are in line with those of a previous study by Hejaz [27], but they considerably shift from the values found in Mohamad's work [22]. Groundwater that exceeds the limit of 300 mg/L is considered to be very hard water [58]. Hard waters can affect water heaters, boilers and cooking utensils, distribution pipes, and well pumps, and more soap is required to wash clothes [59,60]. The Ca^{2+} concentrations in the analysed samples extended from 53 to 135 mg/L, with a mean value of 86.8. The permissible limit of Ca^{2+} in drinking water is 75 mg/L according to the WHO, but 63.3% of the samples exceeded the permissible limit. The Mg^{2+} concentrations within the analysed samples extended from 14.6 to 54 mg/L, with a mean of 32.6; 71.4% of the studied wells exceeded the allowable limit of the WHO (30 mg/L). Since limestone and dolomite are basic components of the rocks in the study area, their dissolution due to rain introduces certain amounts of calcium and magnesium into the groundwater. The Na^+ concentrations in the groundwater wells ranged from 15.5 to 131 mg/L, with a mean of 39.9; these levels are below the permissible limit of PSI standards. According to the concentration of Na^+, which was less than 200 mg/L, the groundwater in the study area is safe to drink [26]. The potassium level in the groundwater spanned from 0.1 to 27.7 mg/L, with an average of 4.2. It is evident that 10.2% of the analysed samples exceeded the permissible limit set by the WHO. In the study area, the natural sources of K^+ in the groundwater are mainly K^+-bearing rocks, which are usually found in small quantities because of their slow weathering rate. However, human activities, such as wastewater discharge and fertilizer use, are considered important sources of K^+ [61]. A recent study revealed that agricultural practices have a significant negative impact on groundwater quality, which is observed in high concentrations of NO_3^-, K^+, and EC [62]. The Cl^- concentrations ranged from 26 to 249 mg/L, with a mean of 71.2. The Cl^- concentrations are within the drinking water regulations' approved levels. The maximum values of Cl^- and Na^+ were recorded in Jenin well ID (18–20/008), which, again, reflects the effect of human activities (e.g., wastewater discharge, cesspit leakage, and fertilizer use) as additional sources of these substances. The SO_4^{2-} levels ranged from 0 to 65 mg/L, with a mean of 25.5. All of the results fall within the acceptable ranges for drinking water. The F^- concentrations ranged from 0 to 0.7 mg/L, with a mean of 0.28. All results are within the permissible limits of drinking water. The NO_3^- levels ranged from 0.5 to 110 mg/L, with a mean of 40.6. It is clear that 38.8% of the samples exceeded the permissible limit set by the WHO and PSI. The high NO_3^- concentration is mainly caused by agricultural activities (e.g., fertilizer leaching), wastewater discharge, and septic tank leakage [1]. Mahmoud [22] reported that there are seasonal differences in the concentrations of NO_3^- and NH_4^{2+} in groundwater, which increase in the dry season and decrease in the wet season. Moreover, he found that the PO_4^{3-} as P in groundwater ranged from 0 to 3 mg/L. The majority of heavy metal concentrations (e.g., chromium, copper, manganese, lead, cadmium, nickel, aluminium, and arsenic concentrations) in all of the examined samples were discovered to be below the detection threshold of the utilised analytical methods. This means that all concentrations were found to be within the drinking water regulations' approved levels; hence, the groundwater in the study area is considered to be free from metal toxicity.

Table 3. Statistical summary of chemical composition of groundwater in the study area.

Parameters/Unit	Mean	Std. Dev.	Min	Max	PSI	WHO
pH	7.47	0.18	7.1	8.3	6.5–8.5	6.5–8.5
EC μS/cm	829.6	175.1	504	1261	-	-
TDS mg/L	477.4	100.2	292	731	1000	1000
NO_3^- mg/L	40.6	27	0.5	110	50	50
F^- mg/L	0.28	0.12	0	0.7	1.5	1.5
SO_4^{2-} mg/L	25.5	12.6	0	65	200	250
Cl^- mg/L	71.2	39.9	26	249	250	250
Ca^{2+} mg/L	86.8	17.9	53	135	100	75
Mg^{2+} mg/L	32.6	6.6	14.6	54	100	30
K^+ mg/L	4.2	5.6	0.1	27.7	10	-
Na^+ mg/L	39.9	21.1	15.5	131	200	200
TH mg/L	358	61.8	260	540	500	500
HCO_3^- mg/L	312.5	46	165	420	-	-

3.1. The Dominant Water Type

A Piper diagram was built and plotted for all analysed wells within the study area, with the aim of classifying the groundwater types (Figure 4). Table 4 shows the groundwater types. In total, 63.3% of the groundwater wells were classified as (Ca-Mg-HCO_3); 34.7% as (Ca-Mg-HCO_3-Cl), (Ca-Mg-Na-HCO_3-Cl); (Ca-Na-HCO3-Cl); and 2% as (Na-Ca-Mg-Cl-HCO_3) (Na-Ca-Cl-HCO_3). Based on the fundamental geochemical characteristics of the constituent ionic concentrations, a Piper trilinear diagram is a useful visual tool for classifying groundwater, and it elaborates the association and variation among the various types of groundwater [1,22,63]. Furthermore, defining the geochemical characteristics and chemical relationships in the groundwater in the study area in more detail is considered widely helpful. Cation and anion concentration categories can be found in different zones known as hydrochemical facies [64]. The extensive use of fertilizer in the area of recharge for the aquifers of different lithological environments shows a modified groundwater hydrochemical composition, causing it to be notably different from what is expected in normal conditions driven by water–rock interactions and other natural processes [1]. Despite the lithological differences, NO_3 contamination homogenises the overall hydrochemistry, making it difficult to perform the geochemical interpretation required for any evaluation study of groundwater resources [65].

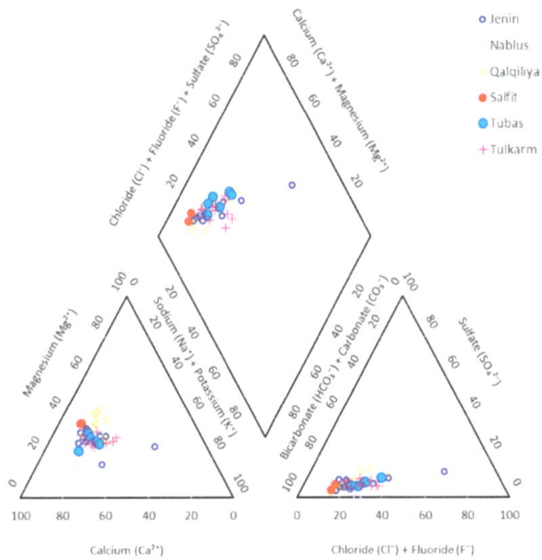

Figure 4. Piper diagram of groundwater sampling area.

Table 4. Categorization of water based on Piper diagram [66,67].

Sl. No.	Water Types	Number and Percentage of Groundwater Source
A	Normal earth alkaline water with prevailing bicarbonate	(31) 63.3%
B	Normal earth alkaline water with prevailing bicarbonate and sulphate or chloride	-
C	Normal earth alkaline water with prevailing sulphate or chloride	-
D	Earth alkaline water with increased portions of alkalis with prevailing bicarbonate	(17) 34.7%
E	Earth alkaline water with increased portions of alkalis with prevailing sulphate and chloride	-
F	Alkaline water with prevailing bicarbonate	-
G	Alkaline water with prevailing sulphate or chloride	(1) 2%

3.2. Water Quality Index in Groundwater

The WQI was classified into five categories, namely excellent, good, poor, very poor, and unsuitable for drinking [47]. Forty-nine groundwater samples and their WQI values are shown in Table 5. The WQI values of the sampled wells ranged from 0.43 to 0.90, of which 77.6% and 22.4% were within the good water and excellent water quality classes, respectively (Table 6). Importantly, all analysed samples were found to be suitable for drinking purposes. Human activities, including the extensive use of fertilizers, wastewater discharge, and cesspit leakage, have affected the groundwater quality of the study area [1,22]. These findings are in line with those of a previous study by Ibrahim and Marei in Jenin in the West Bank [68].

Table 5. Concentration values of physio-chemical analysis of the analysed groundwater samples. WQI assessment results are also provided.

Well_ID	F	NO_3	pH	SO_4	TDS	TH	Cl	Ca	Mg	Na	K	WQI
15–19/017	0.38	68.40	7.45	32.50	558.00	372.00	83.10	87.40	37.33	50.00	19.13	0.82
15–19/018	0.44	72.40	7.36	40.10	626.00	387.00	104.70	90.50	39.16	66.64	24.21	0.90
15–19/046	0.22	63.00	7.54	24.00	545.00	410.00	84.80	106.30	35.10	47.00	11.00	0.76
15–19/006	0.25	71.30	7.27	20.92	543.00	382.80	82.20	93.40	33.95	44.70	10.16	0.75
15–19/001A	0.28	80.00	7.55	29.30	431.00	314.00	55.10	75.64	30.60	30.70	3.34	0.67
15–19/004A	0.29	51.50	7.26	25.90	434.00	373.20	49.70	95.03	33.00	29.70	1.80	0.61
15–20/008	0.40	55.00	7.38	20.50	538.00	411.60	98.88	109.90	33.32	50.36	4.53	0.72
15–19/047A	0.28	32.00	7.60	16.30	400.00	315.80	56.00	79.50	28.50	31.00	0.80	0.51
15–19/028	0.18	43.55	7.69	21.40	320.00	268.10	35.20	66.15	25.00	18.84	0.50	0.48
16–20/005	0.34	26.32	7.35	17.00	387.00	317.60	53.70	85.55	25.25	27.75	1.68	0.50
16–19/002	0.37	40.15	7.24	21.10	455.00	350.50	76.80	94.00	28.12	40.88	2.86	0.60
15–19/010	0.33	48.00	7.61	20.38	510.00	338.60	63.00	91.76	30.21	35.92	0.10	0.61
14–17/043	0.20	59.00	7.31	17.90	460.00	310.00	53.00	72.00	32.00	33.00	1.30	0.59
14–17/044	0.20	65.00	7.32	27.30	466.00	331.00	50.00	80.00	31.00	25.00	0.95	0.62
14–17/001	0.20	110.00	7.50	54.00	556.00	420.00	68.00	90.00	43.00	39.00	2.00	0.84
15–17/012	0.30	45.00	7.57	20.00	443.00	355.00	45.00	71.00	42.00	29.00	2.50	0.59
15–17/004	0.17	63.00	7.70	21.00	499.00	345.00	50.00	65.00	40.00	39.00	6.40	0.66
15–17/007	0.17	31.00	7.40	16.00	415.00	365.00	37.00	90.00	31.00	28.00	1.10	0.52
15–16/003	0.20	51.00	7.50	20.00	415.00	320.00	47.00	65.00	34.50	30.00	3.10	0.57
15–18/004	0.20	70.00	7.55	27.00	415.00	310.00	45.00	57.00	36.00	25.00	4.50	0.62
15–17/019	0.20	19.00	7.50	7.00	370.00	270.00	30.70	53.00	32.00	19.00	3.80	0.43
14–17/021	0.28	60.00	7.30	27.00	496.00	351.00	67.00	85.00	33.80	36.00	1.70	0.64
14–17/034	0.30	82.00	7.40	48.00	658.00	406.00	125.00	97.00	39.00	67.00	3.00	0.82
14–17/052	0.21	91.10	7.50	30.00	550.00	355.20	70.60	83.00	36.00	37.00	27.70	0.89
17–20/050Q	0.70	0.50	7.55	35.00	496.00	670.00	89.00	81.00	35.00	60.00	2.50	0.58
17–19/009	0.42	10.00	7.60	30.00	385.00	335.00	39.00	75.00	31.50	25.00	1.90	0.47
17–20/052J	0.45	3.00	7.30	35.00	478.00	398.00	59.90	95.00	35.00	40.00	3.00	0.53
17–20/033J	0.32	69.00	7.40	52.00	715.00	480.00	160.00	135.00	28.00	100.00	6.00	0.87
17–20/051J	0.37	14.00	7.41	34.00	645.00	499.00	125.00	120.00	46.10	69.00	3.50	0.68
17–20/053	0.31	25.00	7.30	19.00	472.00	395.00	54.00	99.00	30.00	38.00	3.00	0.56
17–21/035	0.17	14.50	7.10	13.00	495.00	449.00	54.00	105.00	39.00	30.00	1.50	0.53
18–20/008	0.21	2.70	8.20	65.00	690.00	295.00	249.00	55.00	25.40	131.00	7.00	0.62
16–19/012	0.35	21.00	7.50	17.00	406.00	349.00	43.00	90.10	25.00	24.00	1.20	0.49
16–20/006	0.20	87.00	7.20	17.00	587.00	481.00	75.00	121.00	38.00	41.00	9.00	0.83

Table 5. *Cont.*

Well_ID	F	NO_3	pH	SO_4	TDS	TH	Cl	Ca	Mg	Na	K	WQI
17–20/051A	0.42	32.00	7.15	23.00	470.00	399.00	55.00	100.00	32.70	35.00	2.20	0.59
18–18/066	0.28	33.90	7.41	23.40	432.00	362.40	59.60	104.20	24.83	32.56	1.83	0.56
18–19/006	0.32	26.06	7.45	39.50	503.00	379.00	98.00	97.60	32.90	52.50	1.68	0.60
18–18/083	0.31	21.22	7.41	19.70	415.00	345.60	64.10	85.55	32.04	30.55	2.02	0.51
Tu_W_001	0.34	23.10	7.71	20.00	393.00	334.70	52.37	81.10	32.10	29.00	3.46	0.52
Tu_W_002	0.35	19.44	7.42	25.70	425.00	333.50	67.73	82.42	31.02	39.79	2.40	0.52
18–18/068	0.31	26.20	7.47	37.90	502.00	381.20	97.52	98.34	33.00	52.71	1.77	0.60
18–18/039	0.25	56.50	7.62	45.00	731.00	540.00	169.00	121.00	54.00	71.00	2.50	0.85
17–17/003	0.00	19.36	7.60	16.20	361.00	264.00	48.00	67.00	24.30	23.50	1.70	0.42
18–18/038	0.17	20.13	7.50	21.00	492.00	324.00	84.00	80.00	30.10	40.50	1.80	0.51
18–18/037	0.16	12.75	7.60	15.00	411.00	280.00	55.00	88.00	14.60	30.70	1.70	0.43
16–18/003!A	0.00	12.87	7.60	0.00	348.00	260.00	45.00	64.00	24.30	20.20	1.60	0.38
16–18/004	0.40	22.74	7.70	8.00	374.00	268.00	52.00	72.00	21.40	23.40	2.10	0.47
Sa_W_001	0.18	9.60	7.70	10.00	292.00	293.00	26.00	65.00	30.00	15.50	1.10	0.39
Sa_W_002	0.55	7.00	7.30	22.00	382.00	350.00	35.50	87.00	38.00	19.00	1.60	0.49

3.3. Human Health Risk Assessment

The calculated values of the non-carcinogenic health risks from ingestion for adults, children, and infants in the study area are presented in Table 6. The HQ_{NO3} values of the samples ranged from 0.014 to 2.99 for adults, with a mean of 1.102; from 0.013 to 2.87 for children, with a mean of 1.057; and from 0.014 to 2.99 for infants, with a mean of 1.102. However, 46.9% of adults, children, and infants are considered to be at high risk. According to the distribution of the hazard quotient HQ_{NO3} values, the western basin of the study area's groundwater is particularly significantly contaminated, with NO_3^- being the result of agricultural practices, cesspit tanks, and insufficiently treated wastewater, raising serious health concerns. In addition to NO_3^-, F^- is another element of concern, which is widely distributed in the Earth's crust. The HQ_F values of the groundwater samples extended from 0 to 0.21 for adults, with a mean of 0.042; from 0 to 0.73 for children, with a mean of 0.297; and from 0 to 0.76 for infants, with a mean of 0.309. HQ_F was found to have the highest values in infants, followed by in children, with the lowest values in adults. However, all values of HQ_F were <1, indicating low or no health risk for F in the analysed groundwater samples.

Table 6. Assessment results of human health risk according to computed HQ and THI of fluoride and nitrate ingestion in the study area.

Well_ID	Adults			Children			Infants		
	HQ_F	HQ_{NO3}	THI	HQ_F	HQ_{NO3}	THI	HQ_F	HQ_{NO3}	THI
15–19/017	0.06	1.86	1.92	0.40	1.78	2.18	0.41	1.86	2.27
15–19/018	0.08	1.97	2.05	0.46	1.89	2.35	0.48	1.97	2.45
15–19/046	0.02	1.71	1.73	0.23	1.64	1.87	0.24	1.71	1.95
15–19/006	0.03	1.94	1.96	0.26	1.86	2.12	0.27	1.94	2.21
15–19/001A	0.03	2.17	2.21	0.29	2.09	2.38	0.30	2.17	2.48
15–19/004A	0.04	1.40	1.44	0.30	1.34	1.64	0.32	1.40	1.71
15–20/008	0.07	1.49	1.56	0.42	1.43	1.85	0.43	1.49	1.93
15–19/047A	0.03	0.87	0.90	0.29	0.83	1.13	0.30	0.87	1.17
15–19/028	0.01	1.18	1.20	0.19	1.14	1.32	0.20	1.18	1.38
16–20/005	0.05	0.72	0.77	0.35	0.69	1.04	0.37	0.72	1.08
16–19/002	0.06	1.09	1.15	0.39	1.05	1.43	0.40	1.09	1.49
15–19/010	0.05	1.30	1.35	0.34	1.25	1.59	0.35	1.30	1.66
14–17/043	0.02	1.60	1.62	0.21	1.54	1.75	0.22	1.60	1.82
14–17/044	0.02	1.77	1.78	0.21	1.69	1.90	0.22	1.77	1.98
14–17/001	0.02	2.99	3.01	0.21	2.87	3.08	0.22	2.99	3.21
15–17/012	0.04	1.22	1.26	0.31	1.17	1.49	0.33	1.22	1.55
15–17/004	0.01	1.71	1.72	0.18	1.64	1.82	0.18	1.71	1.90
15–17/007	0.01	0.84	0.85	0.18	0.81	0.99	0.18	0.84	1.03
15–16/003	0.02	1.39	1.40	0.21	1.33	1.54	0.22	1.39	1.60
15–18/004	0.02	1.90	1.92	0.21	1.82	2.03	0.22	1.90	2.12
15–17/019	0.02	0.52	0.53	0.21	0.50	0.70	0.22	0.52	0.73
14–17/021	0.03	1.63	1.66	0.29	1.56	1.86	0.30	1.63	1.93

Table 6. *Cont.*

Well_ID	Adults			Children			Infants		
	HQ_F	HQ_{NO3}	THI	HQ_F	HQ_{NO3}	THI	HQ_F	HQ_{NO3}	THI
14–17/034	0.04	2.23	2.27	0.31	2.14	2.45	0.33	2.23	2.55
14–17/052	0.02	2.48	2.49	0.22	2.37	2.59	0.23	2.48	2.70
17–20/050Q	0.21	0.01	0.23	0.73	0.01	0.74	0.76	0.01	0.77
17–19/009	0.08	0.27	0.35	0.44	0.26	0.70	0.46	0.27	0.73
17–20/052J	0.09	0.08	0.17	0.47	0.08	0.55	0.49	0.08	0.57
17–20/033J	0.04	1.88	1.92	0.33	1.80	2.13	0.35	1.88	2.22
17–20/051J	0.06	0.38	0.44	0.39	0.36	0.75	0.40	0.38	0.78
17–20/053	0.04	0.68	0.72	0.32	0.65	0.98	0.34	0.68	1.02
17–21/035	0.01	0.39	0.41	0.18	0.38	0.56	0.18	0.39	0.58
18–20/008	0.02	0.07	0.09	0.22	0.07	0.29	0.23	0.07	0.30
16–19/012	0.05	0.57	0.62	0.36	0.55	0.91	0.38	0.57	0.95
16–20/006	0.02	2.36	2.38	0.21	2.27	2.48	0.22	2.36	2.58
17–20/051A	0.08	0.87	0.95	0.44	0.83	1.27	0.46	0.87	1.33
18–18/066	0.03	0.92	0.96	0.29	0.88	1.18	0.30	0.92	1.23
18–19/006	0.04	0.71	0.75	0.33	0.68	1.01	0.35	0.71	1.06
18–18/083	0.04	0.58	0.62	0.32	0.55	0.88	0.34	0.58	0.91
Tu_W_001	0.05	0.63	0.68	0.36	0.60	0.96	0.37	0.63	1.00
Tu_W_002	0.05	0.53	0.58	0.36	0.51	0.87	0.38	0.53	0.91
18–18/068	0.04	0.71	0.75	0.32	0.68	1.01	0.34	0.71	1.05
18–18/039	0.03	1.54	1.56	0.26	1.47	1.73	0.27	1.54	1.81
17–17/003	0.00	0.53	0.53	0.00	0.50	0.50	0.00	0.53	0.53
18–18/038	0.01	0.55	0.56	0.18	0.52	0.70	0.18	0.55	0.73
18–18/037	0.01	0.35	0.36	0.17	0.33	0.50	0.17	0.35	0.52
16–18/003A	0.00	0.35	0.35	0.00	0.34	0.34	0.00	0.35	0.35
16–18/004	0.07	0.62	0.69	0.42	0.59	1.01	0.43	0.62	1.05
Sa_W_001	0.01	0.26	0.27	0.19	0.25	0.44	0.20	0.26	0.46
Sa_W_002	0.13	0.19	0.32	0.57	0.18	0.76	0.60	0.19	0.79

The total hazard (THI) values of the analysed wells varied by the category of age, with the values extending from 0.093 to 3.01 for adults, with a mean of 1.144; from 0.29 to 3.08 for children, with a mean of 1.354; and from 0.302 to 3.21 for infants, with a mean of 1.411 (Table 6). However, the THI results of the groundwater samples for the total non-carcinogenic risk that exceeded the acceptable limit (i.e., THI > 1) were as follows: 46.9% for adults, 61.2% for children, and 67.4% for infants. The order of THI values for the different age categories (i.e., infants > children > adults) was also depicted in the spatial distribution of these values and variations in the health risk of the groundwater wells within the study area (Figure 5a–c). These results confirm that infants are more vulnerable than adults and children to non-carcinogenic risks due to toxicity. This is primarily because they consume more water per unit of body weight than adults and children [69].

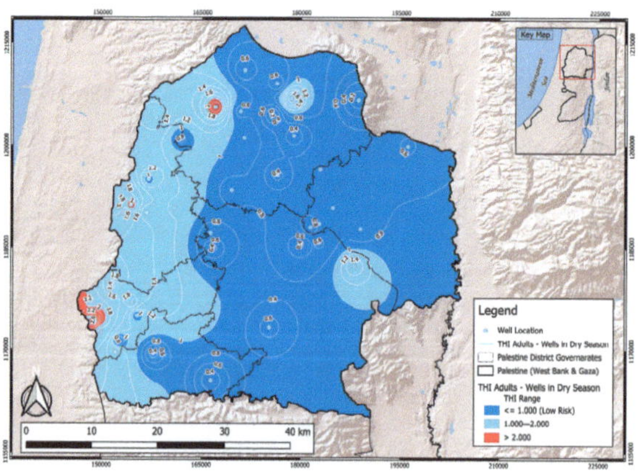

(a)

Figure 5. *Cont.*

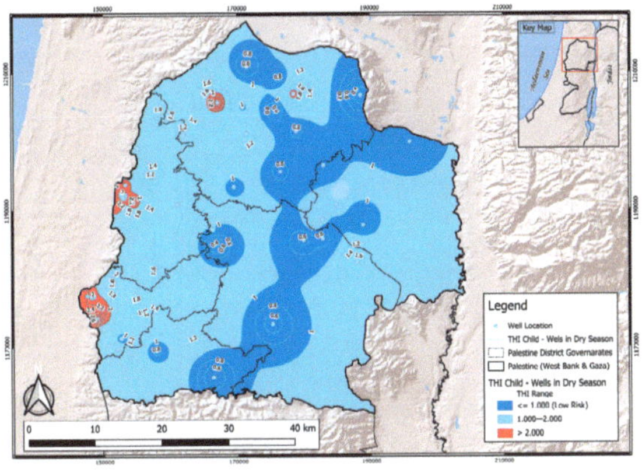

(b)

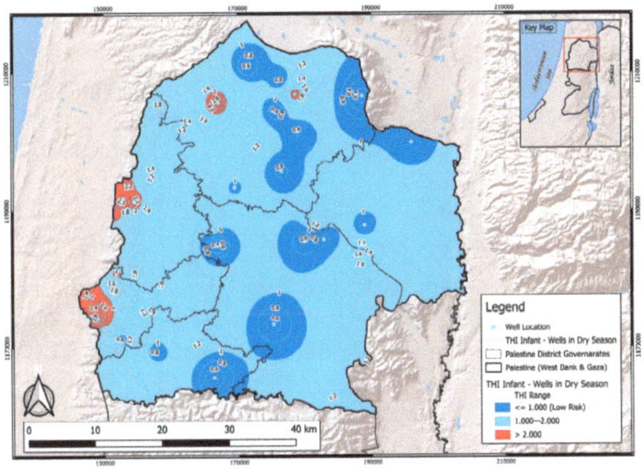

(c)

Figure 5. (**a**) Zoning map of health risk of groundwater for adults. (**b**) Zoning map of health risk of groundwater for children. (**c**) Zoning map of health risk of groundwater for infants.

4. Conclusions

The monitoring of groundwater is crucial, especially in regions grappling with water scarcity. This study utilised three measurement tools (the WQI, HHR, and Piper diagrams) to evaluate groundwater quality. Since each tool covers a specific aspect of water quality, the results offer a broad and comprehensive understanding of the water quality and its impact on the local population.

The main results and conclusions can be summarised as follows:

- The pH values of the groundwater samples were found to vary from 7.1 to 8.3, showing properties that are almost neutral to slightly alkaline. Based on the classification of groundwater according to TDS, 67.3% of the samples fell into the desirable category, while 32.7% were categorised as permissible for drinking. The total hardness values of the groundwater samples ranged from 260 to 540 mg/L; 2% of the groundwater samples exceeded the permissible limit set by the WHO. Moreover, 87.7% were classified as very hard water. The levels of K^+ in the groundwater varied from 0.1 to 27.7 mg/L; notably, 10.2% of the examined samples exceeded the permissible limit set by the WHO. The NO_3^- levels ranged from 0.5 to 110 mg/L, with a mean of 40.6. It is clear that 38.8% of the samples exceeded the permissible limit set by the WHO.

However, the Cl^- and Na^+ concentrations were found to be within the permissible limit set by the WHO.

- The ionic dominance pattern of the groundwater showed the following order for anions: $HCO_3^- > Cl^- > NO_3^- > SO_4^{2-} > F^-$. Moreover, it showed the following order for cations: $Ca^{2+} > Na^+ > Mg^{2+} > K^+$. The predominant water type in this region is fresh water (calcium–magnesium–bicarbonate), which is suitable for drinking. This type can mainly be attributed to the hydrogeological facies of groundwater, determined by natural processes, such as rock weathering, leaching, and evaporation. However, water types such as (Ca-Mg-Na-HCO_3-Cl), (Ca-Na-HCO_3-Cl), (Na-Ca-Mg-Cl-HCO_3), and (Na-Ca-Cl-HCO_3) may be influenced by anthropogenic factors, such as wastewater, cesspits, and agricultural fertilizers. Consequently, these water types may not be suitable for drinking.
- Based on the water quality index (WQI), 77.6% and 22.4% of samples fell into the 'good' and 'excellent' water classes, respectively, indicating that the groundwater is suitable for drinking.
- No health risk was found for fluoride (F^-) in the analysed groundwater samples. The non-carcinogenic risk of NO_3^- ranged from 0.093 to 3.01 for adults, from 0.29 to 3.08 for children, and from 0.302 to 3.21 for infants. Our study indicates that the younger age group is more vulnerable to NO_3^- toxicity than the oldest age group. However, the health risk associated with NO_3^- was 46.9% for all ages. Therefore, more efforts are needed to reduce the nitrate NO_3^- levels in the groundwater of the study area. Moreover, the development of appropriate methods for the protection of groundwater catchments in the West Bank is crucial and could have beneficial implications for groundwater quality and sustainability.

  Recommendation:

Achieving proper groundwater quality cannot be ensured unless effective management of pollution sources is in place. The construction of dams to collect and harvest rainwater and the subsequent injection into the subsurface to replenish groundwater significantly contribute to reducing the concentrations of NO_3^- and other chemicals in the aquifer. Furthermore, mixing groundwater sources with high NO_3^- levels with those that possess low NO_3^- levels has proven to be an efficient strategy for addressing this concern. This method is effective in mitigating the issue by diluting the NO_3^- concentration in the produced groundwater.

Author Contributions: A.Z.: conceptualization, methodology, visualisation, investigation, writing—original draft. L.A. and C.T.G.: supervision, validation, writing—review and editing. All authors have read and agreed to the published version of the manuscript.

Funding: The authors acknowledge the support of a grant, code GUP-2022-065.

Data Availability Statement: The data presented in this study are available on request from the corresponding author.

Conflicts of Interest: The authors declare no conflict of interest.

References

1. Zhang, Q.; Xu, P.; Qian, H. Groundwater Quality Assessment Using Improved Water Quality Index (WQI) and Human Health Risk (HHR) Evaluation in a Semi-arid Region of Northwest China. *Expo. Health* **2020**, *12*, 487–500. [CrossRef]
2. Wu, J.; Wang, L.; Wang, S.; Tian, R.; Xue, C.; Feng, W.; Li, Y. Spatiotemporal variation of groundwater quality in an arid area experiencing long-term paper wastewater irrigation, northwest China. *Environ. Earth Sci.* **2017**, *76*, 460. [CrossRef]
3. Zeidan, B.A. *Groundwater Degradation and Remediation in the Nile Delta Aquifer*; Springer: Berlin/Heidelberg, Germany, 2017; Volume 55, pp. 159–232. [CrossRef]
4. Jonoski, A.; Ahmed, T.; Almasri, M.N.; Abu-Saadah, M. Decision Support System for Sustainable Exploitation of the Eocene Aquifer in the West Bank, Palestine. *Water* **2023**, *15*, 365. [CrossRef]
5. El-Fadel, M.; Quba'a, R.; El-Hougeiri, N.; Hashisho, Z.; Jamali, D. The Israeli Palestinian Mountain Aquifer: A Case Study in Ground Water Conflict Resolution. *J. Nat. Resour. Life Sci. Educ.* **2001**, *30*, 50–61. [CrossRef]

6. Jebreen, H.; Banning, A.; Wohnlich, S.; Niedermayr, A.; Ghanem, M.; Wisotzky, F. The influence of Karst Aquifer mineralogy and geochemistry on groundwater characteristics: West Bank, Palestine. *Water* **2018**, *10*, 1829. [CrossRef]
7. SUSMAQ. A Brief Summary of the Hydrogeology of the West Bank Management. 2002. Available online: http://nora.nerc.ac.uk/id/eprint/504766/1/West_Bank_hg.pdf (accessed on 2 February 2022).
8. Tal-Spiro, O. *Israeli-Palestinian Cooperation on Water Issues*; The Knesset the Research and Information Center, Jerusalem: Jerusalem, Israel, 2011; pp. 1–17. Available online: www.knesset.gov.il/mmm (accessed on 4 March 2022).
9. Adimalla, N. Controlling factors and mechanism of groundwater quality variation in semiarid region of South India: An approach of water quality index (WQI) and health risk assessment (HRA). *Environ. Geochem. Health* **2020**, *42*, 1725–1752. [CrossRef] [PubMed]
10. Kate, S.; Kumbhar, S.; Jamale, P. Water quality analysis of Urun-Islampur City, Maharashtra, India. *Appl. Water Sci.* **2020**, *10*, 1–8. [CrossRef]
11. Hirata, R.; Cagnon, F.; Bernice, A.; Maldaner, C.H.; Galvão, P.; Marques, C.; Terada, R.; Varnier, C.; Ryan, M.C.; Bertolo, R. Nitrate contamination in brazilian urban aquifers: A tenacious problem. *Water* **2020**, *12*, 2709. [CrossRef]
12. Karakuş, C.B. Evaluation of groundwater quality in Sivas province (Turkey) using water quality index and GIS-based analytic hierarchy process. *Int. J. Environ. Health Res.* **2019**, *29*, 500–519. [CrossRef]
13. Wen, X.; Chen, F.; Lin, Y.; Zhu, H.; Yuan, F.; Kuang, D.; Jia, Z.; Yuan, Z. Microbial indicators and their use for monitoring drinkingwater quality-A review. *Sustainability* **2020**, *12*, 2249. [CrossRef]
14. Rajankar, P.N.; Tambekar, D.H.; Wate, S.R. Groundwater quality and water quality index at Bhandara District. *Environ. Monit. Assess.* **2011**, *179*, 619–625. [CrossRef]
15. Pei-Yue, L.; Hui, Q.; Jian-Hua, W. Groundwater quality assessment based on improved water quality index in Pengyang County, Ningxia, Northwest China. *E-J. Chem.* **2010**, *7* (Suppl. S1), S209–S216. [CrossRef]
16. Ketata, M.; Gueddari, M.; Bouhlila, R. Use of geographical information system and water quality index to assess groundwater quality in el khairat deep aquifer (Enfidha, Central East Tunisia). *Arab. J. Geosci.* **2012**, *5*, 1379–1390. [CrossRef]
17. Herschy, R.W. Water quality for drinking: WHO guidelines. In *Encyclopedia of Lakes and Reservoirs*; WHO: Geneva, Switzerland, 2011; pp. 876–883. [CrossRef]
18. Zohud, A.; Alam, L. A Review of Groundwater Contamination in West Bank, Palestine: Quality, Sources, Risks, and Management. *Water* **2022**, *14*, 3417. [CrossRef]
19. WHO. *Guidelines for Drinking-Water Quality*; WHO: Geneva, Switzerland, 2017.
20. Gikas, P.; Manali, A.; Manaroli, X.E.; Makaroglou, G.; Tsamoutsoglou, K. Sustainable Chemical and Environmental Engineering. In Proceedings of the 1st International Conference on Sustainable Chemical and Environmental Engineering, Crete, Greece, 31 August–4 September 2022.
21. Jha, S.K.; Singh, R.K.; Damodaran, T.; Mishra, V.K.; Sharma, D.K.; Rai, D. Fluoride in groundwater: Toxicological exposure and remedies. *J. Toxicol. Environ. Health Part B* **2013**, *16*, 52–66. [CrossRef] [PubMed]
22. Mahmoud, N.; Zayed, O.; Petrusevski, B. Groundwater Quality of Drinking Water Wells in the West Bank, Palestine. *Water* **2022**, *14*, 377. [CrossRef]
23. Abu Jabal, M.S.; Abustan, I.; Rozaimy, M.R.; Al-Najar, H. Fluoride enrichment in groundwater of semi-arid urban area: Khan Younis City, southern Gaza Strip (Palestine). *J. Afr. Earth Sci.* **2014**, *100*, 259–266. [CrossRef]
24. Adimalla, N. Groundwater Quality for Drinking and Irrigation Purposes and Potential Health Risks Assessment: A Case Study from Semi-Arid Region of South India. *Expo. Health* **2018**, *11*, 109–123. [CrossRef]
25. Ghanem, M.; Ahmad, W.; Keilani, Y.; Sawaftah, F.; Schelter, L.; Schuettrumpf, H. Spring water quality in the central West Bank, Palestine. *J. Asian Earth Sci. X* **2021**, *5*, 100052. [CrossRef]
26. Cotruvo, J.A. 2017 WHO guidelines for drinking water quality: First addendum to the fourth edition. *J. Am. Water Work. Assoc.* **2017**, *109*, 44–51. [CrossRef]
27. Hejaz, B.; Al-Khatib, I.A.; Mahmoud, N. Nidal Domestic Groundwater Quality in the Northern Governorates of the West Bank, Palestine. *J. Environ. Public Health* **2020**, *2020*, 6894805. [CrossRef] [PubMed]
28. Daghara, A.; Al-Khatib, I.A.; Al-Jabari, M. Quality of Drinking Water from Springs in Palestine: West Bank as a Case Study. *J. Environ. Public Health* **2019**, *2019*, 8631732. [CrossRef]
29. Todd, D.K. *Groundwater Hydrology*, 2nd ed.; Wiley: Hoboken, NJ, USA, 1980.
30. Arvin, H.S.E.; Bardow, A. Caries affected by calcium and fluoride in drinking water and family income. *Water Health* **2018**, *16*, 49–56. [CrossRef]
31. World Health Organization (WHO). *Potassium in Drinking Water. Background Document for Preparation of WHO Guidelines for Drinking-Water Quality*, 4th ed.; World Health Organization: Geneva, Switzerland, 2009; pp. 1–5. Available online: https://apps.who.int/iris/bitstream/handle/10665/70171/WHO_HSE_WSH_09.01_7_eng.pdf?sequence=1&isAllowed=y (accessed on 8 February 2023).
32. Abbasi, T.; Abbasi, S. *Water Quality Indices*; Elsevier: Amsterdam, The Netherlands, 2012. [CrossRef]
33. Jayarajan, S.K.P.; Kuriachan, L. Exposure and health risk assessment of nitrate contamination in groundwater in Coimbatore and Tirupur districts in Tamil Nadu, South India. *Environ. Sci. Pollut. Res.* **2020**, *28*, 10248–10261. [CrossRef] [PubMed]
34. PWA. *Palestinian Water Authority Annual Status Report on Water Resources, Water Supply, and Wastewater in the Occupied State of Palestine*; PWA: Al-Bireh, Palestine, 2012; pp. 1–94.

35. PCBS. *Palestinians at the End of 2020*; PCBS: Washington, DC, USA, 2020; pp. 1–77. Available online: http://www.pcbs.gov.ps (accessed on 22 October 2022).
36. PWA. *Status Report of Water Resources in the Occupied State of Palestine-2012*; PWA: Al-Bireh, Palestine, 2013; p. 22. Available online: http://www.pwa.ps/userfiles/file/1/WRSTATUSReport-finaldraft2014-04-01.pdf (accessed on 5 June 2023).
37. Abed, A.; Wishahi, S. *Geology of Palestine: The West Bank and Gaza Strip*; Palestinian Hydrology Group: Jerusalem, Israel, 1999.
38. Jebreen, H.; Wohnlich, S.; Banning, A.; Wisotzky, F.; Niedermayr, A.; Ghanem, M. Recharge, geochemical processes and water quality in karst aquifers: Central West Bank, Palestine. *Environ. Earth Sci.* **2018**, *77*, 261. [CrossRef]
39. Sayato, Y. WHO Guidelines for Drinking-Water Quality. *Eisei kagaku* **1989**, *35*, 307–312. [CrossRef]
40. WHO. *Operational Guidance for Maintaining Essential Health Services During an Outbreak*; WHO: Geneva, Switzerland, 2020; pp. 1–10. Available online: https://www.who.int/publications-detail/covid-19-operational-guidance-for-maintaining-essential-health-services-during-an-outbreak (accessed on 2 May 2023).
41. Angela, A. Water Quality Monitoring Protocols for Streams and Rivers. *N. Engl. J. Med.* **2018**, *372*, 2499–2508.
42. APHA (American Public Health Association); American Water Works Association; Water Environment Federation. *Standard Methods for the Examination of Water and Wastewater*; APHA: Washington, DC, USA, 1995; pp. 3–104. Available online: https://www.standardmethods.org/ (accessed on 5 August 2023).
43. CFR. *Guidelines Establishing Test Procedures for the Analysis of Pollutants*; CFR: New York, NY, USA, 1996.
44. APHA. *Standard Methods for the Examination of Water and Wastewater*, 19th ed.; APHA: New York, NY, USA, 1995.
45. Ahamad, A.; Raju, N.J.; Madhav, S.; Gossel, W.; Wycisk, P. Impact of non-engineered Bhalswa landfill on groundwater from Quaternary alluvium in Yamuna flood plain and potential human health risk, New Delhi, India. *Quat. Int.* **2019**, *507*, 352–369. [CrossRef]
46. Palestine Standards Institute (PSI). *The Second Working Draft ofthe Amended Drinking Water Standard*; PSI: Ramallah, Jerusalem, 2004.
47. El Baba, F.D.S.M.; Kayastha, P.; Huysmans, M. Evaluation of the Groundwater Quality Using the Water Quality Index and Geostatistical Analysis in the Dier al-Balah Governorate, Gaza Strip, Palestine. *Water* **2020**, *12*, 262. [CrossRef]
48. USEPA. *Environmental Risk. Your Guide to Analysis and Reducing Risk Rush*; USEPA: Washington, DC, USA, 1991.
49. U.S. Environmental Protection Agency. *Quick Guide to Drinking Water Sample Collection*, 2nd ed.; U.S. Environmental Protection Agency: Washington, DC, USA, 2016; pp. 6–20.
50. USEPA. *Risk Assessment Guidance*; USEPA: Washington, DC, USA, 2014.
51. U.S. Environmental Protection Agency. *Child-Specific Exposure Factors Handbook*; U.S. Environmental Protection Agency: Washington, DC, USA; p. 448.
52. USEPA. *Risk Assessment Guidance for Superfund, Human Health Evaluation Manual*; USEPA: Washington, DC, USA, 2009.
53. Wegahita, N.K.; Ma, L.; Liu, J.; Huang, T.; Luo, Q.; Qian, J. Spatial assessment of groundwater quality and health risk of nitrogen pollution for shallow groundwater aquifer around fuyang city, china. *Water* **2020**, *12*, 3341. [CrossRef]
54. Rao, N.S.; Rao, P.S.; Reddy, G.V.; Nagamani, M.; Vidyasagar, G.; Satyanarayana, N.L.V.V. Chemical characteristics of groundwater and assessment of groundwater quality in Varaha River Basin, Visakhapatnam District, Andhra Pradesh, India. *Environ. Monit. Assess.* **2011**, *184*, 5189–5214. [CrossRef] [PubMed]
55. Ji, Y.; Wu, J.; Wang, Y.; Elumalai, V.; Subramani, T. Seasonal Variation of Drinking Water Quality and Human Health Risk Assessment in Hancheng City of Guanzhong Plain, China. *Expo. Health* **2020**, *12*, 469–485. [CrossRef]
56. Judeh, T.; Bian, H.; Shahrour, I. GIS-based spatiotemporal mapping of groundwater potability and palatability indices in arid and semi-arid areas. *Water* **2021**, *13*, 1323. [CrossRef]
57. Ameen, H.A. Spring water quality assessment using water quality index in villages of Barwari Bala, Duhok, Kurdistan Region, Iraq. *Appl. Water Sci.* **2019**, *9*, 176. [CrossRef]
58. Prakash, K.L.; Somashekar, R.K. Groundwater quality—Assessment on Anekal Taluk, Bangalore Urban district, India. *J. Environ. Biol.* **2006**, *27*, 633–637. [PubMed]
59. Singh, S.; Raju, N.J.; Ramakrishna, C. Evaluation of Groundwater Quality and Its Suitability for Domestic and Irrigation Use in Parts of the Chandauli-Varanasi Region, Uttar Pradesh, India. *J. Water Resour. Prot.* **2015**, *7*, 572–587. [CrossRef]
60. Hem, J.D. *Study and Interpretation of the Chemical Characteristics of Natural Water*; USGS: Reston, VA, USA, 1985; Volume 2254.
61. Saha, S.; Reza, A.H.M.S.; Roy, M.K. Hydrochemical evaluation of groundwater quality of the Tista floodplain, Rangpur, Bangladesh. *Appl. Water Sci.* **2019**, *9*, 198. [CrossRef]
62. Smith, D.N.I.; Ortega-Camacho, D.; Acosta-González, G.; Leal-Bautista, R.M.; Fox, W.E.; Cejudo, E. A multi-approach assessment of land use effects on groundwater quality in a karstic aquifer. *Heliyon* **2020**, *6*, e03970. [CrossRef]
63. Piper, A.M. A graphic procedure in the geochemical interpretation of water-analyses. *Am. Geophys. Union* **1944**, *25*, 914–928. [CrossRef]
64. Agarwal, M.; Singh, M.; Hussain, J. Assessment of groundwater quality with special emphasis on nitrate contamination in parts of Gautam Budh Nagar district, Uttar Pradesh, India. *Acta Geochim.* **2019**, *38*, 703–717. [CrossRef]
65. Menció, A.; Mas-Pla, J.; Otero, N.; Regàs, O.; Boy-Roura, M.; Puig, R.; Bach, J.; Domènech, C.; Zamorano, M.; Brusi, D.; et al. Nitrate pollution of groundwater; all right, but nothing else? *Sci. Total Environ.* **2016**, *539*, 241–251. [CrossRef]
66. Langguth, H.R. *Groundwater Characteristics in Bereiech Des Velberter Sattles*; NRW: Duesseldorf, Germany, 1966; p. 127.
67. Ravikumar, P.; Somashekar, R.; Prakash, K. A comparative study on usage of Durov and Piper diagrams to interpret hydrochemical processes in groundwater from SRLIS river basin, Karnataka, India. *Earth Sci.* **2015**, *80*, 31073–31077.

68. Ibrahim, L.; Marei, A. Water Quality Index (WQI) for Water Resources in Jenin District. Master's Thesis, Al-Quds University-Palestine, Abu Dis, Palestine, 2019; pp. 1–52.
69. Li, P.; He, X.; Guo, W. Spatial groundwater quality and potential health risks due to nitrate ingestion through drinking water: A case study in Yan'an City on the Loess Plateau of northwest China. *Hum. Ecol. Risk Assess. Int. J.* **2019**, *25*, 11–31. [CrossRef]

hydrology

MDPI

Article

Analysis of the Distance between the Measured and Assumed Location of a Point Source of Pollution in Groundwater as a Function of the Variance of the Estimation Error

Ivan Kovač [1], Marko Šrajbek [2,*], Nikolina Klišanin [1] and Gordon Gilja [2]

1 Faculty of Geotechnical Engineering, University of Zagreb, Hallerova aleja 7, 42000 Varaždin, Croatia; ivan.kovac@gfv.unizg.hr (I.K.); nikolina.k11@gmail.com (N.K.)
2 Department of Hydroscience and Engineering, Faculty of Civil Engineering, University of Zagreb, Fra Andrije Kačića-Miošića 26, 10000 Zagreb, Croatia; gordon.gilja@grad.unizg.hr
* Correspondence: marko.srajbek@grad.unizg.hr

Abstract: The localization of pollution sources is one of the main tasks in environmental engineering. For this paper, models of spatial distribution of nitrate concentration in groundwater were created, and the point of highest concentration was determined. This point represents the assumed location of the pollution source and differs from the actual location, so there is a certain distance between the measured and assumed location. This paper puts forward a new hypothesis that the distance between the measured and the assumed location is a function of the variance of the estimation error. The scientific contribution of this paper is based on the fact that the interaction of statistical and geostatistical methods can locate the dominant point source of pollution or narrow down the search area. The above hypothesis is confirmed by the example of the Varaždin wellfield, which was closed due to an excessively high groundwater nitrate concentration. Seven different interpolation methods were used to create spatial distribution models. Each method provides a different model, a different variance of the estimation error, and estimates of the location of the pollution source. The smallest value of variance of the estimation error of 1.65 was obtained for the minimum curvature interpolation method and the largest value of variance (24.49) was obtained for the kriging with logarithmic variogram. Our results show a nonlinear and monotonic relationship between the distance and the variance of the estimation error, so logarithmic and rational quadratic models were fitted to the scatter point data. The models were linearized, a t-test was performed, and the results show that the models can be considered reliable, which is confirmed by the values of the coefficients of determination of the linearized models, which are around 0.50. The obtained results can be used in planning additional research work to determine the measured location of the pollution source. The research methodology we used is universal and can be applied to other locations where high concentrations of certain contaminants have been detected in groundwater in alluvial aquifers.

Keywords: groundwater; nitrate; point pollution source; interpolation; distance; variance; estimation error

Citation: Kovač, I.; Šrajbek, M.; Klišanin, N.; Gilja, G. Analysis of the Distance between the Measured and Assumed Location of a Point Source of Pollution in Groundwater as a Function of the Variance of the Estimation Error. *Hydrology* **2023**, *10*, 199. https://doi.org/10.3390/hydrology10100199

Academic Editors: Roohollah Noori, Augustina Clara Alexander and Pantelis Sidiropoulos

Received: 1 August 2023
Revised: 25 September 2023
Accepted: 5 October 2023
Published: 9 October 2023

1. Introduction

Groundwater is the most abundant source of readily available freshwater in the world, excluding glaciers and polar ice caps, accounting for 97% of all freshwater [1]. For this reason, groundwater is now the most important source of water for various human needs. In the countries of the European Union, groundwater is estimated to meet 55% of domestic water needs [2], and when it comes to drinking water, the percentage increases to 75% [3]. With improvements in quality of life, high quality water is one of the necessary prerequisites for sustainable development [4], but in recent decades, there has been an increased emission of pollutants into the environment due to the increasingly rapid development of industry and agricultural production [5,6]. In this regard, groundwater is no exception, and its degradation is almost inevitable [7–9]. Global water demand has also been steadily increasing over the last three decades, and this trend is expected to continue [10]. In view of this, it is expected that new wellfields will be developed or that the pumping capacity of existing wells will be increased. As recent research has shown, the quality of groundwater can deteriorate when the pumping rate is increased [11].

Until recently, it was believed that the aquifer top deposits would provide sufficient protection to groundwater, but scientific studies and monitoring have shown that pollutants can easily penetrate the unsaturated zone and enter groundwater [12,13]. For this reason, groundwater quality and its changes in space and time have become the subject of numerous scientific studies. The first studies in this regard have shown that an increase in nitrate concentration is positively correlated with an increase in agricultural production on a global scale [14]. Recent studies also show that human activities are the main source of pollution of groundwater with nitrates [15]. Specifically, in order to increase agricultural land and improve conditions for food production, various melioration measures are applied. Studies show that the use of mineral fertilizers in agricultural production increases the nitrate concentration in groundwater [16–19]. For ease of access, agricultural lands are mostly located in the lowland catchment, which is a sensitive ecosystem with slow flow and a high groundwater table [20–22]. Due to these hydrogeological conditions, pollutants can easily penetrate from the surface into the subsurface, where they can persist for a long period of time. This leads to a violation of the natural balance between water and nutrients [23], and a significant correlation between agricultural development and groundwater quality degradation has been found [24–28].

An increased concentration of nitrates in drinking water negatively affects human health [29,30], and the toxicity of nitrates is a consequence of reducing nitrogen to nitrites [31]. Moreover, a high concentration of nitrites in the human body causes vasodilatory and cardiovascular disorders, while even a low concentration of nitrites can lead to methemoglobinemia, a disease which children are especially at risk of [32].

In the total concentration of nitrates in groundwater, one part refers to the natural contribution, the so-called nitrate background level (nbl), while the other part is related to the anthropogenic contribution. In one of the first studies in this field [33], indicative limits for nitrate concentrations in groundwater were presented, and they can be used to determine anthropogenic influence. The results of this particular study show that nitrate concentrations below 1 mg/L NO_3^- can be attributed to the natural nitrates of the aquifer, concentrations in the range of 1–13 mg/L NO_3^- can be characterized as a transitional range, i.e., concentrations that may or may not indicate an anthropogenic influence, and nitrate concentrations above 13 mg/L NO_3^- prove an anthropogenic influence. More recent studies also support these findings. For example, the value of nbl in the area of Rajasthan in India was estimated to be 7.2 mg/L [34], and in the Bono East regions in Ghana, nbl can be expected to be 3.9 mg/L [35]. For the area considered in this paper (a location in Northern Croatia), the Lepeltier method was used to determine nbl values of 8.6 + 2MAD (median absolute deviation) and 8.7 + 2SD (standard deviation) mg/L; thus, the upper limits of the assessment were estimated to be 18.2 mg/L and 24.3 mg/L, respectively [36].

Anthropogenic pollution sources can be divided into point and non-point sources in terms of their dispersion. Non-point sources of pollution represent dispersed sources on the surface. The most common examples of this type of pollution are agricultural lands treated with mineral fertilizers, and the most common type are nitrates, which are very persistent and mobile in groundwater. Although the concentrations in groundwater caused by this type of pollution can be relatively low, such sources can affect a large volume of groundwater due to their dispersion [37]. Point sources of pollution, on the other hand, refer to a localized area from which pollution is emitted. The concentration of pollution near point sources is usually high, but this type of pollution affects a smaller volume of groundwater. As the distance from the point source of pollution increases, the concentration of pollution in space decreases relatively rapidly [37]. Numerous studies have aimed to quantify the sources of pollution. In Koprivnica-Križevci County in Croatia, Nemcic Jurec et al. [38] show the impact of agricultural production as a point and non-point source of nitrate pollution in groundwater. Three different sites were investigated for possible pollution. At one site, where no pollution was expected and which also served as a control group, a very low nitrate concentration of 4.6 mg/L NO_3^- was determined. At the other two sites, an area of intensive agricultural production and a suburban area, the nitrate concentrations were found to be statistically significantly higher, with values of 28.7 and 26.5 mg/L NO_3^-. As point sources of nitrate pollution in groundwater, some studies show increased nitrate concentrations in groundwater caused by poultry and livestock farming [39–41].

Interpolation methods are common techniques used in groundwater resource engineering and management. They can be used very effectively to determine groundwater levels, temperature variations, and the spatial distribution of certain types of pollution. There are several different methods, and the most commonly used are inverse distance weighting (idw), kriging, radial basis function, minimum curvature, and polynomial regression. The use of different interpolation methods can lead to different estimates regarding the parameter values at the interpolation points and, consequently, to the production of different maps presenting changes in the values of the analyzed parameters [42].

Kriging and idw methods were successfully used in the mapping of groundwater tables in the Erbil Basin in Iraq [43]. The authors processed data collected in 2004 and 2022 and concluded that the groundwater level dropped suddenly during this 18-year period. In a very recent study, a spatial interpolation method was used to detect the spatiotemporal dynamic patterns of groundwater level and salinity in the Yellow River delta in China [44]. The results showed that the risk of seawater intrusion into the local shallow aquifers in this region is high, as the mean groundwater level is lower than the local sea level in 42% of the region.

In a very recent study, interpolation methods were successfully used to determine the spatial distribution of temperature in shallow aquifers in Poland in the city of Wroclaw [45]. It was found that the spatial distribution of groundwater temperature for the analyzed period is characterized by higher temperatures in the city center. The best results were obtained using the Ordinary Kriging method. In other studies, interpolation methods based on point measurement data were also used to obtain groundwater temperatures for the analyzed areas [46,47], with particular emphasis on the kriging method [48,49].

As mentioned earlier, interpolation methods are used to map the distribution of pollution in groundwater. The inverse distance weighting method was used for the interpolation of heavy metals in groundwater in the Dehradun district in India [50], while the kriging method was used for the spatial interpolation of the water quality index in the Khyber Pakhtunkhwa province in Pakistan [51].

In this paper, statistical and geostatistical methods are presented to determine point source nitrate pollution in groundwater. Although numerical methods are being increasingly used nowadays, the advantage of the approach presented in this work is the simplicity and speed of the procedure. The development of a reliable numerical model requires detailed field research, which is often not possible due to time or cost constraints. For this reason, statistical and geostatistical methods remain an important tool in groundwater engineering and management. The location that is the focus of this article covers the narrow area of the Varaždin wellfield in the northern part of Croatia. This location was chosen because of the long-standing excessively high nitrate concentration in the groundwater, which is why the wellfield has been closed. Today, groundwater at this site is only used when other wellfields cannot meet the drinking water needs of the Varaždin water supply system. Recent investigations have shown that the majority of the pollution present in the area is attributable to the location being used as an illegal dumpsite for waste derived from poultry farms [41].

The very fact that the location of the dominant point pollution source is known provides an opportunity to test the hypothesis put forward in this work. This paper presents a novel hypothesis: the distance between the measured and assumed location (which can be determined via the use of interpolation methods) of the dominant point source of pollution is a function of the variance of the estimation error. Therefore, the scientific contribution of this paper is based on the fact that the dominant point source of pollution can be located and the fact that the search area can be narrowed down through the interaction of statistical and geostatistical methods. The results obtained have a very practical use, since the search area is narrowed down, which speeds up the search and reduces the associated costs. The method presented in this paper can be applied to other wellfields in alluvial aquifers where high concentrations of certain contaminants have been detected.

2. Materials and Methods

2.1. The Study Area

The study area is located in the north of the Republic of Croatia and includes the area of the Varaždin wellfield (Figure 1), which is located near the town of Varaždin. The Varaždin region is one of the most densely populated and economically developed areas in the Republic of Croatia. Groundwater for the Varaždin regional water supply system is pumped into the Varaždin, Bartolovec and Vinokovščak wellfields from the same alluvial aquifer, which is built of quaternary sand and gravel.

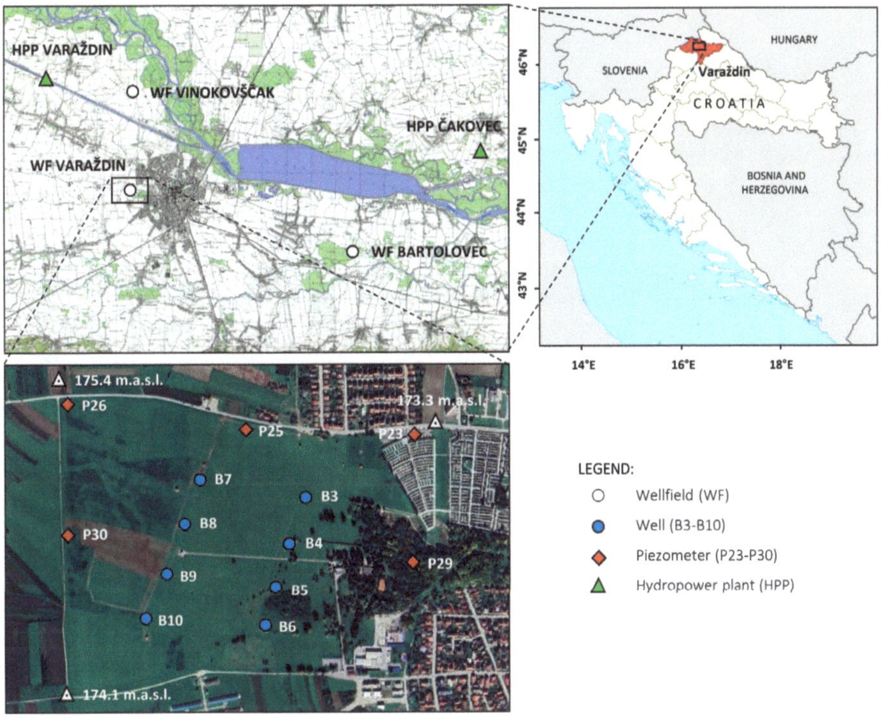

Figure 1. Geographical position of the Varaždin wellfield and spatial arrangement of wells and piezometers.

The Bartolovec and Vinokovščak wellfields are of good quality in terms of nitrate concentration, while the same does not apply for the Varaždin wellfield [52]. Therefore, in the Varaždin regional water supply network, the nitrate concentration was reduced by increasing the pumping rates in the Bartolovec and Vinokovščak wellfields [53]. Due to the high nitrate concentration, the Varaždin wellfield has not been active since 2004. For these reasons, the already existing problem in the water supply will soon become one of the main prerequisites for the sustainable development of the region and improvements to the quality of life of the region's inhabitants.

The research scope of this work covers the period from 1993 to 2003 (until the closure of the Varaždin wellfield in 2004). Groundwater at the wellfield was regularly sampled from eight wells (B3–B10) and five piezometers (P23–P30) (Figure 1) from the upper aquifer layer during the observation period, so all results refer to this layer. A total of 3051 data points were collected, allowing for a reliable statistical analysis. Annual average values were calculated for each sampling point (well and piezometer), and these data were used in subsequent analyses (Table 1).

Table 1. Numerical means regarding nitrate concentration at control points and deviation values obtained via the use of interpolation methods.

	B3	B4	B5	B6	B7	B8	B9	B10	P23	P25	P26	P29	P30	$\sum\Delta^2$	$\overline{\Delta^2}$
Average [mg/L NO_3^-]	74.14	75.35	76.7	80.04	76.71	81.42	85.92	87.80	65.03	71.67	77.4	68.62	87.00		
Kriging–linear	0.06	0.01	−0.05	−0.02	−0.02	0.02	0.10	0.02	−0.03	−0.05	−0.06	−0.04	0.04	0.03	0.00
Kriging–linear nugget	0.06	0.01	−0.05	−0.02	−0.02	0.03	0.10	0.02	−0.03	−0.05	−0.06	−0.04	0.03	0.03	0.00
Kriging–power	0.12	0.03	−0.09	−0.01	−0.05	0.05	0.21	0.08	−0.12	−0.12	−0.14	−0.11	0.11	0.15	0.01
Kriging–power nugget	0.12	0.03	−0.10	−0.03	−0.04	0.06	0.20	0.08	−0.12	−0.11	−0.13	−0.11	0.10	0.14	0.01
Kriging–logarithmic	−0.07	−0.18	−0.28	0.61	−0.40	0.69	2.16	1.43	−2.70	−1.29	−0.56	−1.99	1.90	24.68	1.90
Kriging–logarithmic nugget	−0.02	−0.14	−0.25	0.52	−0.34	0.56	1.86	0.98	−2.20	−1.06	−0.46	−1.68	1.51	16.53	1.27
Minimum curvature	−0.01	−0.10	−0.10	0.06	0.03	−0.04	0.01	0.00	−0.25	0.03	0.08	0.24	0.01	0.15	0.01
Polynomial regression	1.72	−0.24	−1.84	−1.56	−0.33	1.14	2.52	1.25	1.67	−0.21	−2.52	−1.87	0.27	30.92	2.38
Radial basis function	0.00	0.00	−0.01	0.00	0.00	0.00	0.01	0.00	0.00	0.00	0.00	0.00	0.00	0.00	0.00
Inverse distance to a power	−5.50	−7.79	−7.90	−2.87	−0.95	−0.77	1.05	4.46	−9.21	−1.78	4.81	−16.25	3.65	572.63	44.05

The geological structures of the aquifer consist of coarse-grained sediments of gravel and sand, while the underlying sediments consist of silt and clay. The thickness of the aquifer increases from west to east, and at the Varaždin wellfield it is about 75 m (Figure 2). The aquifer consists of two permeable gravel and sand layers, most of which are separated by a semipermeable aquitard of silt and clay [54]. The upper layer of the aquifer is composed of coarse-grained gravel and sand, while the lower part is mainly composed of fine-grained gravel with sand. The groundwater flows in a west–east direction, and it is parallel to the river Drava [55]. As previous investigations have shown, the quality of groundwater in the lower layer of the aquifer in the Bartolovec wellfield is significantly higher [56]. The values of hydraulic conductivity at the Varaždin wellfield were measured to be in the range of 185–241 m/day for the upper layer and 92.6 m/day for the lower layer [55].

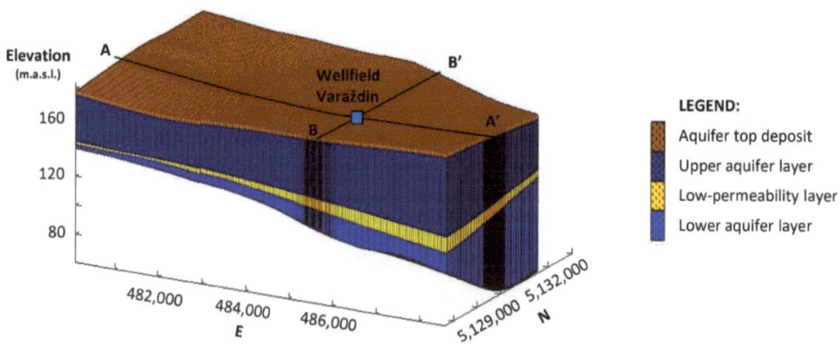

Figure 2. Three-dimensional model of the wider area of the Varaždin wellfield [41].

2.2. Previous Research

For many years, the nitrate concentration in the Varaždin wellfield was higher than the maximum allowed concentration (MAC), according to the Ordinance on conformity parameters and analysis methods, monitoring and water safety plans for human consumption and according to the EU Nitrate Directive [57]. At the same time, the nitrate concentration in the groundwater at the other wellfields in the wider area of the town of Varaždin was much lower [56], which is adequately shown by the time series of the average annual nitrate concentrations of the wellfields (Figure 3). The curves represent the average annual nitrate concentration of all wells in the upper aquifer in each wellfield.

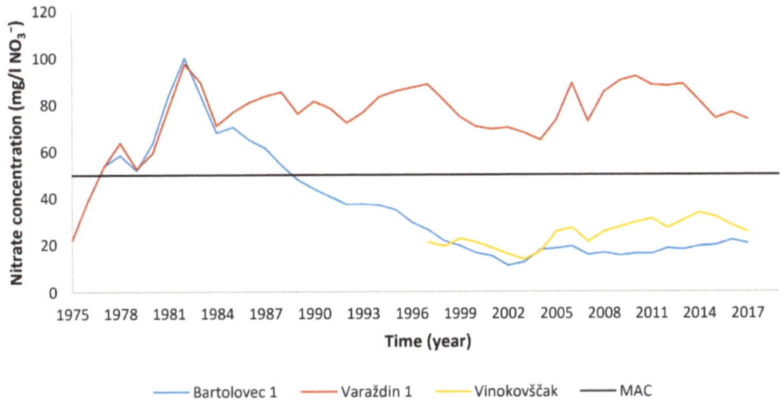

Figure 3. Time series of the annual average nitrate concentrations in the wellfields in Bartolovec, Varaždin, and Vinokovščak.

Several works have been written about the cause of such a high nitrate concentration in the Varaždin wellfield. The first works on this topic highlighted the application of mineral fertilizers as the main cause of such a high nitrate concentration in the groundwater [58]. Although this source of pollution contributes significantly to the total concentration of nitrates, in later works, through the use of numerical modeling methods, it was proved that the main cause of this problem is the illegal disposal of waste material from nearby poultry farms [39,41]. This source of pollution was classified as a point source of pollution.

As can be seen from Figure 3, the nitrate concentration varies significantly in terms of time and space. The construction of the Čakovec and Varaždin hydroelectric power plants (HPP) affected the quality of the groundwater. The influence of these power plants on the nitrate concentration was most evident. After the Varaždin HPP was put into operation in 1975, the groundwater level raised upstream. As a result, the nitrates deposited in the soil were washed away, which contributed to the increase in nitrate concentration in the groundwater [59]. The nitrate concentration in the groundwater increased steadily until 1982, when the reservoir of the Čakovec HPP was filled. The annual average curves at the Varaždin and Bartolovec wellfields then reached their maximum (Figure 3). Thereafter, the nitrate concentration at the wellfields decreased, and a constant decrease was recorded at the Bartolovec wellfield. The initial influence of HPPs gradually decreased, and today, it has completely disappeared [56].

However, as can be seen in Figure 3, the downward trend in the Varaždin wellfield was relatively short lived. The reason for this is the appearance of a new source of groundwater pollution, which had a dominant influence on nitrate concentrations. Since 1984, the average annual nitrate concentrations in the Varaždin wellfield have been in the range of 60 mg/L NO_3^- to 95 mg/L NO_3^-, which is higher than MAC [56,60].

For this reason, the nitrate concentration at all of the groundwater sampling points in the wellfield (wells B3–B10 and piezometers P23–P30) was analyzed, and it was found that there are significant differences in nitrate concentration in the wells in the Varaždin wellfield: in some wells, the concentration is always higher (e.g., well B10), while in others (e.g., B3), it is always lower than the average value of the whole wellfield (Figure 4) [39]. This relationship was observed in all wells throughout the observation period. Since the nitrate concentration in the wells has such a regular relationship, we felt that our use of average nitrate concentrations in subsequent analyses was justified.

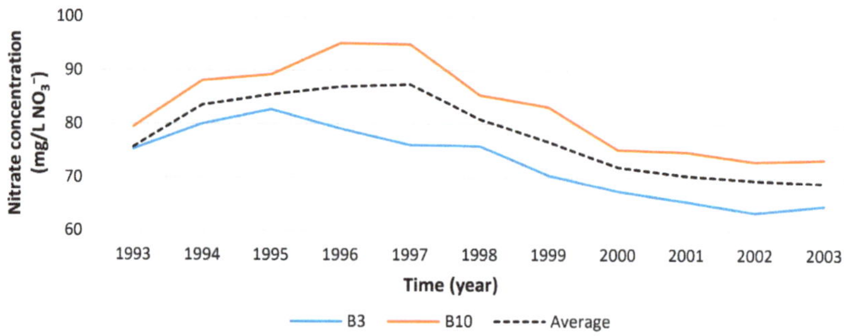

Figure 4. Annual average nitrate concentration curves in groundwater at the Varaždin wellfield in wells B3, B10, and the wellfield annual average.

2.3. *Aim of the Paper*

One of the main tasks of environmental engineering is to determine the main sources of pollution. Therefore, in order to prevent further groundwater degradation, it is necessary to locate and eliminate them. Based on experimental data, it is possible to create a map with iso-lines of pollutant concentration in space by applying various interpolation methods. Such a map represents a 2D model of the spatial distribution of pollutants. The application of different interpolation methods leads to different results. Based on such a map, it is possible to assume the location of the pollution source, i.e., the point where the concentration is highest. However, any assessment may contain some error, i.e., the measured and the assumed location do not completely coincide, and there is a possibility that there is some difference between them. This difference was assumed to be proportional to the variance of the estimation error, and on this basis, the following null hypothesis was established:

The distance between the measured and the assumed location of the pollution source d is a function of the variance of the estimation error σ^2_{error}:

$$H_0: \qquad d = d\left(\sigma^2_{error}\right) \tag{1}$$

That is, the assumed location with the smallest variance of the estimation error will be the closest to the measured location of the pollution source. An alternative hypothesis was also put forward:

$$H_1: \qquad d \neq d\left(\sigma^2_{error}\right) \tag{2}$$

Given that it is possible to obtain different models of the distribution of pollutants in space for the same experimental data, the goals of this work were as follows:

- Based on the value of the variance of the estimation error, choose the interpolation method that gives the best estimate.
- Based on the selected interpolation method, create a model that most closely approximates the actual spatial distribution of groundwater nitrate concentrations.
- Use the created model to assume the location of the pollution source.
- Determine the dependence of the distance between the measured and the assumed location on the variance of the estimation error and prove the hypothesis H_0.

If the null hypothesis is correct, further investigation begin with the assumed location with the smallest error estimation. In this way, it is possible to reduce the area that must be investigated to find the measured location of the pollution source. The above hypothesis was tested at the Varaždin wellfield. In this wellfield, as mentioned above, the location of the point source of pollution has already been determined [41]. Therefore, this location was deemed suitable for testing the presented hypothesis.

2.4. Nitrate Field Analysis

Based on the spatial distribution of sampling locations and the corresponding average nitrate concentrations during the observed period, 2D models of the spatial distribution of nitrate concentration were created using several interpolation methods: (1) kriging, (2) minimum curvature, (3) polynomial regression, (4) radial basis function, (5) inverse distance to a power, (6) nearest neighbor, and (7) moving average. Since the kriging method is very commonly used for such purposes [61–65], an additional experimental variogram was constructed for this interpolation method, in which the following theoretical variograms were fitted: (a) linear, (b) linear with nugget, (c) power, (d) power with nugget, (e) logarithmic, (f) logarithmic with nugget, (g) gaussian, and (h) exponential (Figure 5). Hence, 14 different models were created. For each 2D model, the location of the point source of pollution was assumed, and the distances between the assumed and measured location *d* were determined. Golden Software Surfer 8.0 was used for this purpose.

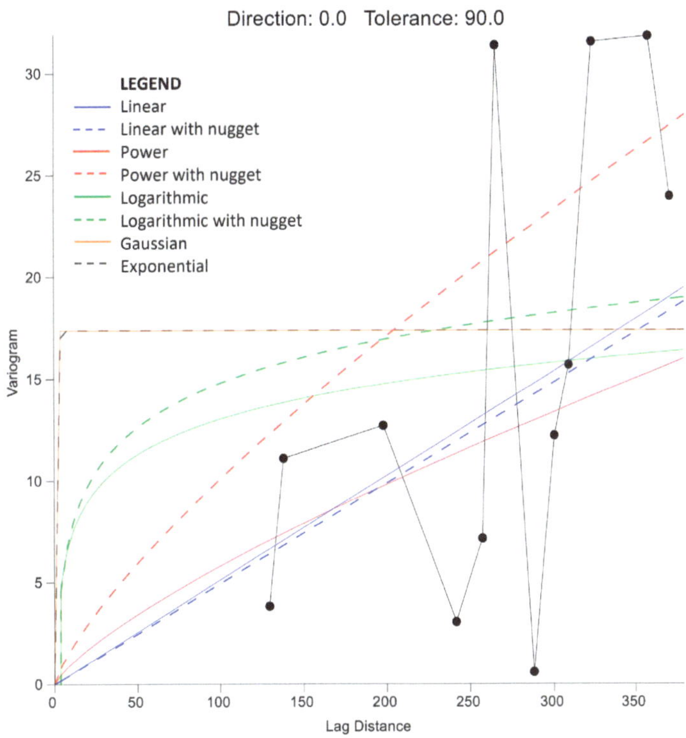

Figure 5. Experimental variograms in kriging interpolation method.

Furthermore, for each interpolation method and theoretical variogram, the variance of the estimation error σ^2_{error}, which represents the average square of the difference between the measured and estimated values, was determined via a cross-validation procedure:

$$\sigma^2_{error} = \frac{1}{N}\sum_{i=1}^{N}(z_i - \hat{z}_i)^2 \tag{3}$$

z_i—*measured i-th value*

$\hat{z}_i$—*estimated i-th value*
N—total number of measured values

For each model, the values of the variance of the estimation error and the corresponding distance between the measured and assumed location of the pollution source were determined. Based on these results, a dispersion diagram was created. With the models previously linearized, mathematical models were incorporated into this diagram and tested using the *t*-test:

$$t = r \cdot \sqrt{\frac{df}{1 - r^2}} \tag{4}$$

r—correlation coefficient for the linearized model
df = *N* − *2, number of degrees of freedom*

The obtained t values were compared with the critical value t_α, which was determined based on the total number of data with a significance level of 5%. If $t > t_\alpha$, the model is considered reliable and the hypothesis H_0 can be accepted:

$$t > t_\alpha \rightarrow H_0 \tag{5}$$

3. Results and Discussion

The arithmetic mean nitrate concentrations at each sampling location in the observed period and the associated deviations obtained from each model are shown in Table 1. The smallest deviations were obtained for the radial basis function, kriging with linear and power variogram, and minimum curvature. Slightly larger deviations were obtained for kriging with logarithmic variogram and polynomial regression, while the largest deviations were obtained for inverse distance to a power.

The distances between the measured location of the point source of nitrate pollution according to [41] and the locations assumed by the interpolation methods used can be clearly seen in Figure 6. The results of all models show an increase in concentration in the northwest–southeast direction. However, for some methods, the highest concentration is within the model domain (kriging–linear variogram, kriging–linear variogram with nugget, kriging–power, kriging–power with nugget, kriging–logarithmic, kriging–logarithmic with nugget and radial basis function), while for others, this point is located at the very edge of the domain or, more precisely, at its origin (minimum curvature and polynomial regression). Moreover, the four methods used did not give good results (kriging–gaussian variogram, kriging–exponential variogram, nearest neighbor, and moving average) and it was not possible to determine the point with the highest nitrate concentration. Therefore, they can be characterized as methods that are not suitable for this type of analysis. For this reason, the results of these methods were not considered in subsequent analyses.

As expected, each of the interpolation methods and tested variograms resulted in a different spatial distribution of nitrate concentration and, accordingly, in a different value of the variance of the estimation error σ^2_{error} and the distance d between the measured and assumed location of the point source of nitrate pollution (Table 2). In other words, it can be recognized that the distance d is a function of the variance of the estimation error σ^2_{error}. As the value of the variance of the estimation error σ^2_{error} increases, the distance d also increases and vice versa. The smallest values of σ^2_{error} and d were obtained for the interpolation method minimum curvature, while the largest values were obtained for the kriging method with logarithmic variogram. In order to prove the hypothesis H_0, a functional dependence $\sigma^2_{error}(d)$ was interpolated using the data from Table 2, and the resulting approximation is shown in Figure 7. For this purpose, the program Statistica 13.5.0.17 TIBCO Software Inc. was used.

Table 2. Values of the variance of the estimation error σ^2_{error} and the distance *d* between the measured and predicted location of the point source of nitrate pollution.

Interpolation Method	σ^2_{error} [(mg/L)^2]	*d* [m]
Kriging–linear variogram	4.13	466.61
Kriging–linear variogram with nugget	4.12	463.41
Kriging–power variogram	5.66	594.50
Kriging–power variogram with nugget	5.62	594.35
Kriging–logarithmic variogram	24.49	596.12
Kriging-logarithmic variogram with nugget	23.44	595.81
Kriging–gaussian variogram	51.87	not applicable
Kriging–exponential variogram	51.87	not applicable
Minimum curvature	1.65	0.00
Polynomial regression	2.38	0.00
Radial basis function	2.94	309.88
Inverse distance to a power	22.33	594.74
Nearest neighbor	22.26	not applicable
Moving average	23.01	not applicable

(1) Kriging
(**a**) linear variogram

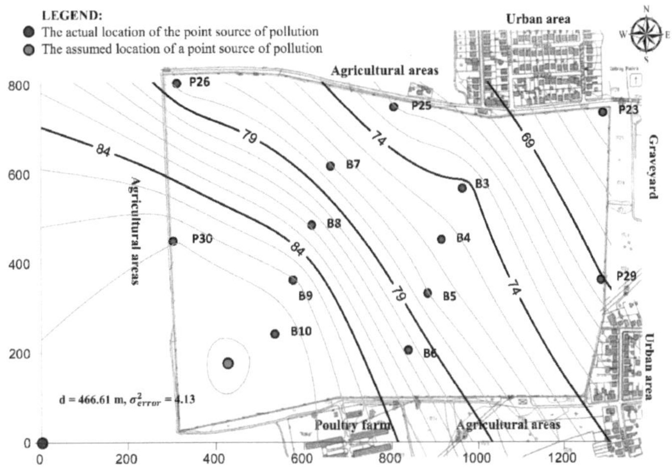

(1) Kriging
(**b**) linear variogram with nugget effect

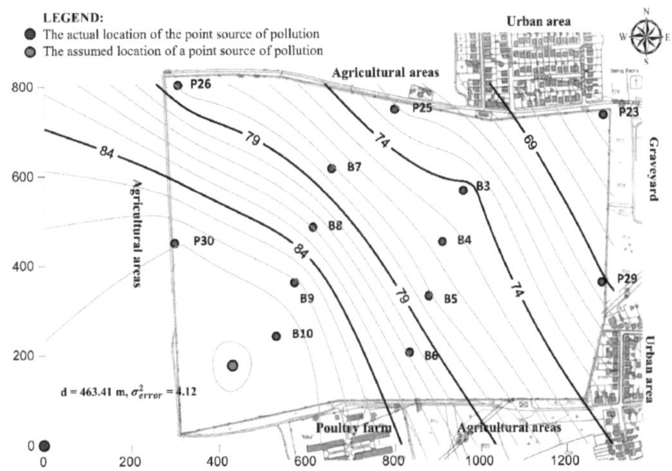

Figure 6. *Cont.*

(1) Kriging
(**c**) power variogram

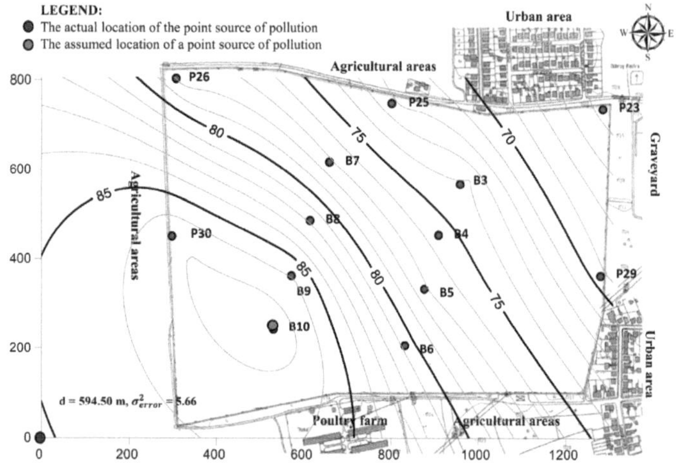

(1) Kriging
(**d**) power variogram with nugget effect

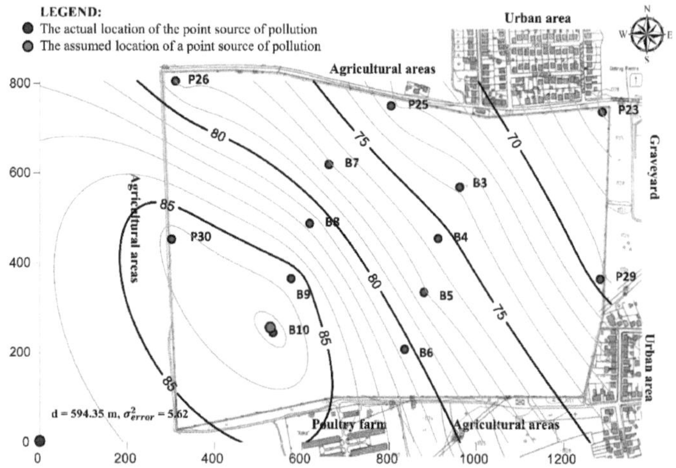

(1) Kriging
(**e**) logarithmic variogram

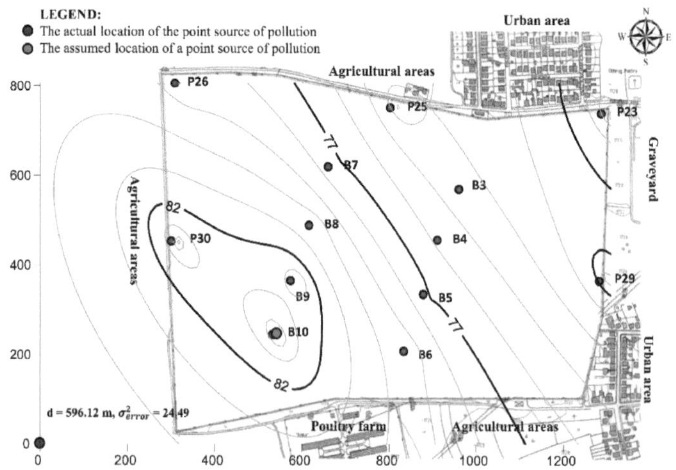

Figure 6. *Cont.*

(1) Kriging
(f) logarithmic with nugget effect

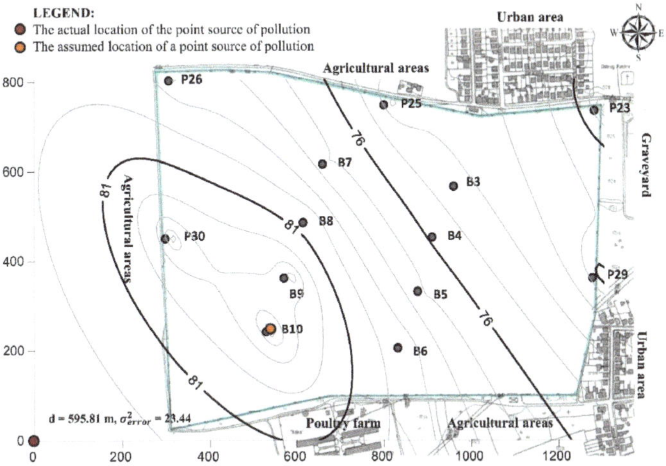

(2) Minimum

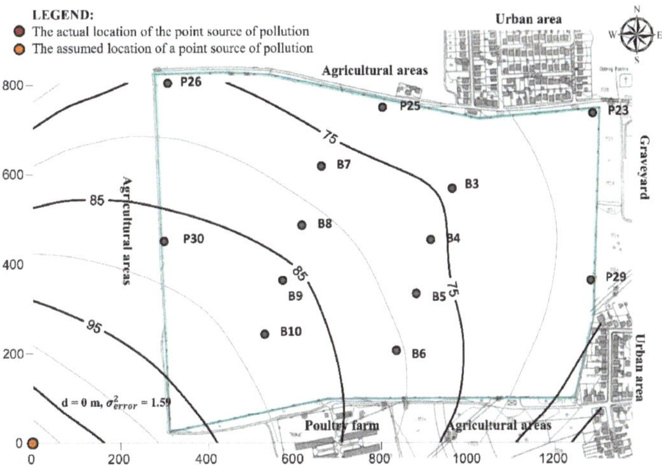

(3) Polynomial regression

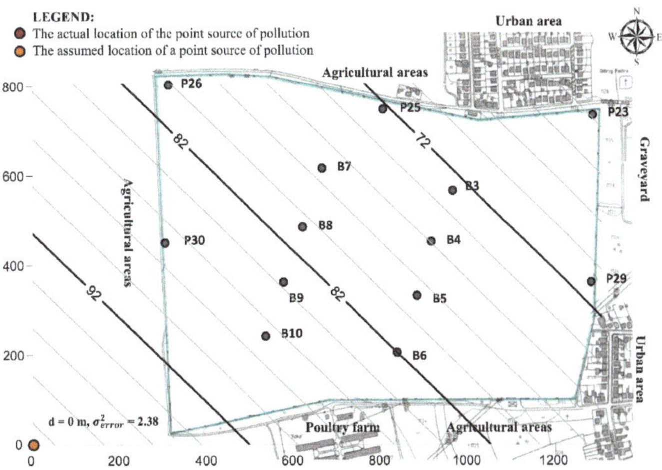

Figure 6. *Cont.*

(4) Radial basis function

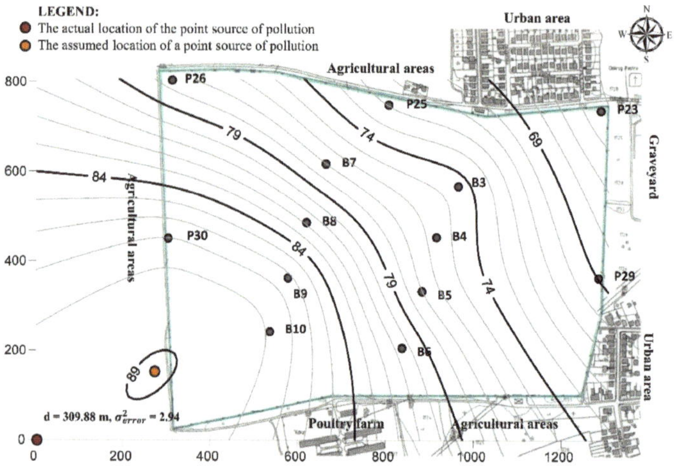

(5) Inverse distance to a power

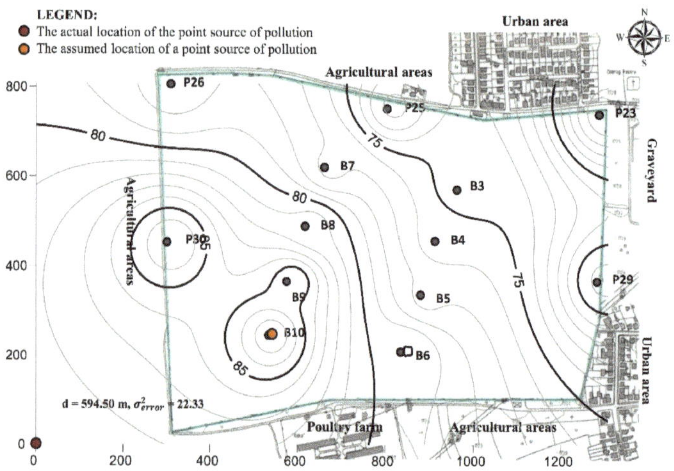

Figure 6. Maps of isolines of nitrate concentration in groundwater made based on various interpolation methods and the assumed location of the source of pollution.

Figure 7a shows the fitted linear model, which calculated a coefficient of determination of only 0.36, indicating a low correlation between the two variables. In fact, from the distribution of the points with coordinates (d, σ^2_{error}), it can be seen that there is a non-linear and monotonic relationship between the mentioned variables. For this reason, the scatter point data (d, σ^2_{error}) were interpolated using the logarithmic model (Figure 7b):

$$d = a\ln\left(\sigma^2_{error}\right) + b \tag{6}$$

with parameters a = 78.76 and b = 188.31, and the modified rational quadratic model (Figure 7c):

$$d = a\frac{\sigma^2_{error}}{b + \sigma^2_{error}} \tag{7}$$

with parameters a = 655.66 and b = 10.81. In order to prove hypothesis H_0 using the t-test, the models given in Equations (6) and (7) need to be linearized. For this purpose, the functional dependency $\sigma^2_{error}(d)$ for the logarithmic model can be expressed as follows:

$$\sigma^2_{error} = \exp\left(\frac{d-b}{a}\right) \tag{8}$$

and for the modified rational quadratic model, the same value can be expressed as follows:

$$\sigma^2_{error} = \sqrt{\frac{db}{a-d}} \tag{9}$$

which yields the linear models illustrated in Figure 8 and characterized by coefficients of determination of R^2 = 0.503 and R^2 = 0.498, respectively.

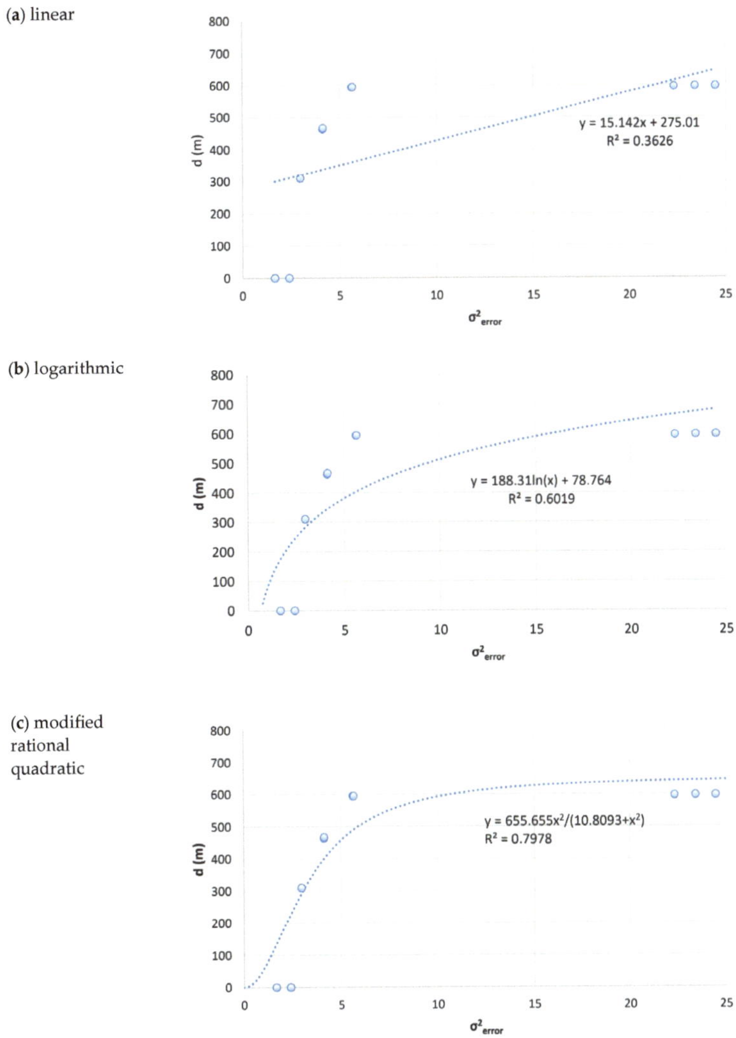

Figure 7. The approximated functional dependency $\sigma^2_{error}(d)$ based on (**a**) a linear model, (**b**) a logarithmic model, and (**c**) a modified rational quadratic model.

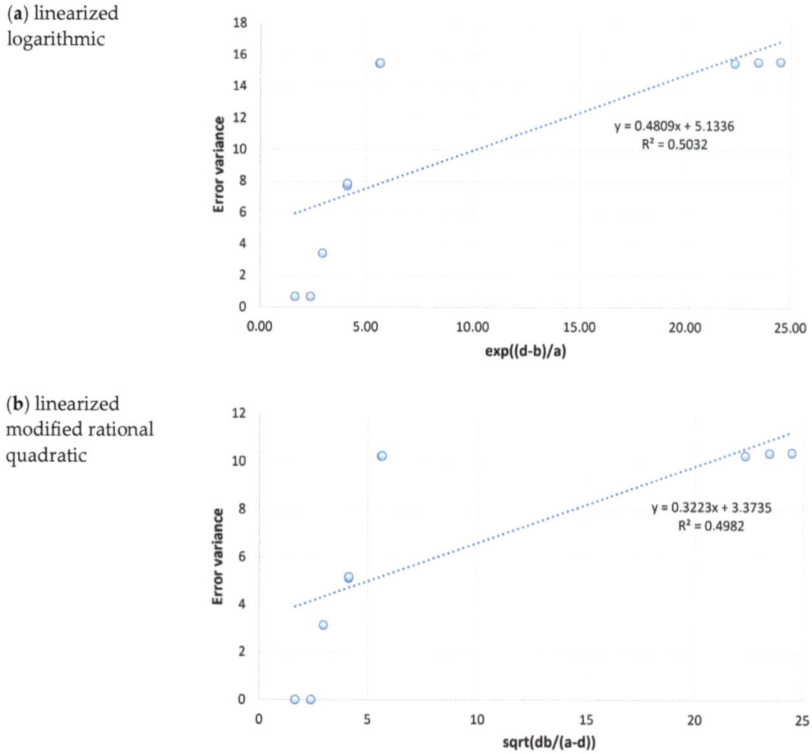

Figure 8. Graphical representation of the linearized logarithmic and modified rational quadratic models given by Equations (8) and (9).

Since the amount of available data points from Table 1 is 10, the number of degrees of freedom in the *t*-test is 8. Accordingly, the *t*-value for the linearized logarithmic model was determined to be 2.846, and for the modified rational quadratic model, the *t*-value was 2.818. These values should be considered with respect to the established critical value t_α = 2.306. Since the obtained *t* values are higher than the critical value t_α, the functional dependence $\sigma^2_{error}(d)$ can be considered reliable for both models, confirming the assumption that the distance *d* between the measured and the predicted location of the point source of pollution is a function of the variance of the estimation error σ^2_{error}. In other words, it can be concluded that the following implication holds:

$$t > t_\alpha \Longrightarrow H_0 : d = f\left(\sigma^2_{error}\right) \tag{10}$$

Interpolation methods are very commonly used to evaluate groundwater level [66–68], temperature variations [45], and the spatial quality of various parameters. Kriging and idw are cited as the most commonly used methods [48,49]. Thus, the idw method has been used for the interpolation of heavy metals in groundwater in the Dehradun district in India [50], while the Kriging method has been used for the spatial interpolation of the water quality index in the Khyber Pakhtunkhwa province in Pakistan [51]. The results in this paper show that different spatial interpolation methods yield different maps from the same data, which is consistent with previous studies [69]. The results of previous studies also indicate that the Kriging interpolation method usually gives better results than the inverse distance to a power [70,71], and the results in this work confirm what has been said, which can clearly be seen from the deviation values shown in Table 1. The radial basis function method is also mentioned as a good method for determining the spatial distribution of nitrogen compounds in groundwater [42], and this is confirmed by the results obtained in this work, where the value of the variance of the estimation error was 2.94, while the deviation values at the control points were minimal.

Based on the smallest value of the variance of the estimation error, the minimum curvature method can be chosen as the best estimation method for a given data set. However, each site should be considered individually, and it is necessary to conduct studies at other sites to determine if this method provides the best results or if other methods would provide better results at other sites. However, the obtained result can be very useful to minimize the area where the source of pollution is suspected. Indeed, it is often the case that this source of pollution is not clearly visible at the surface of the considered aquifer, so the presented methodology can be used to reduce the search area. Since the smallest value of the variance of the estimation error is an indicator of the smallest distance d between the predicted and the measured (but unknown) location of the pollution source, the interpolation model with the smallest variance of the estimation error can be assumed to be the one that most closely resembles the actual spatial distribution of the pollution concentration, and the predicted location of the pollution source (with maximum pollution concentration values) can be used as a starting point for further investigations.

4. Conclusions

Based on measurements of groundwater pollution, our analysis showed that there is a functional dependence between the distance d between the measured and assumed locations of the point source of pollution, which were determined via the use of different interpolation methods, and the variance of the estimation error σ^2_{error}. The smallest value of the variance of the estimation error (1.65) was obtained for the minimum curvature interpolation method, which makes it the best estimate for a given dataset. On the other hand, the largest value, 24.49, was obtained for the kriging method–logarithmic variogram. Through using the three interpolation methods (moving average, nearest neighbor, kriging with gaussian and exponential variogram), it was not possible to determine the assumed location of the point source of pollution; therefore, these methods can be considered unsuitable for the given dataset. Each site should be considered individually because it is possible that, at other sites, a different interpolation method will give better results. To prove this, it is necessary to conduct this type of research at other locations. To verify the established hypothesis, the distance d between the maximum concentration in the model domain (assumed location of the nitrate pollution source) and the measured location of the nitrate pollution source was determined for each model. In addition, the variance of the estimation error σ^2_{error} was calculated for each of the tested methods, and the functional dependencies $\sigma^2_{error}(d)$ were approximated by the logarithmic and modified rational quadratic models. Furthermore, the statistical t-test was used to confirm the established hypothesis. The results of the coefficients of determination of the linearized models are 0.50, and the obtained t values are higher than the critical value t_α, so the models can be considered reliable at a 95% significance level. The results in this paper show that the interaction of statistical and geostatistical methods can successfully locate or reduce the search area of point sources of pollution. The obtained conclusion has a direct practical application. If the source of the groundwater pollution is not known in advance, considering that it is usually difficult to find on the aquifer surface (or even not located on the surface), a procedure to reduce the search area is necessary. The obtained conclusion can be used to define the origin of the search area as the point with the maximum pollutant concentration (obtained via an interpolation method that gives the minimal variance of the estimation error σ^2_{error}). Accordingly, the search area can be narrowed down, and the search time can be shortened, which also leads to a reduction in the associated costs. Regarding future research, it is necessary to determine the influence of anisotropy on the spatial distribution of nitrate concentration in groundwater. If this influence is present, it is reasonable to assume that the variance of the estimation error could be further reduced. Consequently, the distance between the measured and predicted location of the pollution source would be reduced.

Author Contributions: Conceptualization, I.K. and M.Š.; methodology, I.K.; software, M.Š. and N.K.; validation, I.K., M.Š. and G.G.; formal analysis, I.K., M.Š. and N.K.; investigation, I.K. and N.K.; resources, G.G.; data curation, I.K. and M.Š.; writing—original draft preparation, I.K. and M.Š.; writing—review and editing, G.G.; visualization, M.Š.; supervision, I.K. and G.G.; funding acquisition, G.G. All authors have read and agreed to the published version of the manuscript.

Funding: This research received no external funding.

Data Availability Statement: The data used in this study are presented in the article. Additional data are available upon request from the corresponding author.

Conflicts of Interest: The authors declare no conflict of interest.

References

1. Quevauviller, P.; Fouillac, A.M.; Grath, J.; Ward, R. *Groundwater Monitoring*; Wiley: Hoboken, NJ, USA, 2009. [CrossRef]
2. EEA. *Freshwater Quality—The European Environment—State and Outlook 2010*; European Environment Agency: Copenhagen, Denmark, 2010.
3. European Commission. *Groundwater Protection in Europe: The New Groundwater Directive: Consolidating the EU Regulatory Framework*; European Commission: Brussels, Belgium, 2009. [CrossRef]
4. Alcamo, J. Water quality and its interlinkages with the Sustainable Development Goals. *Curr. Opin. Environ. Sustain.* **2019**, *36*, 126–140. [CrossRef]
5. Neves, S.A.; Marques, A.C.; Patrício, M. Determinants of CO_2 emissions in European Union countries: Does environmental regulation reduce environmental pollution? *Econ. Anal. Policy* **2020**, *68*, 114–125. [CrossRef]
6. Wen, J.; Mughal, N.; Zhao, J.; Shabbir, M.S.; Niedbała, G.; Jain, V.; Anwar, A. Does globalization matter for environmental degradation? Nexus among energy consumption, economic growth, and carbon dioxide emission. *Energy Policy* **2021**, *153*, 112230. [CrossRef]
7. Dyvak, M.; Rot, A.; Pasichnyk, R.; Tymchyshyn, V.; Huliiev, N.; Maslyiak, Y. Monitoring and Mathematical Modeling of Soil and Groundwater Contamination by Harmful Emissions of Nitrogen Dioxide from Motor Vehicles. *Sustainability* **2021**, *13*, 2768. [CrossRef]
8. Kurwadkar, S.; Kanel, S.R.; Nakarmi, A. Groundwater pollution: Occurrence, detection, and remediation of organic and inorganic pollutants. *Water Environ. Res.* **2020**, *92*, 1659–1668. [CrossRef]
9. Zhao, X.; Wang, D.; Xu, H.; Ding, Z.; Shi, Y.; Lu, Z.; Cheng, Z. Groundwater pollution risk assessment based on groundwater vulnerability and pollution load on an isolated island. *Chemosphere* **2022**, *289*, 133134. [CrossRef]
10. Plessis, A.D. Water Resources from a Global Perspective. In *South Africa's Water Predicament*; Springer: Cham, Switzerland, 2023; pp. 1–25. [CrossRef]
11. Kovač, I.; Šrajbek, M.; Kranjčević, L.; Novotni-Horčička, N. Nonlinear models of the dependence of nitrate concentrations on the pumping rate of a water supply system. *Geosci. J.* **2020**, *24*, 585–595. [CrossRef]
12. Das, B.; Pal, S.C. Assessment of groundwater vulnerability to over-exploitation using MCDA, AHP, fuzzy logic and novel ensemble models: A case study of Goghat-I and II blocks of West Bengal, India. *Environ. Earth Sci.* **2020**, *79*, 104. [CrossRef]
13. Umar, M.; Khan, S.N.; Arshad, A.; Aslam, R.A.; Khan, H.M.S.; Rashid, H.; Pham, Q.B.; Nasir, A.; Noor, R.; Khedher, K.M.; et al. A modified approach to quantify aquifer vulnerability to pollution towards sustainable groundwater management in Irrigated Indus Basin. *Environ. Sci. Pollut. Res.* **2022**, *29*, 27257–27278. [CrossRef]
14. Addiscott, T.M.; Whitmore, A.P.; Powlson, D.S. *Farming, Fertilizers and the Nitrate Problem*; Oxford University Press: Oxford, UK, 1991.
15. Foster, S.; Hirata, R.; Andreo, B. The aquifer pollution vulnerability concept: Aid or impediment in promoting groundwater protection? *Hydrogeol. J.* **2013**, *21*, 1389–1392. [CrossRef]
16. Almasri, M.N.; Kaluarachchi, J.J. Modular neural networks to predict the nitrate distribution in ground water using the on-ground nitrogen loading and recharge data. *Environ. Model. Softw.* **2005**, *20*, 851–871. [CrossRef]
17. Farhadi, H.; Fataei, E.; Sadeghi, M.K. The Relationship Between Nitrate Distribution in Groundwater and Agricultural Land use (Case study: Ardabil Plain, Iran). *Anthropog. Pollut. J.* **2020**, *4*, 50–56.
18. Srivastav, A.L. Chemical fertilizers and pesticides: Role in groundwater contamination. In *Agrochemicals Detection, Treatment and Remediation*; Elsevier: Amsterdam, The Netherlands, 2020; pp. 143–159. [CrossRef]
19. Yu, L.; Zheng, T.; Yuan, R.; Zheng, X. APCS-MLR model: A convenient and fast method for quantitative identification of nitrate pollution sources in groundwater. *J. Environ. Manag.* **2022**, *314*, 115101. [CrossRef]
20. Müller, L.; Behrendt, A.; Schindler, U. Strukturaspekte der bodendecke und bodeneigenschaften zweier niederungsstandorte in Nordostdeutschland: Structure aspects of the soil landscape and soil properties of two lowland sites in North-East Germany. *Arch. Agron. Soil Sci.* **2004**, *50*, 289–307. [CrossRef]
21. Krause, S.; Bronstert, A.; Zehe, E. Groundwater–surface water interactions in a North German lowland floodplain—Implications for the river discharge dynamics and riparian water balance. *J. Hydrol.* **2007**, *347*, 404–417. [CrossRef]

22. Schmalz, B.; Tavares, F.; Fohrer, N. Assessment of nutrient entry pathways and dominating hydrological processes in lowland catchments. *Adv. Geosci.* **2007**, *11*, 107–112. [CrossRef]
23. Krause, S.; Bronstert, A. An advanced approach for catchment delineation and water balance modelling within wetlands and floodplains. *Adv. Geosci.* **2005**, *5*, 1–5. [CrossRef]
24. Davey, I.R.; Besien, T.J.; Evers, S.; Ward, R. *Nitrates in Groundwater—Agricultural Nitrate Contamination in Groundwater in England and Wales: An Overview*; CRC Press: Boca Raton, FL, USA, 2014.
25. He, X.-S.; Zhang, Y.-L.; Liu, Z.-H.; Wei, D.; Liang, G.; Liu, H.-T.; Xi, B.-D.; Huang, Z.-B.; Ma, Y.; Xing, B.-S. Interaction and coexistence characteristics of dissolved organic matter with toxic metals and pesticides in shallow groundwater. *Environ. Pollut.* **2020**, *258*, 113736. [CrossRef]
26. Razowska-Jaworek, L.; Sadurski, A. *Nitrates in Groundwater—Development of a Groundwater Abstraction Modelling Environment for Drinking Water Supply*; CRC Press: Boca Raton, FL, USA, 2014.
27. Sapek, B.; Sapek, A. *Nitrates in Groundwater—Nitrate in Groundwater as an Indicator of Farmstead Impacts on the Environment*; CRC Press: Boca Raton, FL, USA, 2014.
28. Stockmarr, J.; Nyegaard, P. *Nitrates in Groundwater—Nitrate in Danish Groundwater*; CRC Press: Boca Raton, FL, USA, 2014.
29. Li, P.; He, X.; Guo, W. Spatial groundwater quality and potential health risks due to nitrate ingestion through drinking water: A case study in Yan'an City on the Loess Plateau of northwest China. *Hum. Ecol. Risk Assess. Int. J.* **2019**, *25*, 11–31. [CrossRef]
30. Yu, G.; Wang, J.; Liu, L.; Li, Y.; Zhang, Y.; Wang, S. The analysis of groundwater nitrate pollution and health risk assessment in rural areas of Yantai, China. *BMC Public Health* **2020**, *20*, 437. [CrossRef]
31. Chambers, T.; Douwes, J.; Mannetje, A.; Woodward, A.; Baker, M.; Wilson, N.; Hales, S. Nitrate in drinking water and cancer risk: The biological mechanism, epidemiological evidence and future research. *Aust. N. Z. J. Public Health* **2022**, *46*, 105–108. [CrossRef] [PubMed]
32. Brender, J.D. Human Health Effects of Exposure to Nitrate, Nitrite, and Nitrogen Dioxide. In *Just Enough Nitrogen*; Springer International Publishing: Cham, Switzerland, 2020; pp. 283–294. [CrossRef]
33. Madison, R.J.; Brunett, J.O. Overview of the occurrence of nitrate in ground water in the United States. *US Geol. Surv. Water-Supply Pap.* **1984**, *2275*, 93–105.
34. Rahman, A.; Mondal, N.C.; Tiwari, K.K. Anthropogenic nitrate in groundwater and its health risks in the view of background concentration in a semi arid area of Rajasthan, India. *Sci. Rep.* **2021**, *11*, 9279. [CrossRef] [PubMed]
35. Manu, E.; Afrifa, G.Y.; Ansah-Narh, T.; Sam, F.; Loh, Y.S.A. Estimation of natural background and source identification of nitrate-nitrogen in groundwater in parts of the Bono, Ahafo and Bono East regions of Ghana. *Groundw. Sustain. Dev.* **2022**, *16*, 100696. [CrossRef]
36. Nakić, Z.; Kovač, Z.; Parlov, J.; Perković, D. Ambient Background Values of Selected Chemical Substances in Four Groundwater Bodies in the Pannonian Region of Croatia. *Water* **2020**, *12*, 2671. [CrossRef]
37. Eberts, M.; Thomas, M.A.; Jagucki, M.L. *The Quality of Our Nation's Waters: Factors Affecting Public-Supply-Well Vulnerability to Contamination: Understanding Observed Water Quality and Anticipating Future Water Quality*; U.S. Department of the Interior, U.S. Geological Survey: Reston, VA, USA, 2013.
38. Jurec, J.N.; Mesic, M.; Basic, F.; Kisic, I.; Zgorelec, Z. Nitrate concentration in drinking water from wells at three different locations in Northwest Croatia. *Cereal Res. Commun.* **2007**, *35*, 533–536. [CrossRef]
39. Srajbek, M.; Kovac, I.; Novotni-Horcicka, N.; Kranjcevic, L. Assessment of average contributions of point and diffuse pollution sources to nitrate concentration in groundwater by nonlinear regression. *Environ. Eng. Manag. J.* **2020**, *19*, 95–104. [CrossRef]
40. Kim, H.-R.; Yu, S.; Oh, J.; Kim, K.-H.; Lee, J.-H.; Moniruzzaman, M.; Kim, H.K.; Yun, S.-T. Nitrate contamination and subsequent hydrogeochemical processes of shallow groundwater in agro-livestock farming districts in South Korea. *Agric. Ecosyst. Environ.* **2019**, *273*, 50–61. [CrossRef]
41. Šrajbek, M.; Kranjčević, L.; Kovač, I.; Biondić, R. Groundwater Nitrate Pollution Sources Assessment for Contaminated Wellfield. *Water* **2022**, *14*, 255. [CrossRef]
42. Bronowicka-Mielniczuk, U.; Mielniczuk, J.; Obroślak, R.; Przystupa, W. A Comparison of Some Interpolation Techniques for Determining Spatial Distribution of Nitrogen Compounds in Groundwater. *Int. J. Environ. Res.* **2019**, *13*, 679–687. [CrossRef]
43. Mustafa, J.S.; Mawlood, D.K. Mapping Groundwater Levels in Erbil Basin. *Am. Sci. Res. J. Eng. Technol. Sci.* **2023**, *93*, 21–38.
44. Fan, X.; Min, T.; Dai, X. The Spatio-Temporal Dynamic Patterns of Shallow Groundwater Level and Salinity: The Yellow River Delta, China. *Water* **2023**, *15*, 1426. [CrossRef]
45. Hajnrych, M.; Blachowski, J.; Worsa-Kozak, M. Study of groundwater temperature spatio-temporal variation in the city of Wroclaw. Preliminary results. *IOP Conf. Ser. Earth Environ. Sci.* **2023**, *1189*, 012028. [CrossRef]
46. Lubis, R.F.; Yamano, M.; Delinom, R.; Martosuparno, S.; Sakura, Y.; Goto, S.; Miyakoshi, A.; Taniguchi, M. Assessment of urban groundwater heat contaminant in Jakarta, Indonesia. *Environ. Earth Sci.* **2013**, *70*, 2033–2038. [CrossRef]
47. Taniguchi, M.; Shimada, J.; Fukuda, Y.; Yamano, M.; Onodera, S.-I.; Kaneko, S.; Yoshikoshi, A. Anthropogenic effects on the subsurface thermal and groundwater environments in Osaka, Japan and Bangkok, Thailand. *Sci. Total Environ.* **2009**, *407*, 3153–3164. [CrossRef] [PubMed]
48. Previati, A.; Crosta, G.B. Characterization of the subsurface urban heat island and its sources in the Milan city area, Italy. *Hydrogeol. J.* **2021**, *29*, 2487–2500. [CrossRef]

49. Benz, S.A.; Bayer, P.; Goettsche, F.M.; Olesen, F.S.; Blum, P. Linking Surface Urban Heat Islands with Groundwater Temperatures. *Environ. Sci. Technol.* **2016**, *50*, 70–78. [CrossRef]
50. Nayak, A.; Matta, G.; Uniyal, D.P.; Kumar, A.; Kumar, P.; Pant, G. Assessment of potentially toxic elements in groundwater through interpolation, pollution indices, and chemometric techniques in Dehradun in Uttarakhand State. *Environ. Sci. Pollut. Res.* **2023**. [CrossRef]
51. Khan, M.; Almazah, M.M.A.; Ellahi, A.; Niaz, R.; Al-Rezami, A.Y.; Zaman, B. Spatial interpolation of water quality index based on Ordinary kriging and Universal kriging. *Geomat. Nat. Hazards Risk* **2023**, *14*, 2190853. [CrossRef]
52. Kovač, I.; Kovačev-Marinčić, B.; Novotni-Horčička, N.; Mesec, J.; Vugrinec, J. Komparativna analiza koncentracije nitrata u gornjem i donjem sloju varaždinskog vodonosnika. *Rad. Zavoda Znan. Varaždin* **2017**, *28*, 41–57. [CrossRef]
53. Novotni-Horčička, N.; Šrajbek, M.; Kovač, I. Nitrati u Regionalnom vodovodu Varaždin. In *Voda i Javna Vodoopskrba*; Hrvatski Zavod za Javno Zdravstvo (HZJZ): Zagreb, Croatia, 2010; pp. 123–131.
54. Urumović, K. O kvartnom vodonosnom kopleksu u području Varaždina. *Geološki Vjesn.* **1971**, *43*, 109–118.
55. Larva, O. Aquifer Vulnerability at Catchment Area of Varaždin Pumping Sites. Ph.D Thesis, Faculty of Mining, Geology and Petroleum Engineering, University of Zagreb, Zagreb, Croatia, 2008.
56. Šrajbek, M. Nitrate Pollution Propagation in Groundwater and Wellfield Imapct Assessment. Ph.D Thesis, Faculty of Engineering, University of Rijeka, Rijeka, Croatia, 2021.
57. European Union. Council Directive 91/676/EEC of 12 December 1991 concerning the protection of waters against pollution caused by nitrate from agricultural sources. *Off. J. Eur. Commun.* **1991**, *375*, 1–13.
58. Gjetvaj, G. Identifikacija porijekla nitrata u podzemnim vodama Varaždinske regije. *Hrvat. Vode* **1993**, *1*, 247–252.
59. Grđan, D.; Durman, P.; Kovačev-Marinčić, B. Odnos promjene režima i kvalitete podzemnih voda na crpilištima Varaždin i Bartolovec. *Geološki Vjesn.* **1991**, *44*, 301–308.
60. Kovač, I. Statistical-Variographic Analysis of Ground Water Chemical Composition in Varaždin Region. Ph.D. Thesis, Faculty of Minning, Geology and Petroleum Engineering, University of Zagreb, Zagreb, Croatia, 2004.
61. Arslan, H. Spatial and temporal mapping of groundwater salinity using ordinary kriging and indicator kriging: The case of Bafra Plain, Turkey. *Agric. Water Manag.* **2012**, *113*, 57–63. [CrossRef]
62. Thomas, E.O. Spatial evaluation of groundwater quality using factor analysis and geostatistical Kriging algorithm: A case study of Ibadan Metropolis, Nigeria. *Water Pract. Technol.* **2023**, *18*, 592–607. [CrossRef]
63. Belkhiri, L.; Tiri, A.; Mouni, L. Spatial distribution of the groundwater quality using kriging and Co-kriging interpolations. *Groundw. Sustain. Dev.* **2020**, *11*, 100473. [CrossRef]
64. Gundogdu, K.S.; Guney, I. Spatial analyses of groundwater levels using universal kriging. *J. Earth Syst. Sci.* **2007**, *116*, 49–55. [CrossRef]
65. Rostami, A.A.; Karimi, V.; Khatibi, R.; Pradhan, B. An investigation into seasonal variations of groundwater nitrate by spatial modelling strategies at two levels by kriging and co-kriging models. *J. Environ. Manag.* **2020**, *270*, 110843. [CrossRef]
66. Arkoc, O. Modeling of spatiotemporal variations of groundwater levels using different interpolation methods with the aid of GIS, case study from Ergene Basin, Turkey. *Model. Earth Syst. Environ.* **2022**, *8*, 967–976. [CrossRef]
67. Varouchakis, A.; Hristopulos, D.T. Comparison of stochastic and deterministic methods for mapping groundwater level spatial variability in sparsely monitored basins. *Environ. Monit. Assess.* **2013**, *18*, 1–19. [CrossRef] [PubMed]
68. Xiao, Y.; Gu, X.; Yin, S.; Shao, J.; Cui, Y.; Zhang, Q.; Niu, Y. Geostatistical interpolation model selection based on ArcGIS and spatio-temporal variability analysis of groundwater level in piedmont plains, northwest China. *Springerplus* **2016**, *5*, 425. [CrossRef] [PubMed]
69. Elumalai, V.; Brindha, K.; Sithole, B.; Lakshmanan, E. Spatial interpolation methods and geostatistics for mapping groundwater contamination in a coastal area. *Environ. Sci. Pollut. Res.* **2017**, *24*, 11601–11617. [CrossRef]
70. Ahmad, A.Y.; Saleh, I.A.; Balakrishnan, P.; Al-Ghouti, M.A. Comparison GIS-Based interpolation methods for mapping groundwater quality in the state of Qatar. *Groundw. Sustain. Dev.* **2021**, *13*, 100573. [CrossRef]
71. Mirzaei, R.; Sakizadeh, M. Comparison of interpolation methods for the estimation of groundwater contamination in Andimeshk-Shush Plain, Southwest of Iran. *Environ. Sci. Pollut. Res.* **2016**, *23*, 2758–2769. [CrossRef]

hydrology MDPI

Article

Assessment of Nitrate in Groundwater from Diffuse Sources Considering Spatiotemporal Patterns of Hydrological Systems Using a Coupled SWAT/MODFLOW/MT3DMS Model

Alejandra Correa-González [1,*], Joel Hernández-Bedolla [1], Marco Antonio Martínez-Cinco [2], Sonia Tatiana Sánchez-Quispe [1] and Mario Alberto Hernández-Hernández [3]

1 Facultad de Ingeniería Civil, Universidad Michoacana de San Nicolas de Hidalgo, Morelia 58030, Mexico; joel.hernandez@umich.mx (J.H.-B.); quispe@umich.mx (S.T.S.-Q.)

2 Facultad de Ingeniería Química, Universidad Michoacana de San Nicolas de Hidalgo, Morelia 58030, Mexico; marco.martinez@umich.mx

3 Investigador por México CONAHCYT-Instituto de Geofísica, Universidad Nacional Autónoma de México, Circuito de la Investigación Científica SN, Ciudad Universitaria, Coyoacán 04510, Mexico; malbher@igeofisica.unam.mx

* Correspondence: alejandra.correa@umich.mx; Tel.: +52-425-105-6058

Abstract: In recent years, due to various anthropogenic activities, such as agriculture and livestock, the presence of nitrogen-associated contaminants has been increasing in surface- and groundwater resources. Among these, the main compounds present in groundwater are ammonia, nitrite, and nitrate. However, it is sometimes difficult to assess such effects given the scarcity or lack of information and the complexity of the system. In the current study, a methodology is proposed to assess nitrate in groundwater from diffuse sources considering spatiotemporal patterns of hydrological systems using a coupled SWAT/MODFLOW/MT3DMS model. The application of the model is carried out using a simplified simulation scheme of hydrological and agricultural systems because of the limited spatial and temporal data. The study area includes the Cuitzeo Lake basin in superficial flow form and the Morelia–Querendaro aquifer in groundwater flow form. The results within the methodology are surface runoff, groundwater levels, and nitrate concentrations present in surface- and groundwater systems. The results indicate that the historical and simulated nitrate concentrations were obtained within acceptable values of the statistical parameters and, therefore, are considered adequate.

Keywords: numerical modeling; agriculture; nitrate

Citation: Correa-González, A.; Hernández-Bedolla, J.; Martínez-Cinco, M.A.; Sánchez-Quispe, S.T.; Hernández-Hernández, M.A. Assessment of Nitrate in Groundwater from Diffuse Sources Considering Spatiotemporal Patterns of Hydrological Systems Using a Coupled SWAT/MODFLOW/MT3DMS Model. *Hydrology* **2023**, *10*, 209. https://doi.org/10.3390/hydrology10110209

Academic Editors: Pantelis Sidiropoulos, Augustina Clara Alexander and Peiyue Li

Received: 13 September 2023
Revised: 17 October 2023
Accepted: 29 October 2023
Published: 9 November 2023

1. Introduction

The world's population requires large food resources. Food demand is expected to double by the middle of the century [1]. The growing demand for food has increased pressure on agricultural areas [2,3]. Irrigated and rainfed crops depend on fertilizers for greater efficiency [4]. Moreover, nitrogen fertilizer is a component of food security [5,6]. In agricultural areas, the main contaminants in groundwater are ammonia, nitrite, and nitrate due to the use of nitrogen fertilizers and extensive livestock farming [7]. The presence of nitrate in groundwater is higher than that of nitrite since nitrite is converted to nitrate in the presence of oxygen [8]. In addition, nitrate, being a mobile compound, drains through the subsurface with minimal uptake [9].

Nitrite and nitrate in drinking water can cause hemoglobinemia in infants younger than six months and affect children up to six years of age [10]. In addition, water consumption with high concentrations of nitrite and nitrate has been associated with other types of diseases, such as cancer, thyroid diseases, and congenital disabilities [11,12]. In 2006, the International Agency for Research Cancer (IARC) placed nitrite and nitrate in group 2A as "probably carcinogenic to humans" after several studies [13]. According to several authors, the primary sources of nitrate entering groundwater are agriculture and wastewater discharges, with agriculture being the key factor for the presence of nitrate in water [14–16]. However, the evaluation of nitrogen compounds in subsoil is a complex system with a large number of factors. These include parameters associated with the nitrogen cycle and spatiotemporal hydrological variables such as soil characteristics and water movement through the subsoil and porous media [15,16].

Mathematical modeling in hydrology is a widely used tool for the evaluation of various problems associated with water resources [17–20] and groundwater management [21,22]. Some authors developed surface runoff models to quantify the surface runoff of a basin [23], highlighting the importance of basin-level studies like a fundamental unit for the study of water resources in surface systems [24,25]. Examples of mathematical rainfall runoff models are Témez [26], Sacramento [27], and the Soil Conservation Service Curve Number (SCS-CN) method [28].

The hydrological model of Témez is commonly used in Spain [29,30] for estimating runoff and recharge in the surface system. It is an aggregated continuous simulation model of a few parameters and monthly time scale. It simulates the main water transfer processes in the hydrological cycle considering two storages: soil and aquifer [31]. The Sacramento model is a conceptual daily passage rainfall runoff model developed by the U.S. River Weather Service. This method has been applied to various watersheds worldwide [32,33]. The SCS-CN method is another popular method for obtaining surface runoff [34–36]. It is empirically developed to estimate the amount of runoff according to land type and land use. The SCS-CN method was incorporated in a semi-distributed Soil and Water Assessment Tool (SWAT) [37] and the Hydrologic Modeling System (HEC-HMS) [38]. Other surface runoff models are based on kinematic wave approximation, for example, the KINEmatic Runoff and EROSion Model (KINEROS) [39–43], fully distributed Gridded Surface Subsurface Hydrologic Analysis (GSSHA) [44–46], and fully distributed European Hydrological System Model (MIKE-SHE) [47–50].

For groundwater flow simulation, the finite difference method and the method of eigenvalues and eigenvectors [51] are used to solve the equations that govern groundwater flow. The modular three-dimensional finite difference groundwater flow model, MODFLOW [52], incorporates the finite difference method, while the Aquival model [53] solves the equations of water flow in porous media through the eigenvalue method and eigenvector matrix [54]. Both methods are commonly applied worldwide [55–57].

Furthermore, several water quality models have been formulated and incorporated into surface runoff or groundwater flow models to evaluate various pollutants [37,53,58], as well as the evaluation of nitrate in surface runoff [59,60] and groundwater [61–63]. Over the last decades, several models of nutrient transport have been developed, for example, the export coefficient modelling adopted by the Environmental Protection Agency (EPA) [64–69], Bayesian hierarchical approach [70–72], inverse Bayesian modeling approach [73], GIS mass balance method [74], mass balance coupled modeling approach [75], and lysimeter-based approach [76].

Different methodologies have been developed to obtain the transport of nitrogenous compounds in surface runoff and groundwater. For example, the SWAT model developed by the U.S. Agricultural Research Service, ARS [37], is internationally used [77,78]. The PATRICAL model, a Spanish acronym for precipitation input in integrated network sections with water quality [79], has been used in numerous studies carried out in Spain [80]. The modular three-dimensional transport model, MT3D [81], offers several advanced versions such as the modular three-dimensional multispecies transport model, MT3DMS [82], and the reactive transport in a three-dimensional model, RT3D [83], which are models that have been widely used in combination with MODFLOW to simulate the transport of contaminants in groundwater [62,84,85].

The MT3D, MT3DMS, and RT3D groundwater transport models determine solute transport through the porous media using the sequential iteration approach (SIA). The SIA consists of separately solving the transport and chemical reaction terms [86]. The SIA has been applied to develop several models, such as PHREEQC [87] and PHT3D [88], due to its simplicity compared to the direct solution method. PHREEQC is a one-dimensional transport model designed to perform a wide variety of aqueous geochemical (homogeneous or heterogeneous kinetic reactions, heterogeneous equilibria, and transport of solute) assessments [89] and PHT3D is a computer code for general reactive transport calculations, coupling MODFLOW/MT3DMS for transport and PHREEQC for chemical reactions [58].

Nitrate transport in groundwater is influenced by surface factors associated with hydrologic cycle inputs and subsurface flow; in recent years, several models have been coupled to evaluate the previous factors [90,91]. The conjunctive use of the SWAT and MODFLOW models has been one of the most common [92,93]. This model coupling has also been used with the MT3DMS or RT3D models to assess nitrate transport in groundwater [94–96]. The present study proposes nitrate assessment in groundwater using a coupled SWAT/MODFLOW/MT3DMS model in the Morelia–Querendaro aquifer (central Mexico) with limited information. The objective of multiple coupling is to implement a robust model to calculate nitrate in groundwater, which involves surface factors and reflects its spatial and temporal variability. In developing countries, there is no information on the use of nitrogen fertilizers in agricultural areas. Therefore, it is necessary to develop a reliable methodology capable of generating results for the study area.

2. Materials and Methods

2.1. Study Area

The study area is located in west-central Mexico (Figure 1). The study area includes the Cuitzeo Lake basin (CLB) on the surface and the Morelia–Querendaro aquifer (MQA). The CLB is endorheic and has an area of 4000 km^2 [97]. According to CONAGUA [98], the MQA has an extension of 3507 km^2. The study area defines geomorphological features resulting from very intense tectonic forces and volcanic activity. To the southwest, there are mountains and ridges made up predominantly of extrusive igneous rocks of Tertiary age, and in the southern portion of the area, andesites and basalts of the Paleogene Neogene, which reach elevations above 3000 m above sea level and have a very steep relief, with steep fronts and steep slopes. In the northeast, there are valleys of tectonic and fluvial origin. The plains located in the central portion of the zone are formed by the floodplains of the rivers and streams that flow into Lake Cuitzeo [69]. The average annual temperature, precipitation, and potential evaporation are 17.4 °C, 797 mm, and 1810 mm, respectively, with a rainfall regime mostly from June to September [99].

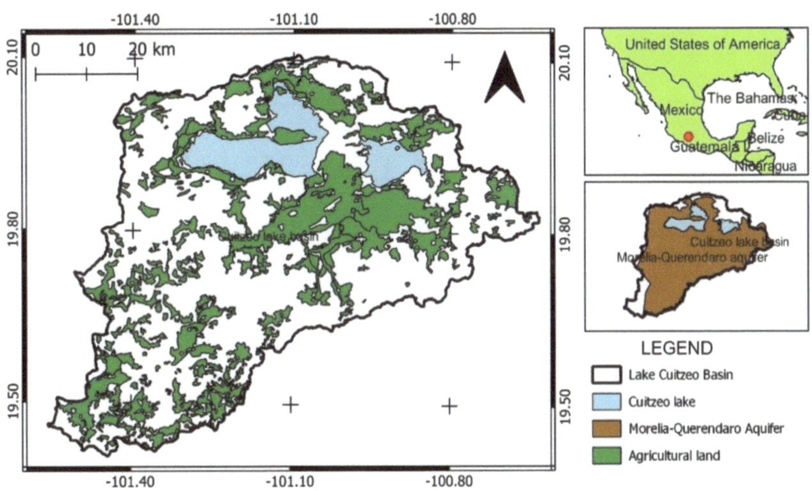

Figure 1. Agricultural land distribution in the basin.

Agriculture is an activity that is carried out in 39.65% of the basin, which accounts for 53.69% of the extraction destined for the aquifer. Most of the agricultural area is concentrated in the lowland areas of the basin (Figure 1) and its production is presented in summer with activities beginning in February–March and ending in September–October. The primary agricultural products in the region are corn, grain, alfalfa, vegetable and fodder sorghum, oats, and wheat (Figure 2). In addition, depending on the crop and municipality, the type and amount of mineral fertilizer may vary. The main fertilizers are ammonium sulfate and nitrates with an applied amount of 200 to 600 kg/ha [100].

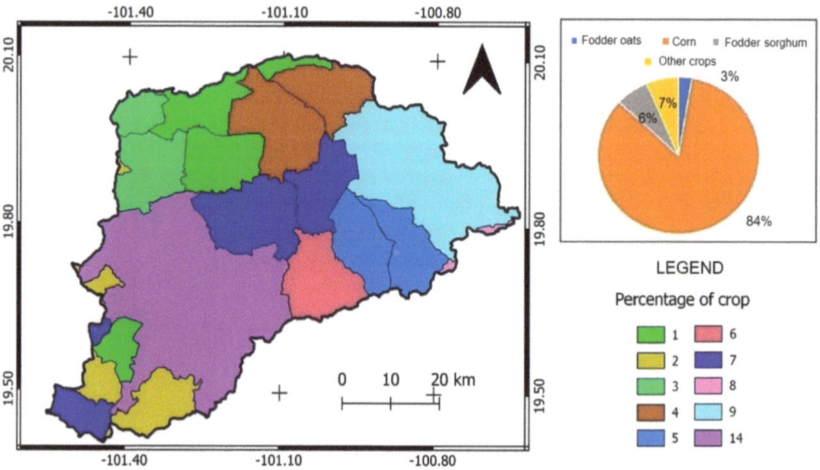

Figure 2. Percentage of crops of total agriculture in LCB—main crops.

The predominant soil types in CLB are Andosols, Luvisols, Acrisols, Vertisols, and Planosols. To the southwest are Andosols and Acrisols. The Andosols occur at altitudes of 2300 to 3100 masl, with a medium texture and hydraulic conductivity of 10 to 12 mm/h. Luvisols occur at altitudes of 2000 to 2350 masl, with a medium texture and hydraulic conductivity of 1.5 to 12 mm/h. In the plains, there are Vertisols and Planosols, with fine and medium texture, respectively. The Vertisol and Planosol soils have hydraulic conductivities ranging from 0.05 to 2.5 mm/h and from 1.5 to 12 mm/h, respectively.

MQA is heterogeneous and anisotropic, generally unconfined, with local semi-confinement conditions. The aquifer is located in a tectonic depression and is made up, in its upper portion, of clastic sediments of varied granulometry as well as clayey sediments that were deposited in the lower part of the basin where Lake Cuitzeo is located; the lower portion is made up of volcanic rocks. The granular and fractured environments form a single hydrogeological unit that typically has medium to high permeability and an average thickness of 300 to 400 m, and depth to the groundwater levels varies from 10 to 200 m. The different units that make up the aquifer show variations: to the west, it is mainly hosted by pumiceous tuffs with lithic and clayey horizons and welded tuffs; towards the plain, it is mainly made up of clastic sediments of the size of gravels and sands; and towards the area adjacent to Lake Cuitzeo by clayey sediments, andesitic breccias, basalts, as well as basaltic breccias and ashes (Figure 3). The groundwater balance in the MQA shows a yearly deficit of 20.95 cubic hectometers (hm^3) [101].

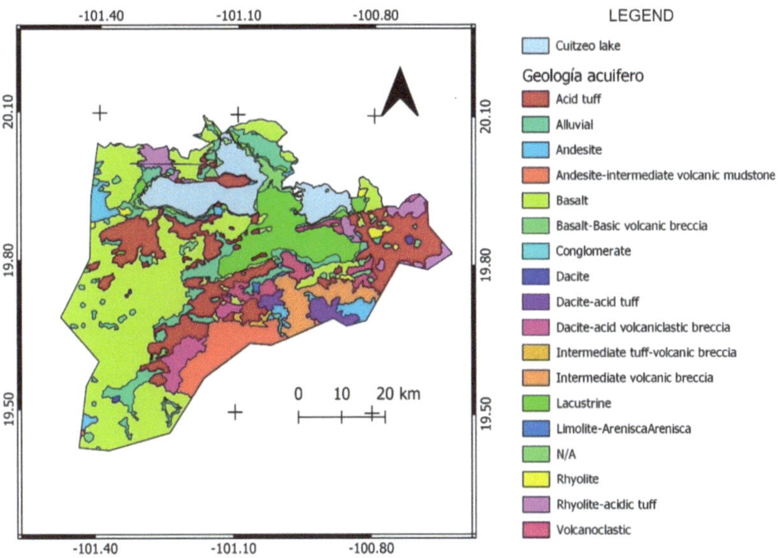

Figure 3. Geology of the Morelia–Querendaro aquifer, obtained from National Institute of Statistics and Geography (INEGI; https://www.inegi.org.mx/temas/; accessed on 28 October 2023).

2.2. Coupled SWAT/MODFLOW/MT3DMS Model

An integrated modeling framework was used to assess the concentration of nitrate in groundwater by quantifying the spatiotemporal patterns of groundwater–surface water interaction with limited information. The first stage is related to the hydrological surface modeling, used to calculate the spatial and temporal aquifer recharge. The second stage consists of using the same model to estimate the nitrate concentration that enters through subsurface percolation and later reaches the saturated zone as recharge. The third stage evaluates the groundwater levels of the aquifer. The last stage assesses nitrate transport within the aquifer. The coupling of the models was carried out manually. The information obtained in the hydrological modeling was used for the implementation of the modeling in MODFLOW/MT3DMS. Figure 4 shows the stages of model coupling, as well as the quantitative and qualitative inputs and outputs of each model.

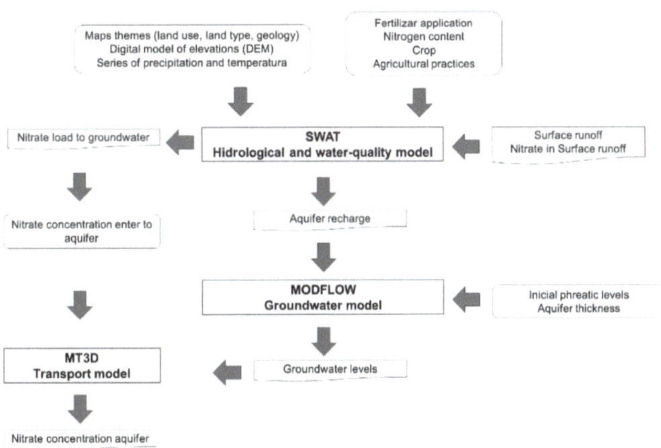

Figure 4. Coupling of mathematical models. Quantitative and qualitative inputs and outputs of each model.

SWAT (the Soil and Water Assessment Tool) [20], developed by the U.S. Agricultural Research Service (ARS), is a sub-aggregated, physically based, daily time-lagged model. It incorporates watershed equations of the hydrological balance and nitrogen cycle. Although SWAT simulates two aquifers in each subbasin (shallow and deep aquifer), it does not currently print groundwater height in the output files [102]. SWAT considers five different forms of nitrogen in the soil: two are inorganic forms of nitrogen (NH_4 and NO_3), while the other three forms are organic (fresh, stable, and active organic nitrogen). The model contemplates immobilization, mineralization, nitrification, denitrification, fixation, and leaching reactions.

MODFLOW [52] simulates three-dimensional, multilayer-type groundwater flow through a porous medium using a finite difference method. The three-dimensional movement of constant density groundwater through the porous earth material is calculated using the partial differential equation governed by the law of Darcy.

The coupling of these models allows us to evaluate the spatiotemporal patterns according to integrated water and balance, which influence the presence of nitrate in groundwater. Therefore, the transport of nitrate in the aquifer system is evaluated using inputs and transformations of different nitrogen compounds from the surface system that, when coupled in MODFLOW, recharge to the aquifer (Equation (1)).

$$NO3_{rcharg,i} = \left(1 - exp\left[\frac{-1}{\delta_{gw}}\right]\right) NO3_{perc} + exp\left[\frac{-1}{\delta_{gw}}\right] NO3_{rcharg,i-1} \tag{1}$$

where $NO3_{rchrg,i}$ is the amount of nitrate in the recharge entering the aquifer at day i (kg N/ha), δgw is the lag time or drainage time of the overlying geological formations (days), $NO3_{perc}$ is the total amount of nitrate leaving the bottom of the soil profile on day i (kg N/ha), $NO3_{rchrg,i-1}$ is the amount of nitrate in the recharge entering the aquifer at day $i-1$ (kg N/ha).

The linkage between the models is performed by means of hydrologic response units (HRU) defined in SWAT and the numerical grid made in the aquifer in MODFLOW. Once the concentrations of nitrate enter the aquifer with the recharge, its transport is simulated with the MT3DMS model. Advection and hydrodynamic dispersion were simulated (Equation (2)) considering nitrate as a conservative compound, once it enters the aquifer.

$$\frac{\partial(\theta C_k)}{\partial t} = \frac{\partial}{\partial x_i}\left[\theta D_{ij}\frac{\partial C_K}{\partial x_j}\right] - \frac{\partial}{\partial x_i}(\theta v_i C_K) + q_s C_s^k + \sum R_n \tag{2}$$

where θ is the porosity of the geological media (dimensionless), C_k is the dissolved concentration of species k (ML^{-3}), t is time (t), x (i,j) is the distance along the respective coordinate axis (L), D_{ij} is the tensor of the hydrodynamic dispersion coefficient (L/t), v_i is the infiltration velocity or linear pore water velocity (Lt^{-1}), q_s is the volumetric flow rate per volume unit of the aquifer representing fluid sources and sinks (t^{-1}), C_s^k is the concentration of source or sink flow for species k (ML^{-3}), and $\sum R_n$ is the chemical reaction term (ML^{-3} t^{-1}).

2.3. Watershed Model

The hydrological modeling was performed by means of ten climatological stations. The precipitation and temperature series cover the years between 1960 and 2010. This information was obtained from the CLICOM database [103].

Observed surface runoff point information was collected from hydrometric stations permanently established within the drainage network. There are three hydrometric stations in the study area, with a variable period according to each station (1960–1989, 1960–2002, and 2000–2002, respectively). This information was obtained from the BANDAS database [104].

The thematic georeferenced geographic Information was obtained from INEGI (https://www.inegi.org.mx/temas/; accessed on 28 October 2023), including the digital elevation model (DEM), land use, and soil type. The DEM was entered with a resolution of 500 by 500 m. The information about soil type was provided by thematic georeferenced geographic information. This information includes basic soil characteristics, such as dominant surface texture and chemical (salt, sodium) or physical (rock, tepetate, stoniness). The reclassification of the type of use map was carried out according to the SWAT database. The modeling period was from 1960 to 2010 (50 years). The basin was delimited into 73 subbasins with a threshold of 300 ha. Potential evapotranspiration was calculated using the Penman–Monteith method. A global agricultural scheme was proposed for the entire study area, in which it was considered seasonal, starting in February–March and ending in September–October. The planted crop was corn, given that it is the predominant crop in the region (Table 1). The modeled actions were planting, fertilizing, and cultivating without leaving residues in the fields. The proposed fertilizer application is 200 to 600 kg/ha, with a mineral nitrogen content of 0 to 46% or an organic nitrogen content of 0.5 to 3.7%.

Table 1. Soil characteristics calibrated in the study area.

Soil Type	Hydraulic Conductivity (mm/h)	Hydrogeological Group	Soil Thickness (mm)	Base Flow Factor (1/d)	Porosity Fraction	Organic Carbon Content (%)
Vertic luvisol	15	A	3500	0.01	0.005	0.05
Dystric andosol	5	B	1400	0.06	1	0.7
Fine leptic skeletal	2	D	300	0.01	0.005	0
Eutric regosol	2	D	1000	0.06	0.7	0.05
Leptic planosol dístric	10	A	1000	0.048	0.8	0
Phaeozem skeletal	10	A	1500	0.01	1	1
Lytic dystric leptosol	10	A	1000	0.01	1	0

To simulate the nitrate passing to the subsoil, the amount of nitrate runoff and infiltration was considered equal in the model. This is the simple model used by Perez [79]. Since around 1970 in LCB, agricultural activities started intensively due to the enlargement of the hydraulic structure, the initial reserve of ammonia in the soil, and the initial concentrations of the aquifer being considered zero. Nitrogen consumption by plants is according to soil use.

The model parameters that were calibrated cover mainly watershed and soil characteristics, the relationship with groundwater, and agricultural practices; these were hydraulic conductivity, hydrogeological group, thickness depth, and base flow factor; in the case of nitrate, it is the porosity fraction from which anions are excluded and the organic carbon content. The values obtained are shown in Table 1.

2.4. Groundwater Flow Model

The groundwater flow modeling was carried out in a single unconfined layer, considering anisotropy in the hydraulic properties, according to geological materials present in the area. The configuration of initial groundwater levels for the year 1970 proposed are according to the study carried out by CONAGUA [105]. Since no additional information on this aquifer is available, the vertical limits of the aquifer are proposed based on the topographic limits depicted using the DEM, with an average thickness of 400 m. The lateral boundary conditions of the aquifer consist of cells of constant heights and constant flow. The lower boundary in the domain was assumed as the impermeable boundary. On the margin of the lake, limits of imposed potential were established. For the model, the discretization of the study area was carried out in 26 columns by 30 rows in cells of 3000 by 3000 m. The 3507 km^2 area of the aquifer was represented by 407 active cells with 343 inactive cells. There are 40 cells defining the Cuitzeo Lake, assigning them a constant level of 1835 masl. Inflows to the aquifer are composed of lateral groundwater channels, vertical and induced recharge, and outflows are represented by well extraction, groundwater channels, spring discharge, and evapotranspiration. The modeling was performed in transient states for the period 1970 to 2010. The calibration period was divided into 480 stress periods where each stress period corresponds to one month.

Inflows and outflows from groundwater side channels, induced recharge, well extraction, groundwater channels, spring discharge, and evapotranspiration were used as annual values with constant values. The induced recharge and lateral groundwater were 38.5 and 88 hm^3, respectively [101]. Extraction by wells were 162.2 hm^3, groundwater channels were 3.3 hm^3, springs were 60.267 hm^3, and evapotranspiration was 81.8 hm^3 [71]. Recharge from precipitation was 182 hm^3, which has a spatial and temporal distribution according to the results obtained in the SWAT modeling (Figure 5).

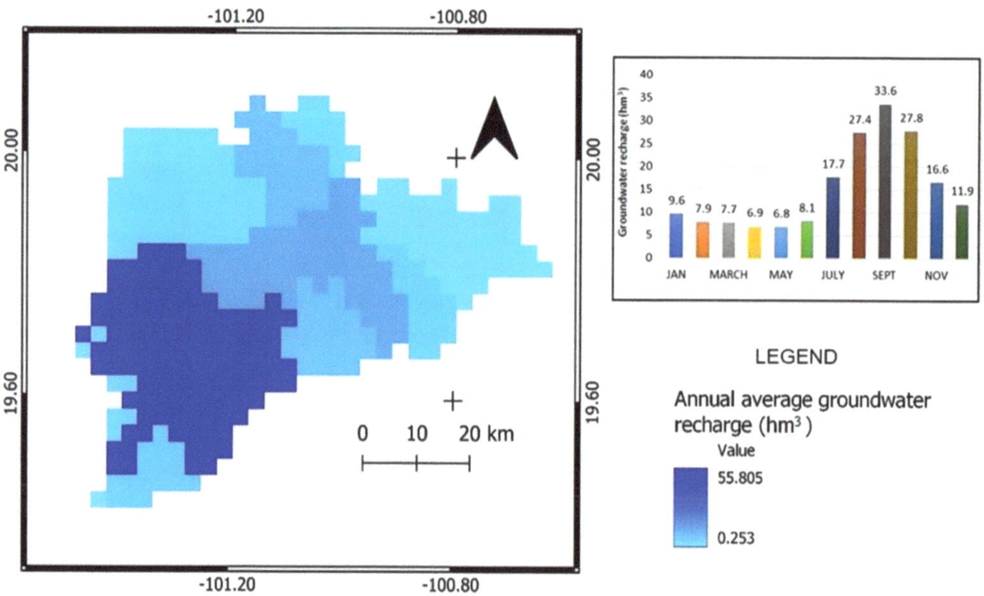

Figure 5. Monthly average recharge of the Morelia–Querendaro aquifer; 1 hm^3 = 1,000,000 m^3.

2.5. Nitrate Transport Model

The groundwater levels obtained from MODFLOW were used for the nitrate transport model. Porosity and longitudinal dispersivity were assigned depending on the soil material. The calibration of the transport model was carried out manually by trial and error. Only advection and longitudinal dispersion phenomena were considered in the modeling. We obtained a longitudinal dispersion value of 0.3–1.2 m. The daily average nitrate concentrations are calculated from the nitrate recharge and nitrate load, both monthly averages.

The monthly nitrate load entering the aquifer is considered with a spatial and temporal distribution according to the results of the SWAT model (Figure 6). April and May demonstrate a greater load of nitrates entering the aquifer, which coincides with the fertilizer placement in the soil.

The calibration was performed at thirteen calibration points. Three surface points to compare runoff and nitrate concentrations; ten points in the aquifer; five points with elevation information; and five nitrate concentration points (Figure 7). The surface points are located permanently. The stream network runs from southwest to northeast, stations S1 and S3 are located within the main network and S2 is located in a minor tributary. The surface nitrate calibration points are the only ones for which information is available. The groundwater level measurement points were selected because they have a greater amount of information.

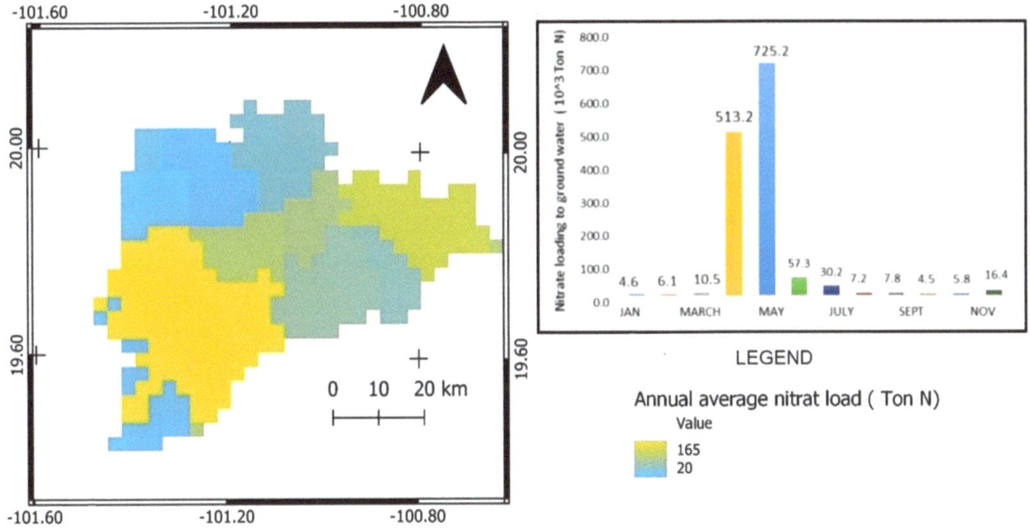

Figure 6. Annual and monthly average nitrate load entering the Morelia–Querendaro aquifer. Amounts are shown in 10^3 tons of nitrogen.

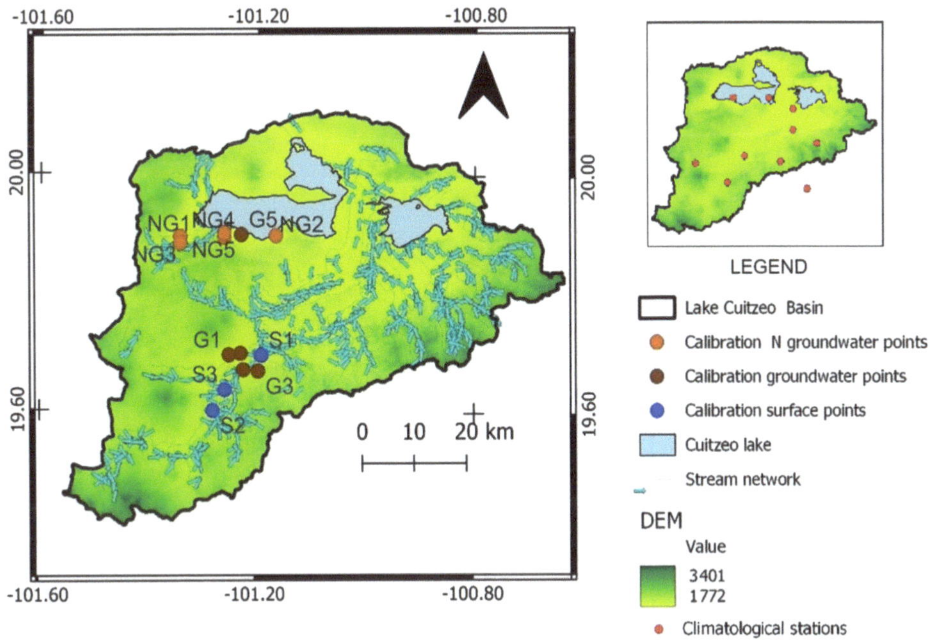

Figure 7. Quantitative and qualitative inputs and outputs of each model.

3. Model Calibration and Validation

Calibration and validation are performed using six statistical parameters to quantify the best fit of the models, distinguishing between the surface system and the aquifer system. These parameters are the mean absolute error (*MAE*; Equation (3)), the root mean square error (*RMSE*; Equation (4)), the correlation coefficient (*R*; Equation (5)), the Nash–

Sutcliffe efficiency index (NSE; Equation (6)), standard deviation of measured data (RSR; Equation (7)), and the percentage bias ($PBIAS$; Equation (8)).

$$MAE = \frac{\sum_{i=1}^{n}|H_i - M_i|}{n} \quad (3)$$

$$RMSE = \sqrt{\frac{\sum_{i=1}^{n}(H_i - M_i)^2}{n}} \quad (4)$$

$$R = \frac{\sum_{i=1}^{n}(H_i - H_m)(P_i - M_m)}{\sqrt{\sum_{i=1}^{n}(H_i - H_m)^2(M_i - M_m)^2}} \quad (5)$$

$$NSE = 1 - \frac{\sum_{i=1}^{n}(H_i - M_i)^2}{\sum_{i=1}^{n}(H_i - H_m)^2} \quad (6)$$

$$RSR = \frac{\sqrt{\sum_{i=1}^{n}(H_i - M_i)^2}}{\sqrt{\sum_{i=1}^{n}(H_i - M_m)^2}} \quad (7)$$

$$PBIAS = \frac{\sum_{i=1}^{n}H_i - M_i}{\sum_{i=1}^{n}H_i} \times 100 \quad (8)$$

where H_i is the historical value for event i, M_i is the modeled value for event i, H_m is the mean of historical values, M_m is the mean of modeled values, n is the number of events.

For the surface system, the parameters R, RSR, NSE, and $PBIAS$ are used to evaluate according to four levels of adjustment [106,107] for the quantification of the flow and transport of nitrogen compounds. In the case of R and NSE, values of one represent a perfect fit and values of zero for RSR and $PBIAS$ obtain the best results [59].

In the case of the aquifer system, MAE, R, and $RMSE$ parameters are used, which are widely used to evaluate the groundwater level fit [7,108,109]. For MAE and $RMSE$, values close to zero provide the best fit [110]. R has been used because it is recommended by several guidelines in hydrological and groundwater modeling [106,107,111,112]. $RMSE$ is one of the most widely used statistical parameters in groundwater SWAT/MODFLOW coupling [94,108,113–115].

3.1. Watershed Model

The calibration was carried out based on the surface runoff's monthly values. Two points, S1 and S2, were used to calibrate and point S3 was used to validate the series.

Calibration of nitrate concentrations was performed based on daily values for 2008 and 2009.

According to the statistical parameters, values ranging from "poor" to "very good" were obtained. It is observed that the values of the historical mean and the modeled mean are similar. The statistical parameter with the best fit is R, and the one with the poorest fit is RSR (Table 2).

Table 2. SWAT model fit of flow rate and nitrate concentrations according to statistical parameters.

Type	Point	M.M.	M.H.	*R*	*NSE*	*RSR*	*PBIAS*
Flow	S1	0.40	0.34	0.58	0.33	0.82	16.17
	S2	2.14	2.36	0.80	0.51	0.74	−15.18
	S3	0.33	0.37	0.84	0.42	0.76	−8.31
Nitrate	S3	0.29	0.335	0.45	0.18	0.89	−16.29

M.M. = modeled mean; M.H. = historical mean.

At S1, where $R = 0.60$ and $PBIAS = 16.17$, a "good" level was reached. $NSE = 0.33$ had an "acceptable" level, while $RSR = 0.82$ was a "poor" level. In S2, $R = 0.80$ obtained a "very good" level. Both $PBIAS = -15.18$ and $NSE = 0.51$ showed a "good" classification, while $RSR = 0.74$ reached a "poor" level. At point S3, a "very good" level was presented, with $R = 0.84$. $PBIAS = 8.31$ had an "excellent" level, while $NSE = 0.42$ and $RSR = 0.76$ were a "poor" level. At S3 for nitrate. The $PBIAS$ of -16.29 was reached, which is considered "good", while the R with a value of 0.45 is "acceptable". The parameters NSE and RSR are a "poor" fit.

3.2. Groundwater Flow Model

Results of the groundwater level estimations results were calibrated with five wells, which are located in the central part of the aquifer, where most observed data are available. In both cases, similar values were observed in the monthly mean and correlation coefficient, $R > 0.61$. These values are considered "acceptable" (Table 3).

Table 3. MODFLOW and MT3DMS model fits according to statistical parameters.

Type	Point	M.M. [1]	M.H. [1]	*MAE*	*RMSE*	*R*
Groundwater	G1	1888.54	1885.46	0.35	4.64	1
	G2	1875.71	1869.98	0.64	6.9	0.98
	G3	1909.23	1914.87	0.93	8.46	0.85
	G4	1887.45	1890.69	0.08	1.2	0.96
	G5	1820.71	1820.18	0.02	0.99	0.61

M.M. = modeled mean; M.H. = historical mean. [1] The units of M.M and M.H. are m for groundwater level.

In G1, there is a difference between the modeled and historical mean groundwater level of 3.08 m, where $MAE = 0.35$, $RMSE = 4.64$, and $R = 1$. G2 has the largest difference between the modeled and historical mean groundwater level of 5.73 m, where $MAE = 0.64$, $RMSE = 6$, and $R = 0.98$. In G3, elevations of 1909.23 and 1914.87 were obtained, where $MAE = 0.93$, $RMSE = 8.46$, and $R = 0.85$. Overall, the best modeling results were acquired in G4, although there is a difference between the mean modeled and historical groundwater level of 3.24 m, with an $R = 0.96$, $MAE = 0.08$, and $RMSE = 1.2$. In G5, similar results were obtained for modeled and historical mean groundwater level of 1877.31 and 1877.60 masl, respectively. The results $R = 0.16$, $MAE = 0.01$, and $RMSE = 0.89$ were obtained.

3.3. Nitrate Transport Model

Calibration of nitrate concentrations in groundwater was performed over a 4-month period (March, April, October 2013 and February 2014). The comparison of historical and modeled means, as well as the results of the statistical parameters, are presented in Table 4. The value of the correlation coefficient according to the calibration data could not be calculated. In NG1, the modeled and historical mean values are almost equal, with $MAE = 0.004$, $RMSE = 0.0004$. In NG2, the simulated and historical mean values are 0.0079 and 0.0074, respectively, $MAE = 0.0007$, $RMSE = 0.008$. In NG3, the modeled and historical means are 0.0044 and 0.0046, respectively, while $MAE = 0.0098$, $RMSE = 0.0107$. In NG4, the simulated and historical mean values are 0.003 and 0.0061, respectively, with $MAE = 0.0031$, $RMSE = 0.0034$. Finally, in NG5, the modeled and historical means are 0.00288 and 0.0041, respectively, while $MAE = 0.0012$, $RMSE = 0.001$.

Table 4. MT3DMS models fit according to statistical parameters.

Type	Point	M.M. [1]	M.H. [1]	*MAE*	*RMSE*
Nitrate	NG1	0.0115	0.0113	0.0004	0.0004
	NG2	0.0079	0.0074	0.0007	0.0008
	NG3	0.0144	0.0046	0.0098	0.0107
	NG4	0.0030	0.0061	0.0031	0.0034
	NG5	0.0028	0.0041	0.0012	0.001

M.M. = modeled mean; M.H. = historical mean. [1] The units of M.M and M.H. are mg/L for nitrate concentrations.

4. Results

4.1. Watershed Model

The results presented in the SWAT hydrologic modeling show an underestimation of peak flows. At the S1, modeling was performed for the period from 1960 to 1989; the historical and modeled mean obtained were 0.34 and 0.40 m^3/s, respectively (Figure 8A). At the S2, modeling was performed for the period from 1960 to 2002. The historical and modeled mean obtained were 2.36 and 2.14 m^3/s, respectively (Figure 8B). At the S3, the modeling was validated over a period from 2000 to 2002; the historical and modeled mean obtained were 0.37 and 0.33 m^3/s, respectively (Figure 8C).

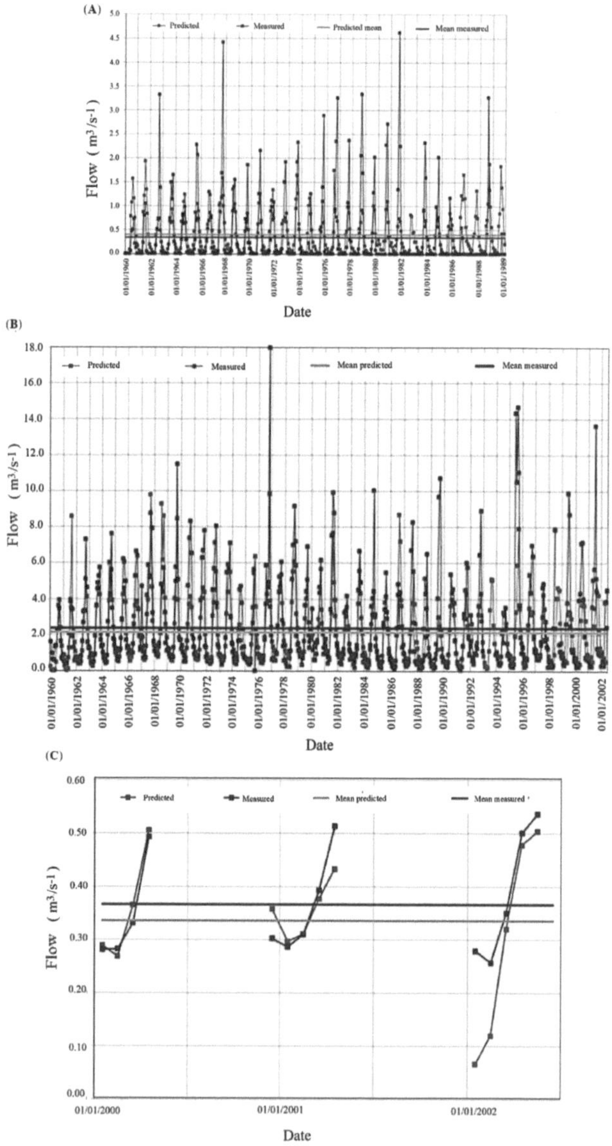

Figure 8. Historical and modeled flow series, calibrated. (**A**) Point S1, period 1960–1989. (**B**) Point S2, period 1960–2002. (**C**) Point S3. Validation, period 2000–2002. 1 hm^3 = 1,000,000 m^3.

The calibration of nitrate concentrations was performed based on daily values for 2008 and 2009. The historical and modeled means obtained were 0.29 and 0.335 mg N/L, respectively (Figure 9).

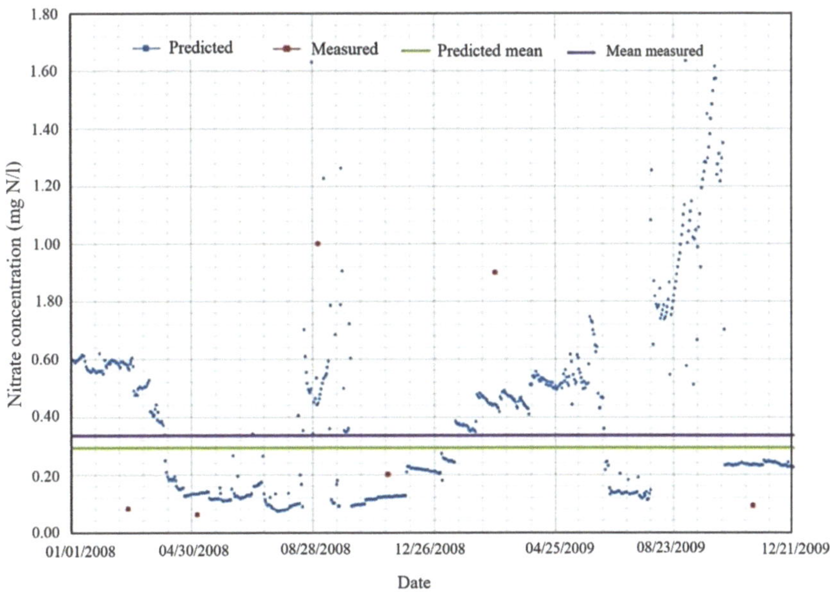

Figure 9. Point S3, calibrated. Historical and modeled nitrate concentrations, period 2008–2009.

The monthly series of nitrate concentration in the river was obtained. Therefore, the values of nitrate concentrations range from 0 to 3.42 mg N/L, with a monthly average of 0.43 mg N/L.

4.2. Groundwater Flow Model

The modeling results of nitrate transport in the aquifer were calibrated at six points, all located on the southern margin of Lake Cuitzeo, which is considered the discharge area for the aquifer system. The historical and modeled means obtained were 1876.328 masl and 1876.236 masl, respectively. At point G4, there is a significant drop in water table levels due to the fact that it is located in an urban area (Figure 10A). At point G5, on the other hand, being on the shore of Lake Cuitzeo, the groundwater level remains practically constant over time (Figure 10B).

4.3. Nitrate Transport Model

Results from the estimation of nitrate concentrations in groundwater show that observed and historical information indicates low concentrations according to the recommended limits for human use and consumption [116]. Since the results obtained with SWAT were calibrated using points in the southwest and the calibration of MT3DMS was performed using points near the lake, the calibration performed in MT3DMS shows a difference between modeled and historical values.

It is observed in Figure 11B–D that at points NG3, NG4, NG5, respectively, there is a graphical difference. At points NG4 and NG5, it is observed that the modeled mean (0.0030 and 0.0028, respectively) remains below the historical mean (0.0061 and 0.0041, respectively), and there is a difference of 0.0031 and 0.0013, respectively. These two points are observed to be close; therefore, it is considered that this underestimation of concentrations occurs in this area, in particular. In addition, at point NG3, there is an underestimation of the historical and modeled mean (0.0144 and 0.0046, respectively).

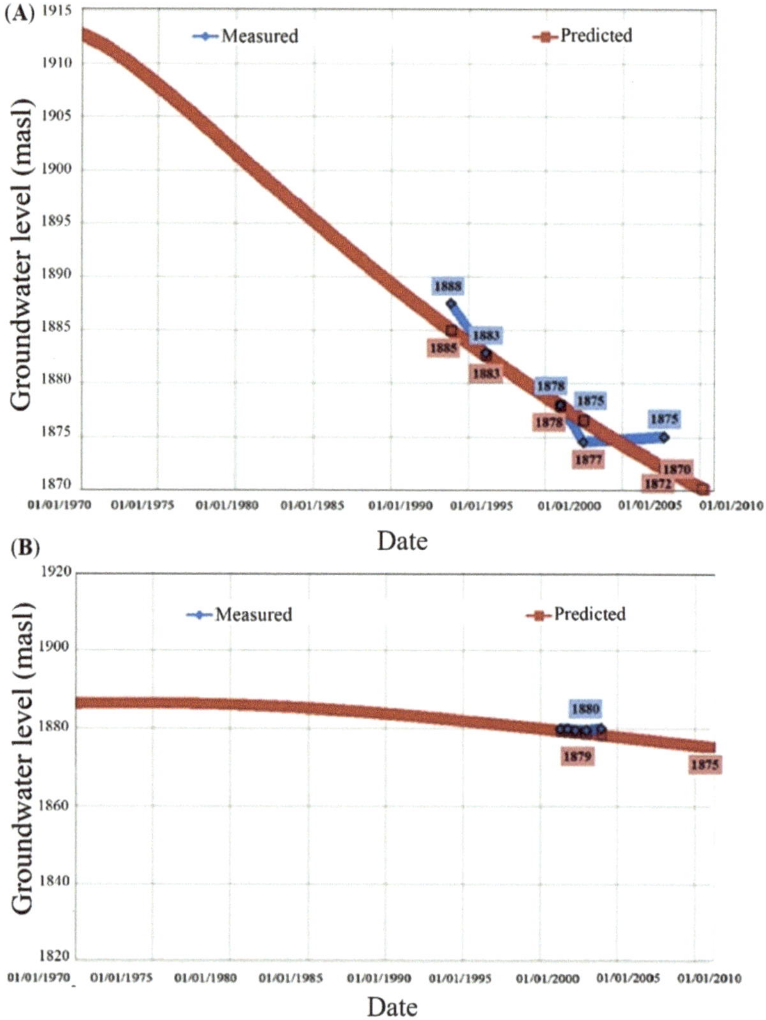

Figure 10. Historical and simulated groundwater level. (**A**) Point G4, period 1970–2010. (**B**) Point G5, period 1970–2010.

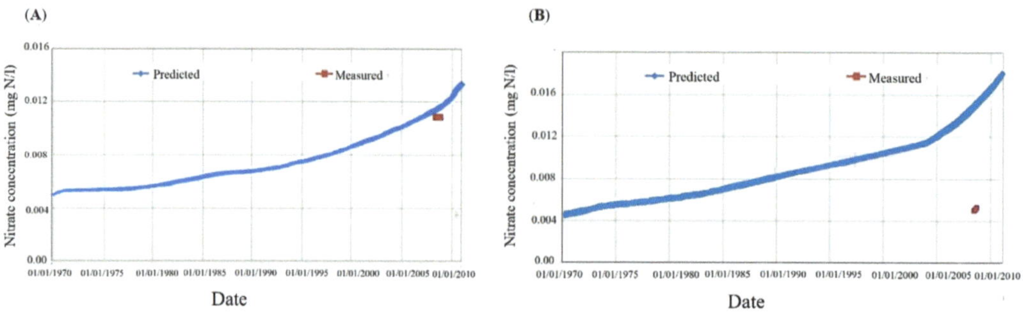

Figure 11. *Cont.*

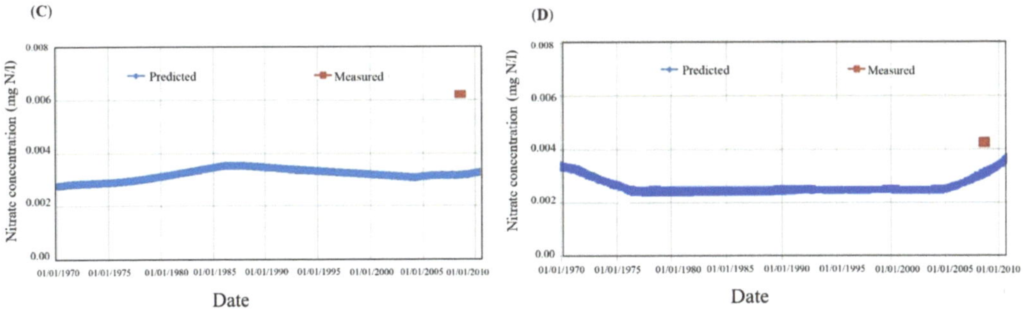

Figure 11. Nitrate concentrations in groundwater. (**A**) Point NG1. (**B**) Point NG3. (**C**) Point NG4. (**D**) Point NG5. Period simulated: 1970–2010.

5. Discussion

The following proposed method was used to assess nitrate concentration in groundwater by quantifying the spatiotemporal patterns of groundwater–surface water interaction with limited information. It is presented as a methodology, since according to studies [93,94,117–121], the coupling of SWAT/MODFLOW/MT3DMS and SWAT/MODFLOW models is performed by calibrating parameters associated with recharge and the passage of water through the subsurface (plant available soil water capacity, curve number, surface runoff lag coefficient, deep aquifer percolation fraction, and others). However, in the present study, in addition to these parameters associated with recharge, it is calibrated using the characteristics of a single soil type.

In recent years, the importance of quantifying the influence of multiple variables associated with the presence of nitrogen in groundwater has become evident [122–124]. One of the most relevant is the diffuse sources associated with agricultural areas [125–127]. However, this is not a variable that can be directly quantified, since it depends on soil characteristics, agricultural practices, hydrological factors, as well as factors inherent to the nitrogen cycle [128,129].

Over the years, several methodologies have been developed for the assessment of nitrogen compounds in groundwater. They mainly focus on the modeling of the groundwater system (aquifer nitrate) [130–132], leaving aside the interrelation of surface variables of great influence on the presence of nitrogen compounds. Furthermore, the spatial and temporal variation in these variables is not considered.

In recent years, studies at the basin level have been proposed for the management of water resources and their adequate management in the conjunctive surface–groundwater system [133,134]. Therefore, the coupling of the basin–aquifer system by means of mathematical models has been used more frequently in recent years, highlighting the importance of basin-level studies and their interrelationship with the subsurface system [59,96], as it allows for the assessment of climatic variables and soil characteristics that influence surface runoff and important spatiotemporal patterns that influence groundwater, as well as conceptualizing and quantifying their interrelationships [135,136].

However, the application of a mathematical model as a tool to assess the presence of nitrogenous compounds in groundwater requires information that is not equally available in all regions. Therefore, this methodology aims to take advantage of spatiotemporal patterns within the interaction of the hydrological system, surface–subsurface. The implementation of the SWAT model, as a tool to evaluate this interrelation of the hydrological system, presents a great advantage in water resources management. In addition, as observed in previous works [115,136], the calibrated parameters are related to the interrelationship that occurs in the subsoil [117,118,137], towards the recharge in order to make the coupling with MODFLOW. However, in this study, in addition to calibrating these parameters, soil characteristics were calibrated, as shown in Table 1.

The use of various statistical parameters reinforces the calibration of each stage with the comparison of the historical and modeled data, unlike other articles that present a smaller number of parameters used in the calibration or the absence of these in some stages of the calibration [118,137,138]. The values obtained in the statistical parameters performed using SWAT, with the SWAT/MODFLOW coupling, are presented with values similar to those of several authors [94,115,139] and the mean values for hydrological modeling and water quality in the calibration [59,106,107,140] are acceptable.

The lack of a better fit is due to the limited information and the uncertainty associated with the use of the model as a third stage, which did not allow for better results. The calibration performed in MODFLOW and MT3DMS presented a lower error than the present study [117,131,137,138], with acceptable results [132].

In addition, this methodology evaluates diffuse sources through a global scheme of agriculture in the study area to quantify the presence of nitrate in groundwater considering variables of the water system in a spatiotemporal way and presenting a first approximation of the interrelation of the surface water–groundwater system. This can be used in regions with limited information and prone to the presence of nitrogen compounds in groundwater, whose main source is agriculture, to be used as a tool to evaluate diverse schemes of practices within the water system and its interrelationships. One of the main challenges in the evaluation of nitrogen compounds in the study area is the variables associated with soil characteristics. Therefore, as a first step, the present study takes the equal partitioning of nitrate to runoff and infiltration, as is performed in some models [79], in addition to considering nitrate in the subsoil as a conservative being a possible limiting factor. Therefore, for future studies, there may be an area of opportunity to evaluate the transformations that nitrate may present in subsoil, as well as studies associated with the factors that govern the dynamics of infiltration–runoff.

6. Conclusions

This paper presented an assessment of nitrate in groundwater from diffuse sources considering spatiotemporal patterns of hydrological systems using a coupled SWAT/MODFLOW/MT3DMS model involving the variability of different factors that directly or indirectly influence nitrate transport to groundwater.

Groundwater recharge was calculated based on climatological variables, such as temperature and precipitation, topography, geology, hydraulic parameters, and soil information. Subsequently, a nitrogen balance in the subsoil was performed to obtain the amount of nitrogen that can drain or run off from the unsaturated zone to the aquifer. The implementation of a surface hydrological model such as SWAT, for this stage, can present a challenge in sites with limited information, but the present study was adapted to work with the available information within a simplified conceptual scheme.

In order to strengthen the methodology and validate the results quantitatively, statistical parameters were used to compare the results obtained against historical data. According to the results of the runoff simulation in SWAT, it was observed that the *R* and *PBIAS* present acceptable values in mathematical terms. Since they presented values close to one in R and small values of *PBIAS*, there was a coherent trend with the measured and modeled values. For the *NSE* parameter, intermediate values were obtained, considering an intermediate consistency. The *RSR* parameter resulted in the lowest level because the coincidence between historical and modeled maximum expenditures was reduced. The statistical results for the groundwater levels of the aquifer simulated in MODFLOW and compared with the observed values were considered with a good approximation, presenting high values in the correlation coefficients. The mean absolute error and mean square error values were regarded as acceptable values. Regarding nitrate transport, despite the scarce information on its presence in groundwater, the results are acceptable and valid since they comply with the restrictions of the statistical parameters. It was observed that, in the historical and modeled means, the values are similar, with high R values and low *MAE* and *RMSE* values close to zero.

The results obtained show that the concentrations present in the groundwater are small compared to those recommended by the WHO (50 mg/L as nitrate ion) [116]. In addition, it was observed that, according to the results obtained in the conjunctive simulation, agriculture and diffuse sources of nitrate concentration largely explain the concentrations observed in the groundwater of the study area.

One of the most important limitations of the methodology is the lack of data for groundwater management in time and space, which is a limitation to reducing the uncertainty of the estimated distributed values of groundwater elevation and nitrate concentrations, which we believe can be achieved by incorporating local stakeholders and water users in order to promote continuous monitoring of hydrological and chemical variables. In addition, this study does not quantify nitrogen transformations in the surface and groundwater system, considering physical and nonchemical factors in its transport.

It is worth mentioning that the methodology is recommended for hydrological basins in which there is great affectation by surface factors due to industrial and urban discharges or by diffuse contamination from agriculture. An adequate conceptualization of the study area is necessary, for example, identifying the main inputs and their associated processes, since the present methodology does not consider other sources of nitrogen compounds, such as wastewater discharges and livestock areas, among others.

Author Contributions: Conceptualization, J.H.-B. and A.C.-G.; methodology, J.H.-B. and M.A.H.-H.; software, M.A.M.-C. and J.H.-B.; validation, S.T.S.-Q., A.C.-G. and J.H.-B.; formal analysis, M.A.H.-H. and A.C.-G.; investigation, J.H.-B. and S.T.S.-Q.; resources, M.A.H.-H.; data curation, M.A.H.-H.; writing—original draft preparation, S.T.S.-Q. and J.H.-B.; writing—review and editing, M.A.H.-H. and S.T.S.-Q.; visualization, M.A.M.-C.; supervision, S.T.S.-Q.; project administration, S.T.S.-Q.; funding acquisition, M.A.M.-C.; and S.T.S.-Q. All authors have read and agreed to the published version of the manuscript.

Funding: This research received no external funding.

Data Availability Statement: Not applicable.

Acknowledgments: We thank the Catedras CONAHCYT program, the anonymous reviewers, and the editor for their constructive comments on the manuscript.

Conflicts of Interest: The authors declare no conflict of interest.

References

1. Valin, H.; Sands, R.D.; van der Mensbrugghe, D.; Nelson, G.C.; Ahammad, H.; Blanc, E.; Bodirsky, B.; Fujimori, S.; Hasegawa, T.; Havlik, P.; et al. The Future of Food Demand: Understanding Differences in Global Economic Models. *Agric. Econ.* **2014**, *45*, 51–67. [CrossRef]
2. Narain-Ford, D.M.; Bartholomeus, R.P.; Raterman, B.; van Zaanen, I.; ter Laak, T.T.; van Wezel, A.P.; Dekker, S.C. Shifting the Imbalance: Intentional Reuse of Dutch Sewage Effluent in Sub-Surface Irrigation. *Sci. Total Environ.* **2021**, *752*, 142214. [CrossRef] [PubMed]
3. Salmoral, G.; Viñarta Carbó, A.; Zegarra, E.; Knox, J.W.; Rey, D. Reconciling Irrigation Demands for Agricultural Expansion with Environmental Sustainability—A Preliminary Assessment for the Ica Valley, Peru. *J. Clean. Prod.* **2020**, *276*, 123544. [CrossRef]
4. Fazel, S.; Kakhki, F.; Sharifian, A.R.; Beikzadeh, N. Evaluation of Effect of Different Growing Media and Nitrogen Fertilizer on Some Morphological Traits in *Spathiphyllum wallisii* L. *J. Ornam. Plants* **2020**, *10*, 49–58.
5. Rosa, L.; Gabrielli, P. Energy and Food Security Implications of Transitioning Synthetic Nitrogen Fertilizers to Net-Zero Emissions. *Environ. Res. Lett.* **2023**, *18*, 014008. [CrossRef]
6. Wang, X.; Xu, M.; Lin, B.; Bodirsky, B.L.; Xuan, J.; Dietrich, J.P.; Stevanović, M.; Bai, Z.; Ma, L.; Jin, S.; et al. Reforming China's Fertilizer Policies: Implications for Nitrogen Pollution Reduction and Food Security. *Sustain. Sci.* **2023**, *18*, 1063–1076. [CrossRef]
7. He, J.; Yao, J.; Li, A.; Tan, Z.; Xie, G.; Shi, H.; Zhang, X.; Sun, W.; Du, P. Potential Impact of Water Transfer Policy Implementation on Lake Eutrophication on the Shandong Peninsula: A Difference-in-Differences Approach. *Hydrol. Res.* **2020**, *51*. [CrossRef]
8. Health Canada. *Guidelines for Canadian Drinking Water Quality: Guideline Technical Document—Nitrate and Nitrite*; Minister of Health: Ottawa, ON, Canada, 2013; Volume 30, ISBN 9781100229997.
9. Zimnicki, T.; Boring, T.; Evenson, G.; Kalcic, M.; Karlen, D.L.; Wilson, R.S.; Zhang, Y.; Blesh, J. On Quantifying Water Quality Benefits of Healthy Soils. *Bioscience* **2020**, *70*, 343–352. [CrossRef]
10. Vitoria, I.; Maraver, F.; Sánchez-Valverde, F.; Armijo, F. Contenido En Nitratos de Aguas de Consumo Público Españolas. *Gac. Sanit.* **2015**, *29*, 217–220. [CrossRef]

11. García Torres, E.; Pérez Morales, R.; González Zamora, A.; Ríos Sánchez, E.; Olivas Calderón, E.H.; de Alba Romero, J.J.; Calleros Rincón, E.Y. Consumption of Water Contaminated by Nitrate and Its Deleterious Effects on the Human Thyroid Gland: A Review and Update. *Int. J. Environ. Health Res.* **2022**, *32*, 984–1001. [CrossRef]
12. Van Maanen, J.M.S.; van Dijk, A.; Mulder, K.; de Baets, M.H.; Menheere, P.C.A.; van der Heide, D.; Mertens, P.L.J.M.; Kleinjans, J.C.S. Consumption of Drinking Water with High Nitrate Levels Causes Hypertrophy of the Thyroid. *Toxicol. Lett.* **1994**, *72*, 365–374. [CrossRef] [PubMed]
13. Grosse, Y.; Baan, R.; Straif, K.; Secretan, B.; El Ghissassi, F.; Cogliano, V. Carcinogenicity of Nitrate, Nitrite, and Cyanobacterial Peptide Toxins. *Lancet Oncol.* **2006**, *7*, 628–629. [CrossRef] [PubMed]
14. Chowdary, V.M.; Rao, N.H.; Sarma, P.B.S. Decision Support Framework for Assessment of Non-Point-Source Pollution of Groundwater in Large Irrigation Projects. *Agric. Water Manag.* **2005**, *75*, 194–225. [CrossRef]
15. Holt, M.S. Sources of Chemical Contaminants and Routes into the Freshwater Environment. *Food Chem. Toxicol.* **2000**, *38*, S21–S27. [CrossRef] [PubMed]
16. Lake, I.R.; Lovett, A.A.; Hiscock, K.M.; Betson, M.; Foley, A.; Sünnenberg, G.; Evers, S.; Fletcher, S. Evaluating Factors Influencing Groundwater Vulnerability to Nitrate Pollution: Developing the Potential of GIS. *J. Environ. Manag.* **2003**, *68*, 315–328. [CrossRef] [PubMed]
17. Debele, B.; Srinivasan, R.; Parlange, J.Y. Coupling Upland Watershed and Downstream Waterbody Hydrodynamic and Water Quality Models (SWAT and CE-QUAL-W2) for Better Water Resources Management in Complex River Basins. *Environ. Model. Assess.* **2008**, *13*, 135–153. [CrossRef]
18. Jayakrishnan, R.; Srinivasan, R.; Santhi, C.; Arnold, J.G. Advances in the Application of the SWAT Model for Water Resources Management. *Hydrol. Process* **2005**, *19*, 749–762. [CrossRef]
19. Ostad-Ali-Askari, K.; Shayannejad, M. Quantity and Quality Modelling of Groundwater to Manage Water Resources in Isfahan-Borkhar Aquifer. *Environ. Dev. Sustain.* **2021**, *23*, 15943–15959. [CrossRef]
20. Srinivasan, R.; Arnold, J.G. Integration of a Basin-Scale Water Quality Model with GIS. *J. Am. Water Resour. Assoc.* **1994**, *30*, 453–462. [CrossRef]
21. Ireson, A.; Makropoulos, C.; Maksimovic, C. Water Resources Modelling under Data Scarcity: Coupling MIKE BASIN and ASM Groundwater Model. *Water Resour. Manag.* **2006**, *20*, 567–590. [CrossRef]
22. Refsgaard, J.C.; Højberg, A.L.; Møller, I.; Hansen, M.; Søndergaard, V. Groundwater Modeling in Integrated Water Resources Management—Visions for 2020. *Ground Water* **2010**, *48*, 633–648. [CrossRef] [PubMed]
23. Devia, G.K.; Ganasri, B.P.; Dwarakish, G.S. A Review on Hydrological Models. *Aquat. Procedia* **2015**, *4*, 1001–1007. [CrossRef]
24. Cheng, G.D.; Li, X. Integrated Research Methods in Watershed Science. *Sci. China Earth Sci.* **2015**, *58*, 1159–1168. [CrossRef]
25. He, C.; Harden, C.P.; Holden, J.; Mdee, A. Watershed Science: Coupling Hydrological Science and Water Resources Management. *Hydrol. Process* **2023**, *37*, e14889. [CrossRef]
26. Témez, J.R. Modelo Matemático de Trasformación "Precipitación-Escorrentía". *Asoc. De Investig. Ind. Eléctrica* **1977**, *39*.
27. Burnash, R.J.C.; Ferral, R.L.; McGuire, R.A. *A Generalized Streamflow Simulation System–Conceptual Modeling for Digital Computers*; US Department of Commerce: Washington, DC, USA; National Weather Service: Silver Spring, MA, USA; State of California, Department of Water Resources: Sacramento, CA, USA, 1973.
28. *Soil Conservation Service (SCS) National Engineering Handbook, Section 4*; Hydrology, Soil Conservation Service, US Department of Agriculture: Washington, DC, USA, 1972.
29. Oñate-Valdivieso, F.; Bosque-Sendra, J.; Sastre-Merlin, A.; Miguel Ponce, V. Calibration, Validation and Evaluation of a Lumped Hydrological Model in a Mountain Area in Southern Ecuador. *Agrociencia* **2015**, *50*, 945–963.
30. Potenciano De Las Heras, A.; Villaverde Valero, J.J. Implementación del Modelo Hidrológico de Témez Para la Evaluación de Recursos Hídricos Con Grass GIS: Fase Superficial y Subterránea; 2009. II Jornadas de Sig Libre. Universitat de Girona. ISBN 978-48-691-9409-6. Available online: https://dugi-doc.udg.edu/bitstream/handle/10256/1387/C18.pdf?sequence=1&isAllowed=y; (accessed on 1 November 2023).
31. CEDEX. *Cálculo Hidrometeorológico de Aportaciones y Crecidas. Manual CHAC*; CEDEX: Madrid, Spain, 2013.
32. Gallart, F.; Delgado, J.; Beatson, S.J.V.; Posner, H.; Llorens, P.; Marcé, R. Analysing the Effect of Global Change on the Historical Trends of Water Resources in the Headwaters of the Llobregat and Ter River Basins (Catalonia, Spain). *Phys. Chem. Earth* **2011**, *36*, 655–661. [CrossRef]
33. Katsanou, K.; Lambrakis, N. Modelleren van de Helleense Karststroomgebieden Met Het Sacramento Soil Moisture Accounting Model. *Hydrogeol. J.* **2017**, *25*, 757–769. [CrossRef]
34. Ajmal, M.; Waseem, M.; Kim, D.; Kim, T.-W. A Pragmatic Slope-Adjusted Curve Number Model to Reduce Uncertainty in Predicting Flood Runoff from Steep Watersheds. *Water* **2020**, *12*, 1469. [CrossRef]
35. Młyński, D.; Wałęga, A.; Książek, L.; Florek, J.; Petroselli, A. Possibility of Using Selected Rainfall-Runoff Models for Determining the Design Hydrograph in Mountainous Catchments: A Case Study in Poland. *Water* **2020**, *12*, 1450. [CrossRef]
36. Psomiadis, E.; Soulis, K.X.; Efthimiou, N. Using SCS-CN and Earth Observation for the Comparative Assessment of the Hydrological Effect of Gradual and Abrupt Spatiotemporal Land Cover Changes. *Water* **2020**, *12*, 1386. [CrossRef]
37. Arnold, J.G.; Srinivasan, R.; Muttiah, R.S.; Williams, J.R. Large Area Hydrologic Modeling and Assessment. Part I: Model Development. *Am. Water Resour. Assoc.* **1998**, *34*, 73–89. [CrossRef]

38. Feldman, A.D.; Hydrologic Engineering Center (U.S.). *Hydrologic Modeling System HEC-HMS Technical Reference Manual*; US Army Corps of Engineers: Washington, DC, USA, 2000.
39. Muñoz-Carpena, R.; Parsons, J.E.; Gilliam, J.W. Modeling Hydrology and Sediment Transport in Vegetative Filter Strips. *J. Hydrol.* **1999**, *214*, 111–129. [CrossRef]
40. Ibáñez, J.F.; Valentín, M.G.; Jang, D. Application of the Kineros 2 Model to Natural Basin for Estimation of Erosion. *Appl. Sci.* **2021**, *11*, 9320. [CrossRef]
41. Hernandez, M.; Miller, S.N.; Goodrich, D.C.; Goff, B.F.; Kepner, W.G.; Edmonds, C.M.; Jones, K.B. Modeling Runoff Response to Land Cover and Rainfall Spatial Variability in Semi-Arid Watersheds. *Environ. Monit. Assess.* **2000**, *64*, 285–298. [CrossRef]
42. Miller, S.N.; Kepner, W.G.; Mehaffey, M.H.; Hernandez, M.; Miller, R.C.; Goodrich, D.C.; Devonald, K.K.; Heggem, D.T.; Miller, W.P. Integrating Landscape Assessment and Hydrologic Modeling for Land Cover Change Analysis. *Proc. J. Am. Water Resour. Assoc.* **2002**, *38*, 915–929. [CrossRef]
43. Michaud, J.; Sorooshian, S. Comparison of Simple versus Complex Distributed Runoff Models on a Midsized Semiarid Watershed. *Water Resour. Res.* **1994**, *30*, 593–605. [CrossRef]
44. Downer, C.W.; Ogden, F.L. GSSHA: Model to Simulate Diverse Stream Flow Producing Processes. *J. Hydrol. Eng.* **2004**, *9*, 161–174. [CrossRef]
45. Sith, R.; Nadaoka, K. Comparison of SWAT and GSSHA for High Time Resolution Prediction of Stream Flow and Sediment Concentration in a Small Agricultural Watershed. *Hydrology* **2017**, *4*, 27. [CrossRef]
46. Ogden, F.L.; Raj Pradhan, N.; Downer, C.W.; Zahner, J.A. Relative Importance of Impervious Area, Drainage Density, Width Function, and Subsurface Storm Drainage on Flood Runoff from an Urbanized Catchment. *Water Resour. Res.* **2011**, *47*. [CrossRef]
47. Lagos, M.S.; Muñoz, J.F.; Suárez, F.I.; Fuenzalida, M.J.; Yáñez-Morroni, G.; Sanzana, P. Investigating the Effects of Channelization in the Silala River: A Review of the Implementation of a Coupled MIKE-11 and MIKE-SHE Modeling System. *Wiley Interdiscip. Rev. Water* **2023**. [CrossRef]
48. Sandholt, I.; Rasmussen, K.; Andersen, J. A Simple Interpretation of the Surface Temperature/Vegetation Index Space for Assessment of Surface Moisture Status. *Remote Sens. Environ.* **2002**, *79*, 213–224. [CrossRef]
49. Refsgaard, J.C. Parameterisation, Calibration and Validation of Distributed Hydrological Models. *J. Hydrol.* **1997**, *198*, 69–97. [CrossRef]
50. Golmohammadi, G.; Prasher, S.; Madani, A.; Rudra, R. Evaluating Three Hydrological Distributed Watershed Models: MIKE-SHE, APEX, SWAT. *Hydrology* **2014**, *1*, 20–39. [CrossRef]
51. Andreu, J.; Sala, J.D. *La Modelación en la Planificación Hidráulica: Modelos de Simulación y Síntesis de Hidrología Superficial*; Servicios de Publicación UPV: Valencia, Spain, 1986.
52. McDonald, M.; Harbaugh, A.W. A Modular Three-Dimensional Finite Difference Ground-Water Flow Model. In *Techniques of Water-Resources Investigations, Book 6*; US Geological Survey: Reston, VA, USA, 1988; p. 588.
53. Andreu, J.; Capilla, J.; Sanchís, E. AQUATOOL, a Generalized Decision-Support System for Water-Resources Planning and Operational Management. *J. Hydrol.* **1996**, *177*, 269–291. [CrossRef]
54. Sahuquillo, A.; Andreu, J. Eigenvalue Simulation of Aquifers and River Interaction for Conjunctive Use. In Proceedings of the Congress of the International Association of Hydraulic Research, IAHR, San Francisco, CA, USA, 10–15 August 1997; Volume C.
55. Flores-Fernández, G.C.; García-Romero, L.; Sánchez-Quispe, S.T. Assessing Climate Change Impacts on Hydrology: Application to Zacapu and Pastor Ortiz Aquifers (Mexico). *J. Water Clim. Change* **2022**, *13*, 2129–2142. [CrossRef]
56. Guzman, J.A.; Moriasi, D.N.; Gowda, P.H.; Steiner, J.L.; Starks, P.J.; Arnold, J.G.; Srinivasan, R. A Model Integration Framework for Linking SWAT and MODFLOW. *Environ. Model. Softw.* **2015**, *73*, 103–116. [CrossRef]
57. Restrepo, J.I.; Montoya, A.M.; Obeysekera, J. A Wetland Simulation Module for the MODFLOW Ground Water Model. *Ground Water* **1998**, *36*, 764–770. [CrossRef]
58. Appelo, C.A.J.; Rolle, M. PHT3D: A Reactive Multicomponent Transport Model for Saturated Porous Media. *Ground Water* **2010**, *48*, 627–632. [CrossRef]
59. Parajuli, P.B.; Nelson, N.O.; Frees, L.D.; Mankin, K.R. Comparison of AnnAGNPS and SWAT Model Simulation Results in USDA-CEAP Agricultural Watersheds in South-Central Kansas. *Hydrol. Process* **2009**, *23*, 748–763. [CrossRef]
60. Yang, C.C.; Prasher, S.O.; Wang, S.; Kim, S.H.; Tan, C.S.; Drury, C.; Patel, R.M. Simulation of Nitrate-N Movement in Southern Ontario, Canada with DRAINMOD-N. *Agric. Water Manag.* **2007**, *87*, 299–306. [CrossRef]
61. Barkow, I.S.; Oswald, S.E.; Lensing, H.-J.; Munz, M. Seasonal Dynamics Modifies Fate of Oxygen, Nitrate, and Organic Micropollutants during Bank Filtration—Temperature-Dependent Reactive Transport Modeling of Field Data. *Environ. Sci. Pollut. Res.* **2021**, *28*, 9682–9700. [CrossRef] [PubMed]
62. Liu, P.; Wang, G.; Shang, M.; Liu, M. Groundwater Nitrate Bioremediation Simulation of In Situ Horizontal Well by Microbial Denitrification Using PHREEQC. *Water Air Soil. Pollut.* **2021**, *232*, 356. [CrossRef]
63. Ranjbar, F.; Jalali, M. Measuring and Modeling Ammonium Adsorption by Calcareous Soils. *Environ. Monit. Assess.* **2013**, *185*, 3191–3199. [CrossRef] [PubMed]
64. Johnes, P.J. Evaluation and Management of the Impact of Land Use Change on the Nitrogen and Phosphorus Load Delivered to Surface Waters: The Export Coefficient Modelling Approach. *J. Hydrol.* **1996**, *183*, 323–349. [CrossRef]
65. De Girolamo, A.M.; Spanò, M.; D'Ambrosio, E.; Ricci, G.F.; Gentile, F. Developing a Nitrogen Load Apportionment Tool: Theory and Application. *Agric. Water Manag.* **2019**, *226*, 105806. [CrossRef]

66. Bell, C.D.; Tague, C.L.; McMillan, S.K. Modeling Runoff and Nitrogen Loads From a Watershed at Different Levels of Impervious Surface Coverage and Connectivity to Storm Water Control Measures. *Water Resour. Res.* **2019**, *55*, 2690–2707. [CrossRef]
67. Han, B.; Reidy, A.; Li, A. Modeling Nutrient Release with Compiled Data in a Typical Midwest Watershed. *Ecol. Indic.* **2021**, *121*, 107213. [CrossRef]
68. Povilaitis, A.; Stalnacke, P.; Vassiljev, A. Nutrient Retention and Export to Surface Waters in Lithuanian and Estonian River Basins. *Hydrol. Res.* **2012**, *43*, 359–373. [CrossRef]
69. Zhang, H.; Hiscock, K.M. Modelling the Effect of Forest Cover in Mitigating Nitrate Contamination of Groundwater: A Case Study of the Sherwood Sandstone Aquifer in the East Midlands, UK. *J. Hydrol.* **2011**, *399*, 212–225. [CrossRef]
70. Strickling, H.L.; Obenour, D.R. Leveraging Spatial and Temporal Variability to Probabilistically Characterize Nutrient Sources and Export Rates in a Developing Watershed. *Water Resour. Res.* **2018**, *54*, 5143–5162. [CrossRef]
71. Chen, D.; Dahlgren, R.A.; Shen, Y.; Lu, J. A Bayesian Approach for Calculating Variable Total Maximum Daily Loads and Uncertainty Assessment. *Sci. Total Environ.* **2012**, *430*, 59–67. [CrossRef] [PubMed]
72. Li, X.; Feng, J.; Wellen, C.; Wang, Y. A Bayesian Approach of High Impaired River Reaches Identification and Total Nitrogen Load Estimation in a Sparsely Monitored Basin. *Environ. Sci. Pollut. Res.* **2017**, *24*, 987–996. [CrossRef]
73. Lu, J.; Gong, D.; Shen, Y.; Liu, M.; Chen, D. An Inversed Bayesian Modeling Approach for Estimating Nitrogen Export Coefficients and Uncertainty Assessment in an Agricultural Watershed in Eastern China. *Agric. Water Manag.* **2013**, *116*, 79–88. [CrossRef]
74. Yang, G.; Best, E.P.H.; Whiteaker, T.; Teklitz, A.; Yeghiazarian, L. A Screening-Level Modeling Approach to Estimate Nitrogen Loading and Standard Exceedance Risk, with Application to the Tippecanoe River Watershed, Indiana. *J. Environ. Manag.* **2014**, *135*, 1–10. [CrossRef]
75. Yan, R.; Brunner, P.; Gao, J. Simulation of Nitrogen Dynamics in Lowland Polders Using a New Coupled Modelling Approach: Insights into Management. *J. Clean. Prod.* **2021**, *313*, 127753. [CrossRef]
76. Wriedt, G.; Bronstert, A.; Geistlinger, H.; Rode, M. Modelling of Nitrogen Transport and Turnover during Soil and Groundwater Passage in a Small Lowland Catchment of Northern Germany. Ph.D. Thesis, Universität Potsdam, Potsdam, Germany, 2004.
77. Akhavan, S.; Abedi-Koupai, J.; Mousavi, S.-F.; Afyuni, M.; Eslamian, S.-S.; Abbaspour, K.C. Application of SWAT Model to Investigate Nitrate Leaching in Hamadan–Bahar Watershed, Iran. *Agric. Ecosyst. Environ.* **2010**, *139*, 675–688. [CrossRef]
78. Schilling, K.E.; Wolter, C.F. Modeling Nitrate-Nitrogen Load Reduction Strategies for the Des Moines River, Iowa Using SWAT. *Environ. Manag.* **2009**, *44*, 671–682. [CrossRef]
79. Pérez, M.Á. Modelo Distribuido de Simulación Del Ciclo Hidrológo y Calidad Del Agua, Integrado a Sistemas de Información Geográfica, Para Grandes Cuencas. Aportación La Análisis de Presiones e Impactos de La Directiva Marco Del Agua. Ph.D. Thesis, Universidad Politécnica de Valencia, Valencia, Spain, 2005.
80. Muñoz Del Prado, F.C.; Velarde Rodríguez, A. Actualización y Mejora Del Modelo Hidrológico Patrical En La Demarcación Hidrográfica Del Júcar. Ph.D. Thesis, Universidad Politécnica de Valencia, Valencia, Spain, 2016.
81. Zheng, C. *MT3D: A Modular Three-Dimensional Transport Model for Simulation of Advection, Dispersion and Chemical Reaction of Contaminants in Groundwater Systems*; SS Papadopulos & Associates: Bethesda, MD, USA, 1990.
82. Zheng, C.; Wang, P.P. *MT3DMS: A Modular Three-Dimensional Multispecies Transport Model for Simulation of Advection, Dispersion, and Chemical Reactions of Contaminants in Groundwater Systems*; Documentation and User's Guide; The University of Alabama: Tuscaloosa, AL, USA, 1999.
83. Clement, T.P. *A Modular Computer Code for Simulating Reactive Multi-Species Transport in 3-Dimensional Groundwater Systems*; Pacific Northwest National Lab.(PNNL): Richland, WA, USA, 1999.
84. Haerens, B.; Prommer, H.; Lerner, D.N.; Dassargues, A. Reactive Transport Modeling of a Groundwater Contamination by Ammoniacal Liquor. 2006. Available online: https://orbi.uliege.be/bitstream/2268/3409/1/publi152-2006.pdf (accessed on 27 October 2023).
85. Park, S.; Jeong, J.; Motter, E.; Bailey, R.T.; Green, C.H.M. Introducing APEXMOD—A QGIS Plugin for Developing Coupled Surface-Subsurface Hydrologic Modeling Framework of APEX, MODFLOW, and RT3D-Salt. *Environ. Model. Softw.* **2023**, *165*, 105723. [CrossRef]
86. Saaltink, M.W.; Carrera, J.; Ayora, C. On the Behavior of Approaches to Simulate Reactive Transport. *J Contam Hydrol.* **2001**, *48*, 213–235. [CrossRef] [PubMed]
87. Parkhurst, D.L.; Appelo, C.A.J. User's Guide to PHREEQC (Version 2)-a Computer Program for Speciation, Batch-Reaction, One-Dimensional Transport, and Inverse Geochemical Calculations. *Water-Resour. Investig. Rep.* **1999**, *99*, 312.
88. Prommer, H.; Barry, D.A.; Zheng, C. MODFLOW/MT3DMS-Based Reactive Multicomponent Transport Modeling. *Ground Water* **2003**, *41*, 247–257. [CrossRef] [PubMed]
89. Steefel, C.I.; Appelo, C.A.J.; Arora, B.; Jacques, D.; Kalbacher, T.; Kolditz, O.; Lagneau, V.; Lichtner, P.C.; Mayer, K.U.; Meeussen, J.C.L.; et al. Reactive Transport Codes for Subsurface Environmental Simulation. *Comput. Geosci.* **2015**, *19*, 445–478. [CrossRef]
90. Elçi, A.; Karadaş, D.; Fıstıkoğlu, O. The Combined Use of MODFLOW and Precipitation-Runoff Modeling to Simulate Groundwater Flow in a Diffuse-Pollution Prone Watershed. *Water Sci. Technol.* **2010**, *62*, 180–188. [CrossRef] [PubMed]
91. Styczen, M.; Storm, B. Modelling of N-Movements on Catchment Scale—A Tool for Analysis and Decision Making. *Fertil. Res.* **1993**, *36*, 1–6. [CrossRef]

92. Chunn, D.; Faramarzi, M.; Smerdon, B.; Alessi, D. Application of an Integrated SWAT–MODFLOW Model to Evaluate Potential Impacts of Climate Change and Water Withdrawals on Groundwater–Surface Water Interactions in West-Central Alberta. *Water* **2019**, *11*, 110. [CrossRef]
93. Molina-Navarro, E.; Bailey, R.T.; Andersen, H.E.; Thodsen, H.; Nielsen, A.; Park, S.; Jensen, J.S.; Jensen, J.B.; Trolle, D. Comparison of Abstraction Scenarios Simulated by SWAT and SWAT-MODFLOW. *Hydrol. Sci. J.* **2019**, *64*, 434–454. [CrossRef]
94. Ehtiat, M.; Jamshid Mousavi, S.; Srinivasan, R. Groundwater Modeling Under Variable Operating Conditions Using SWAT, MODFLOW and MT3DMS: A Catchment Scale Approach to Water Resources Management. *Water Resour. Manag.* **2018**, *32*, 1631–1649. [CrossRef]
95. Wei, X.; Bailey, R.T.; Records, R.M.; Wible, T.C.; Arabi, M. Comprehensive Simulation of Nitrate Transport in Coupled Surface-Subsurface Hydrologic Systems Using the Linked SWAT-MODFLOW-RT3D Model. *Environ. Model. Softw.* **2019**, *122*, 104242. [CrossRef]
96. Wei, X.; Bailey, R.T. Evaluating Nitrate and Phosphorus Remediation in Intensively Irrigated Stream-Aquifer Systems Using a Coupled Flow and Reactive Transport Model. *J. Hydrol.* **2021**, *598*, 126304. [CrossRef]
97. Granados, E.L.; Mendoza, M.; Acosta, A. Cambio de Cobertura Vegetal y Uso de La Tierra. El Caso de La Cuenca Endorreica Del Lago de Cuitzeo, Michoacán. *Gac. Ecol.* **2002**, *64*, 19–34.
98. CONAGUA. *Actualización de la Disponibilidad Media Anual de Agua en el Acuífero Morelia-Queréndaro (1602), Estado de Michoacán*; CONAGUA: Mexico City, México, 2020.
99. CONAGUA. *Actualización de la Disponibilidad Media Anual de Agua en el Acuífero Morelia-Queréndaro (1602), Estado de Michoacán*; CONAGUA: Mexico City, México, 2009.
100. Cram, S.; Galicia, L.; Israde-Alcantara, I. *Atlas de la Cuenca del lago de Cuitzeo: Analisis de su Geografia y Entorno Socioambiental: National or Regional Atlas Mexico*; Universidad Nacional Autonoma de Mexico: Mexico City, Mexico, 2010; ISBN 9786070218309.
101. CONAGUA. *Actualización Hidrogeológica de los Acuíferos: Maravatío-Contepec-Epitacio Huerta, Zacapu, Morelia-Queréndaro y Pastor Ortiz, en el Estado de Michoacán*; CONAGUA: Mexico City, México, 2007.
102. Neitsch, S.L.; Arnold, J.G.; Kiniry, J.R.; Williams, J.R. *Soil and Water Assessment Tool Theoretical Documentation Version 2009*; College of Agriculture and Life Sciences: College Station, TX, USA, 2011.
103. Serratos, L.; Douglas, A.; Douglas, A. NAME: Mexico Climatological Station Network Data. Version 1.0. UCAR/NCAR—Earth Observing Laboratory. 2007. Available online: https://data.eol.ucar.edu/dataset/82.175 (accessed on 27 October 2023).
104. CONAGUA. *Banco Nacional de Datos de Aguas Superficiales (BANDAS)*; CONAGUA: Mexico City, Mexico, 2016.
105. CONAGUA. *Estudio de Diagnóstico de las Condiciones Geohidrológicas Actuales y Análisis de Alternativas de Operación del Acuífero de Morelia Queréndaro, Michoacán*; CONAGUA: Mexico City, México, 1993.
106. Moriasi, D.N.; Gitau, M.W.; Pai, N.; Daggupati, P. Hydrologic and Water Quality Models: Performance Measures and Evaluation Criteria. *Trans. ASABE* **2015**, *58*, 1763–1785. [CrossRef]
107. Moriasi, D.N.; Arnold, J.G.; Van Liew, M.W.; Bingner, R.L.; Harmel, R.D.; Veith, T.L. Model Evaluation Guidelines for Systematic Quantification of Accuracy in Watershed Simulations. *Trans. ASABE* **1983**, *50*, 885–900. [CrossRef]
108. Huang, X.; Gao, L.; Crosbie, R.S.; Zhang, N.; Fu, G.; Doble, R. Groundwater Recharge Prediction Using Linear Regression, Multi-Layer Perception Network, and Deep Learning. *Water* **2019**, *11*, 1879. [CrossRef]
109. Sahoo, M.; Das, T.; Kumari, K.; Dhar, A. Space–Time Forecasting of Groundwater Level Using a Hybrid Soft Computing Model. *Hydrol. Sci. J.* **2017**, *62*, 561–574. [CrossRef]
110. Duan, R.; Fedler, C.B.; Borrelli, J. Field Evaluation of Infiltration Models in Lawn Soils. *Irrig. Sci.* **2011**, *29*, 379–389. [CrossRef]
111. Zheng, C.; Hill, M.C.; Cao, G.; Ma, R. MT3DMS: Model use, calibration, and validation. *Trans ASABE* **2012**, *55*, 1549. [CrossRef]
112. Hill, M.C. *Methods and Guidelines for Effective Model Calibration Water-Resources Investigations Report 98-4005 with Application to: UCODE, a Computer Code for Universal Inverse Modeling, and MODFLOWP, a Computer Code for Inverse Modeling with MODFLOW*; U.S. Geological Survey, Branch of Information Services: Denver, CO, USA, 1998.
113. Tolera, M.B.; Chung, I.-M. Integrated Hydrological Analysis of Little Akaki Watershed Using SWAT-MODFLOW, Ethiopia. *Appl. Sci.* **2021**, *11*, 6011. [CrossRef]
114. Yifru, B.A.; Chung, I.-M.; Kim, M.-G.; Chang, S.W. Assessment of Groundwater Recharge in Agro-Urban Watersheds Using Integrated SWAT-MODFLOW Model. *Sustainability* **2020**, *12*, 6593. [CrossRef]
115. Guevara-Ochoa, C.; Medina-Sierra, A.; Vives, L. Spatio-Temporal Effect of Climate Change on Water Balance and Interactions between Groundwater and Surface Water in Plains. *Sci. Total Environ.* **2020**, *722*, 137886. [CrossRef]
116. WHO. *Nitrate and Nitrite in Drinking-Water Background Document for Development of WHO Guidelines for Drinking-Water Quality*; WHO: Geneva, Switzerland, 2016.
117. Yifru, B.A.; Chung, I.M.; Kim, M.G.; Chang, S.W. Assessing the Effect of Urbanization on Regional-Scale Surface Water-Groundwater Interaction and Nitrate Transport. *Sci. Rep.* **2022**, *12*, 12520. [CrossRef]
118. Line Conan, C.; Bouraoui, F.; Turpin, N.; De Marsily, G.; Bidoglio, G. Modeling Flow and Nitrate Fate at Catchment Scale in Brittany (France). *J. Environ. Qual.* **2003**, *32*, 2026–2032. [CrossRef]
119. Aliyari, F.; Bailey, R.T.; Tasdighi, A.; Dozier, A.; Arabi, M.; Zeiler, K. Coupled SWAT-MODFLOW Model for Large-Scale Mixed Agro-Urban River Basins. *Environ. Model. Softw.* **2019**, *115*, 200–210. [CrossRef]
120. Bailey, R.T.; Wible, T.C.; Arabi, M.; Records, R.M.; Ditty, J. Assessing Regional-Scale Spatio-Temporal Patterns of Groundwater–Surface Water Interactions Using a Coupled SWAT-MODFLOW Model. *Hydrol. Process* **2016**, *30*, 4420–4433. [CrossRef]

121. Kim, N.W.; Chung, I.M.; Won, Y.S.; Arnold, J.G. Development and Application of the Integrated SWAT-MODFLOW Model. *J. Hydrol.* **2008**, *356*, 1–16. [CrossRef]
122. Power, J.F.; Schepers, J.S. Nitrate Contamination of Groundwater in North America. *Agric. Ecosyst. Environ.* **1989**, *26*, 165–187. [CrossRef]
123. Spalding, R.F.; Exner, M.E. Occurrence of Nitrate in Groundwater—A Review. *J. Environ. Qual.* **1993**, *22*, 392–402. [CrossRef]
124. Wick, K.; Heumesser, C.; Schmid, E. Groundwater Nitrate Contamination: Factors and Indicators. *J. Environ. Manag.* **2012**, *111*, 178–186. [CrossRef]
125. Benes, V.; Pěkný, V.; Skorepa, J.; Vrba, J. Impact of Diffuse Nitrate Pollution Sources on Groundwater Quality--Some Examples from Czechoslovakia. *Environ. Health Perspect.* **1989**, *83*, 5–24. [CrossRef]
126. Meinardi, C.R.; Beusen, A.H.W.; Bollen, M.J.S.; Klepper, O.; Willems, W.J. Vulnerability to Diffuse Pollution and Average Nitrate Contamination of European Soils and Groundwater. *Water Sci. Technol.* **1995**, *31*, 159–165. [CrossRef]
127. Agrawal, G.D.; Lunkad, S.K.; Malkhed, T. Diffuse Agricultural Nitrate Pollution of Groundwaters in India. *Water Sci. Technol.* **1999**, *39*, 67–75. [CrossRef]
128. Arauzo, M.; Martínez-Bastida, J.J. Environmental Factors Affecting Diffuse Nitrate Pollution in the Major Aquifers of Central Spain: Groundwater Vulnerability vs. Groundwater Pollution. *Environ. Earth Sci.* **2015**, *73*, 8271–8286. [CrossRef]
129. Krause, S.; Jacobs, J.; Voss, A.; Bronstert, A.; Zehe, E. Assessing the Impact of Changes in Landuse and Management Practices on the Diffuse Pollution and Retention of Nitrate in a Riparian Floodplain. *Sci. Total Environ.* **2008**, *389*, 149–164. [CrossRef]
130. Ameur, M.; Aouiti, S.; Hamzaoui-Azaza, F.; Cheikha, L.B.; Gueddari, M. Vulnerability Assessment, Transport Modeling and Simulation of Nitrate in Groundwater Using SI Method and Modflow-MT3DMS Software: Case of Sminja Aquifer, Tunisia. *Environ. Earth Sci.* **2021**, *80*, 1–16. [CrossRef]
131. Eryiğit, M.; Engel, B. Spatiotemporal Modelling of Groundwater Flow and Nitrate Contamination in An Agriculture-Dominated Watershed. *J. Environ. Inform.* **2022**, *39*, 125–135. [CrossRef]
132. Karlović, I.; Posavec, K.; Larva, O.; Marković, T. Numerical Groundwater Flow and Nitrate Transport Assessment in Alluvial Aquifer of Varaždin Region, NW Croatia. *J. Hydrol. Reg. Stud.* **2022**, *41*, 101084. [CrossRef]
133. Macian-Sorribes, H.; Tilmant, A.; Pulido-Velazquez, M. Improving Operating Policies of Large-scale Surface-groundwater Systems through Stochastic Programming. *Water Resour. Res.* **2017**, *53*, 1407–1423. [CrossRef]
134. Taie Semiromi, M.; Koch, M. Analysis of Spatio-Temporal Variability of Surface–Groundwater Interactions in the Gharehsoo River Basin, Iran, Using a Coupled SWAT-MODFLOW Model. *Environ. Earth Sci.* **2019**, *78*, 201. [CrossRef]
135. Kollet, S.J.; Maxwell, R.M. Capturing the Influence of Groundwater Dynamics on Land Surface Processes Using an Integrated, Distributed Watershed Model. *Water Resour. Res.* **2008**, *44*. [CrossRef]
136. Wittenberg, H.; Sivapalan, M. Watershed Groundwater Balance Estimation Using Streamflow Recession Analysis and Baseflow Separation. *J. Hydrol.* **1999**, *219*, 20–33. [CrossRef]
137. Szymkiewicz, A.; Potrykus, D.; Jaworska-Szulc, B.; Gumuła-Kawecka, A.; Pruszkowska-Caceres, M.; Dzierzbicka-Głowacka, L. Evaluation of the Influence of Farming Practices and Land Use on Groundwater Resources in a Coastal Multi-Aquifer System in Puck Region (Northern Poland). *Water* **2020**, *12*, 1042. [CrossRef]
138. Pulido-Velazquez, M.; Peña-Haro, S.; García-Prats, A.; Mocholi-Almudever, A.F.; Henriquez-Dole, L.; Macian-Sorribes, H.; Lopez-Nicolas, A. Integrated Assessment of the Impact of Climate and Land Use Changes on Groundwater Quantity and Quality in the Mancha Oriental System (Spain). *Hydrol. Earth Syst. Sci.* **2015**, *19*, 1677–1693. [CrossRef]
139. Wang, Y.; Chen, N. Recent Progress in Coupled Surface–Ground Water Models and Their Potential in Watershed Hydro-Biogeochemical Studies: A Review. *Watershed Ecol. Environ.* **2021**, *3*, 17–29. [CrossRef]
140. ASABE. *Guidelines for Calibrating, Validating, and Evaluating Hydrologic and Water Quality (H/WQ) Models American Society of Agricultural and Biological Engineers*; ASABE: St. Joseph, MI, USA, 2017.

hydrology

MDPI

Article

An Integrated Framework to Assess the Environmental and Economic Impact of Fertilizer Restrictions in a Nitrate-Contaminated Aquifer

Ilias Siarkos *, Zisis Mallios and Pericles Latinopoulos

School of Civil Engineering, Aristotle University of Thessaloniki, GR54124 Thessaloniki, Greece; zmallios@civil.auth.gr (Z.M.); latin@civil.auth.gr (P.L.)

* Correspondence: isiarkos@aegean.gr; Tel.: +30-6986-505775

Abstract: Groundwater nitrate contamination caused by the excessive use of nitrogen-based fertilizers has been widely recognized as an issue of significant concern in numerous rural areas worldwide. To mitigate nitrate contamination, corrective management practices, such as regulations on fertilizer usage, should be implemented. However, these measures often entail economic consequences that impact farmers' income, and thus should be properly assessed. Within this context, an integrated framework combining the environmental and economic assessment of fertilization restrictions through multi-criteria decision analysis is presented in an effort to efficiently manage groundwater nitrate contamination in rural areas. For this task, various scenarios involving reductions (10%, 20%, 30%, 40% and 50%) in fertilizer application were investigated, evaluated and ranked in order to determine the most suitable option. The environmental assessment considered occurrences of nitrates in groundwater, with a specific emphasis on nitrate concentrations in water-supply wells, as obtained by a nitrate fate and transport model, while the economic analysis focused on the losses experienced by farmers due to the reduced fertilizer usage. Our case-study implementation showed that a 30% reduction in fertilization is the most appropriate option for the area being studied, highlighting the importance of adopting such an approach when confronted with conflicting outcomes among alternatives.

Keywords: groundwater resources management; agricultural activities; nitrate contamination; numerical modeling; decision analysis; Nea Moudania aquifer

Citation: Siarkos, I.; Mallios, Z.; Latinopoulos, P. An Integrated Framework to Assess the Environmental and Economic Impact of Fertilizer Restrictions in a Nitrate-Contaminated Aquifer. *Hydrology* **2024**, *11*, 8. https://doi.org/10.3390/hydrology11010008

Academic Editor: Peiyue Li

Received: 5 December 2023
Revised: 30 December 2023
Accepted: 2 January 2024
Published: 4 January 2024

1. Introduction

Nitrate is generally considered one of the most widespread and persistent pollutants found in groundwater, originating from multiple nitrogen sources, primarily related to human activities, and to a lesser extent, to natural processes [1–5]. Anthropogenic sources mainly include agricultural activities such as fertilizer and manure applications and animal feedlot operations, as well as industrial and sewage discharges. Natural sources like decomposing organic matter or atmospheric nitrogen deposition also contribute to nitrate occurrence in groundwater, but anthropogenic sources tend to have a more pronounced impact due to their scale and intensity [1,6–10]. Among various human nitrate sources, agricultural activities, and especially the use of nitrogen-rich fertilizers for enhancing crop productivity, have been recognized as the main cause of groundwater nitrate contamination worldwide, posing a serious threat to rural areas [8,11–14].

Concerns over nitrate contamination further increase in regions where groundwater is used for human consumption [4,15,16], since elevated nitrate levels in drinking water have been linked to adverse health issues, such as methemoglobinemia in infants and stomach cancer in adults [8,17–20], although this connection remains a topic of controversy [21,22]. As a result, the Drinking Water Directive (98/83/EC) and the World Health Organization (WHO) set a maximum allowable nitrate concentration limit of 50 mg/L, along with an indicative threshold of 25 mg/L, defined as the "guidance value", in order to prevent potential health consequences resulting from the prolonged consumption of nitrate-contaminated water [19,23]. Beyond concerns for public health, excessive nitrate concentrations can also induce serious environmental harm particularly within natural ecosystems by leading to the eutrophication of interconnected surface water bodies [3,5,24,25].

In this context, it is notably important to adopt and implement corrective management practices aimed at preventing nitrate contamination caused by the excessive usage of agricultural fertilizers, which also aligns with all relevant European legislation, such as the Nitrates Directive (91/676/EEC), the Water Framework Directive (2000/60/EC), and the Groundwater Directive (2006/118/EC), as well as with the objectives outlined in the UN Sustainable Development Goals (SDGs), and especially those of SDG 6. According to SDG 6, access to safe and affordable drinking water for everyone by 2030 has to be ensured by reducing water pollution, which involves, among other things, the mitigation of groundwater nitrate contamination [26–28]. In general, imposing restrictions on the usage of nitrogen-based fertilizers—which is actually translated to reducing fertilizer application rate and thus limiting nitrogen input into the soil and nitrate leaching into groundwater—is a common and effective practice applied in the effort of controlling nitrate contamination in rural areas [8,29–32]; apparently, the higher the reduction in the rate of fertilizer application is, the greater the decrease in nitrate contamination levels might be, thus increasing the likelihood of reaching or maintaining compliance with established groundwater quality standards.

However, implementing effective management alternatives requires a comprehensive consideration of the groundwater system, along with a proper assessment of their efficiency in achieving the desired goals [32–34]. In this context, numerical modelling is usually implemented to investigate and evaluate the effectiveness of management practices on nitrate occurrences in groundwater. Numerical models, by simulating how nitrates move and behave in groundwater over space and time, provide insight into the potential outcomes of protection measures in advance, and they serve as a tool to support science-based policy decisions (i.e., nitrate abatement policies) [25,35–38]. Several studies have implemented numerical modeling through the development of nitrate fate and transport models to simulate and assess the impact of fertilization regulation practices on nitrate contamination levels (e.g., [14,17,31,32,35,38–40]). Other studies (e.g., [25,29,41–43]) have expanded their scope beyond the environmental benefits of fertilization regulations by incorporating the potential economic consequences stemming from their implementation into their analysis. Fertilization reduction practices, by having an adverse impact on crop productivity due to decreased yields or hindered crop growth, can result in negative economic ramifications for farmers, noticeably affecting the local economy in rural areas [6,25,35,43]; apparently, the higher the reduction in fertilizer application rate is, the greater the farmers' economic losses might be.

Based on the preceding analysis, it becomes clear that fertilizer regulations implicitly involve conflicting objectives. These practices actually seek to reduce nitrate concentrations to address environmental concerns, while, at the same time, aiming to minimize the economic losses resulting from their implementation [6,29]. These conflicting objectives can lead to different prioritization schemes when tasked with choosing among various scenarios involving fertilization reduction, i.e., scenarios featuring varying rates of fertilization reduction, since no single scenario, i.e., no certain rate of reduction, can be deemed universally optimal from both environmental and economic perspectives. This complexity necessitates the use of a multi-criteria decision analysis to prioritize the applied fertilization reduction scenarios [2,6]. However, to date, few studies (e.g., [29,43,44]) have undertaken such analysis, generally leading to a lack of decision-support tools for identifying and selecting the most appropriate solutions when designing nitrate abatement policies [23,45].

In this context, the main objective of this work is to develop an integrated decision-support framework aimed at efficiently managing groundwater nitrate contamination caused by fertilizer use in rural areas. More specifically, the framework proposes the most suitable fertilizer regulation policy (i.e., fertilizer reduction scenario) that minimizes the income losses in agriculture resulting from the reduction in nitrogen fertilizer application, while, at the same time, adhering to certain environmental constraints, particularly nitrate contamination levels. The proposed methodology integrates results from a nitrate fate and transport model, and an economic model within a decision analysis framework developed through the Analytic Hierarchy Process (AHP) method, in an effort to establish a connection between fertilization patterns and nitrate concentrations, as well as farmers' income. This comprehensive analysis can assist policymakers and stakeholders in making informed decisions regarding agricultural practices and regulations, while considering both environmental and economic aspects.

2. Materials and Methods

2.1. Study Area

The hydrological basin of Nea Moudania, located in the south-western part of Chalkidiki peninsula, Northern Greece, (south-east of the city of Thessaloniki), was chosen as the study area. It is a coastal basin, directly connected to the sea in the south, where it is surrounded by the Thermaikos Gulf (Figure 1). The basin occupies an area of approximately 127 km^2, with a mean topographic elevation of about 176 m above sea level and a mean slope of about 12%. The climate of the study area, which is hydrologically divided into two sub-regions (Figure 1), the lowland area in the south (53.7%) and the hilly area in the north (46.3%), is classified as semi-arid to humid, typically Mediterranean, with an average annual precipitation of 417 mm for the lowland area and 504 mm for the hilly one.

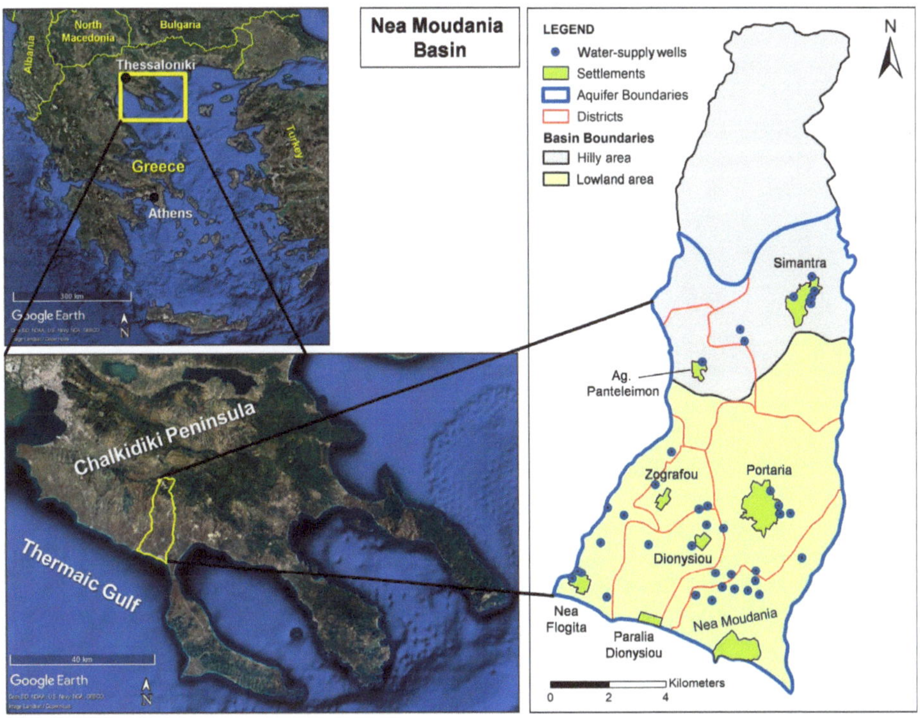

Figure 1. Geographical location and boundaries of Nea Moudania basin (also showing the lowland and hilly areas delineated based on local hydrological conditions), along with the boundaries of the aquifer under study, the several administrative districts and settlements within the region and the water-supply wells (39 in total) used as critical receptors for observing nitrate concentrations.

In terms of geology, most of the region belongs to the Peonia geologic zone and, more specifically, to the Moudania geologic formation, which is part of the Neogene deposits detected in western Chalkidiki. In the Moudania formation, which mostly consists of alternated beds of sandstones, sands, silts and red to brick red clays, the main aquifer system of the study area developed (Figure 1) in the form of successive water-bearing layers, separated by lenses of semi-permeable or impermeable materials [46,47]. The Nea Moudania aquifer serves as the primary exploitable aquifer and the sole source of freshwater in the region, used to meet irrigation, domestic and livestock needs.

The study area, as part of the main agricultural area of Chalkidiki (referred to as the "Kalamaria plains") is a typical rural area, where agriculture holds a dominant role in the local economy. The main crops are wheat, olive and apricot cultivations (about 96% of the total cultivated area), while other crops include barley, vegetables (e.g., tomatoes), pistachios and vineyards. Livestock activities are also present in the region, including cattle, sheep, goat, pig and poultry breeding, but make a relatively small contribution. Regarding the land use allocation, nearly 76% of the Nea Moudania basin is designated as agricultural land, with woodland covering 20% (mostly in the northern part) and the remaining 4% accounting for urban and touristic development. Especially for the latter case, eight settlements are located in the study area, on the basis of which the region is divided into seven administrative districts (two settlements belong to the same district) (Figure 1) [48]. In Table 1, information on the settlements' total populations (including both permanent and seasonal residents), along with the corresponding number of wells used to provide water to each settlement is contained. In the approach followed in the current study, these wells served as critical receptors for observing nitrate concentrations.

Table 1. The total population (including both permanent and seasonal residents), the corresponding number of wells used for water supply and the wastewater sources in each settlement [46,48].

Settlements	Total Population	Water-Supply Wells	Wastewater Sources
Nea Moudania	25,042	10	Wastewater network
Nea Flogita	4595	7	Septic systems
Dionysiou Paralia Dionysiou	13,152	6	Septic systems Wastewater network
Portaria	2983	5	Septic systems
Zografou	903	2	Septic systems
Ag. Panteleimon	762	3	Septic systems
Simantra	4931	6	Septic systems

In general, the Nea Moudania aquifer is characterized by severe quantitative degradation, since a substantial decline in groundwater levels is observed as a result of excessive abstraction of groundwater resources, especially for irrigation purposes. In addition, the study area has also experienced groundwater quality deterioration due to seawater intrusion and nitrate contamination, both linked to the intensification of agricultural activities [46,49]. Nitrate contamination has been well-documented in the study area. Latinopoulos (2003) [46], who carried out a groundwater sampling campaign in 34 wells during November 2001, found that, locally, nitrate concentrations have reached values up to 180 mg/L, exceeding the allowable limit of 50 mg/L. The study revealed that higher nitrate values are primarily related to medium-to-low-depth wells (<50 m), and they are mainly detected in the north and central part of the reference area, where intensive agricultural activities and fertilization practices are observed. Siarkos (2015) [48] drew similar conclusions regarding the spatial distribution of nitrate concentrations, attributing higher values to increased agricultural activities taking place in the aforementioned part of the study area. The analysis was based on nitrate concentration measurements in 12 deep wells (>100 m) designated for domestic use between 2011 and 2013. Specifically, it was found that nitrate concentrations in the sampled water-supply wells situated within agricultural areas in the north and central part of the aquifer were close to or exceeded (in one case) the "guidance value" of

25 mg/L, indicating the potential for further increases in concentrations in the coming years. Furthermore, it is worth mentioning that the broader region of the "Kalamaria plains", part of which is the Nea Moudania basin, has been identified as a "Nitrate Vulnerable Zone (NVZ)", in line with Nitrates Directive.

Based on the preceding analysis, it can be concluded that nitrate contamination has evolved into a matter of significant concern for the study area, posing a threat not only to the availability of groundwater reserves but also to the local population, thus raising public health issues. In this context, ensuring the protection of groundwater resources' quality becomes of utmost importance, necessitating the implementation of protective measures, while properly assessing their efficiency. As previously stated, agricultural activities and, more specifically, the application of nitrogen-based fertilizers, have been recognized as the main source of groundwater nitrate contamination in the region, and therefore management practices should target this certain source. However, there are several other potential on-ground nitrogen sources that might contribute to the presence of nitrate in groundwater, including irrigation with nitrogen-contaminated groundwater, leakage from septic systems and wastewater networks (as listed in Table 1 for each settlement), livestock breeding (assumed to be mainly situated within settlements due to the relatively small number of livestock, primarily catering to household needs) and atmospheric deposition. All these sources were considered when calculating the on-ground nitrogen loadings for the nitrate fate and transport model developed in this study.

2.2. General Methodological Framework

The methodology followed in the present study is aimed at investigating and evaluating the effect of fertilization reduction as a protective measure against groundwater nitrate contamination with the ultimate goal of providing the best management option, that is the most effective fertilization reduction scenario, taking into account both environmental and economic aspects. More specifically, in terms of the environmental perspective, the study examines the presence of nitrate in local groundwater resources, especially emphasizing nitrate concentrations in water-supply wells, thus accounting for potential risks associated with the consumption of poor-quality water. With respect to the economic point of view, the analysis focuses on estimating the costs of groundwater protection in the agricultural sector, specifically by assessing the productivity losses incurred by farmers as a result of the reduced use of fertilizers. A general view of the proposed methodology is depicted in Figure 2, while its various parts are briefly described in the following:

- Step 1: Building a nitrate fate and transport model to study the spatial and temporal variations in nitrate concentrations under various scenarios, i.e., do-nothing and fertilization reduction scenarios, as well as to obtain the nitrate values at water-supply wells, for each applied scenario.
- Step 2: Combining crop production functions with agro-economic data to estimate expected crop-yield losses under different fertilization reduction scenarios, while conducting an economic analysis based on cropping patterns and agro-economic data to calculate the agricultural income for both the do-nothing and fertilization reduction scenarios. The difference in agricultural income between the do-nothing scenario and any scenarios involving fertilization restrictions (fertilization reduction scenarios) is considered as the potential income losses associated with the applied scenario.
- Step 3: Performing a multi-criteria decision analysis subject to a set of criteria derived from both the environmental (from Step 1) and economic (from Step 2) assessment in order to prioritize the proposed fertilization reduction options and eventually select the most suitable that best meets the predefined decision criteria.

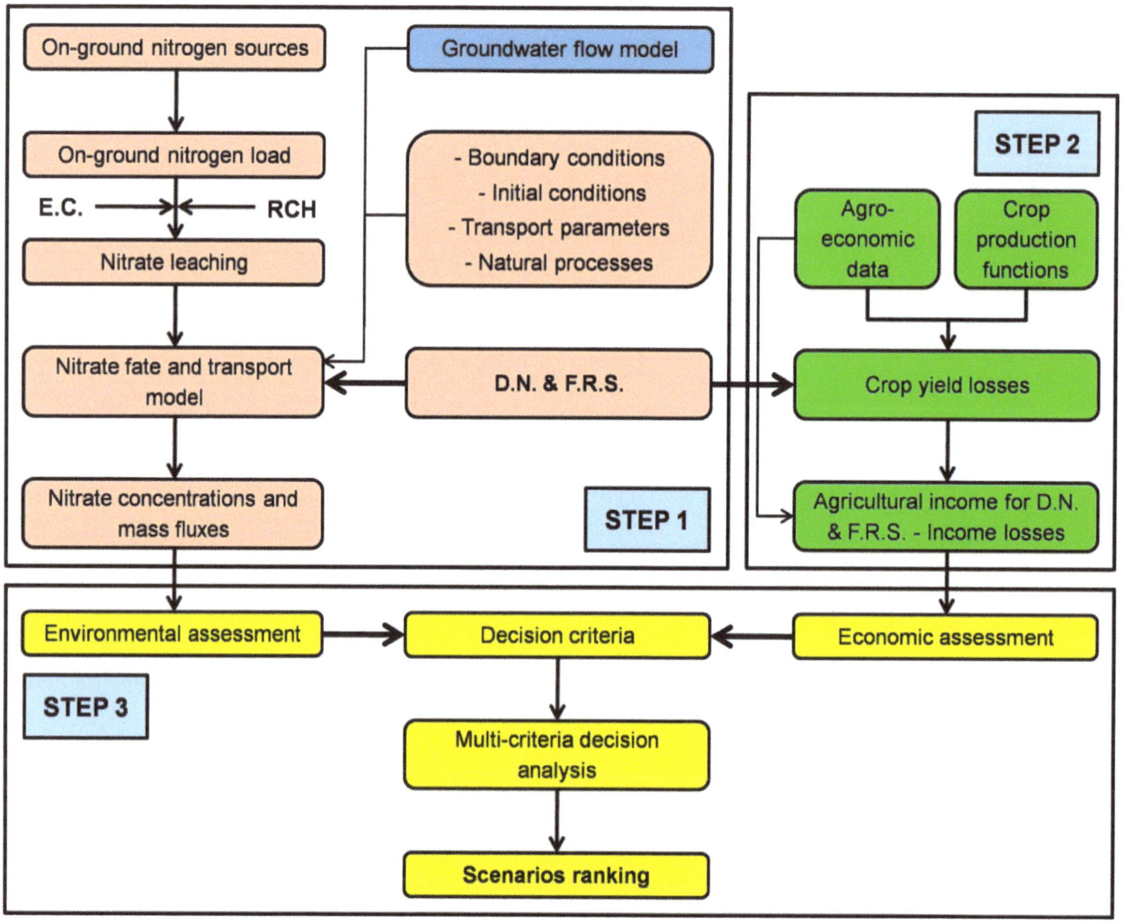

Figure 2. Schematic representation of the methodological framework followed in the present study (E.C.—Export coefficients; RCH—Groundwater recharge; D.N.—Do-nothing scenario; F.R.S.—Fertilization reduction scenarios).

2.3. *Numerical Modeling*

In general, solute mass transport simulation in groundwater requires the simulation of a groundwater flow system as the first step. This is considered essential for determining the hydraulic head distribution and groundwater velocity field throughout the model's domain, as well as for calculating the fluxes across grid cell interfaces in all directions [19,25,29,50]. In the present study, the development of the nitrate fate and transport model employed to investigate the spatio-temporal variation in nitrate concentrations was based on an existing calibrated/validated transient groundwater flow and transport model [49,51], originally created for simulating seawater intrusion (i.e., chloride transport) in the area of interest. This coupled flow and transport model was built by applying the SEAWAT code [52], thus taking into consideration the effects of density variations on groundwater flow caused by changes in chloride concentrations. The newly formed nitrate fate and transport model was developed by applying the MT3DMS code [53].

Given that the existing model was initially designed to simulate seawater intrusion, several adjustments, primarily associated with solute transport, are required in order to account for the new simulation conditions involving the modeling of nitrate contamination. These modifications are mainly related to the initial and boundary conditions of the transport model, as well as the type of pollution sources and the way the pollutants enter the aquifer. However, certain aquifer parameters linked to solute transport and affecting advection and dispersion of chemical constituents in groundwater, such as effective porosity (n_e) and longitudinal and transverse dispersivities (α_L, α_T), were not subject to modifications due to the prior calibration/validation of the existing model. In the subsequent sections, an overview of the existing groundwater model, mainly focusing on the part related to groundwater flow simulation, is provided, along with a thorough description of the development and calibration of the newly formed nitrate fate and transport model, which is then used to simulate various fertilizer application scenarios.

2.3.1. Overview of the Existing Groundwater Model

Detailed information on the overall modelling procedure, including the construction and calibration/validation of the existing seawater intrusion model, can be found in Siarkos and Latinopoulos (2016) [49] and Siarkos et al. (2017) [51]. However, in the following, an outline of the essential components of the existing groundwater model, especially referring to the aquifer conceptual model and the flow simulation, is provided [49]:

i. The boundaries of the aquifer under study coincide with those of the Nea Moudania basin, with the exception of the northern boundary, which was defined based on an isopiezometric contour map (year 2001) realized by Latinopoulos (2003) [46] by applying the Kriging method (Figure 1);
ii. Together, the various successive permeable stratigraphic layers form a single, unified aquifer system with a uniform thickness of 250 m, thus resulting in a vertically integrated two-dimensional areal model comprising one layer in the z-direction;
iii. The aquifer's eastern and western boundaries are treated as no-flow boundaries based on the overall arrangement of flow lines on a regional scale [46], whereas the southern and northern boundaries are assigned as constant head boundary (CHB, h = 0 m) and general head boundary (GHB, h = 150 m), respectively (Figure 3a). Regarding the southern boundary, the hydraulic head values remain constant over time, while concerning the northern boundary, the hydraulic head values gradually decrease over time following the overall decline in groundwater levels observed in the region;
iv. The study area is divided into six distinct hydro-geological zones (HP zones, Figure 3a) based on a number of pumping tests carried out in individual wells [46]; each of these zones is assigned a different value with regard to various aquifer parameters, such as hydraulic conductivity (K), specific storage (S_S), specific yield (S_y) and effective porosity (n_e). However, longitudinal and transverse dispersivities (α_L, α_T) are considered to be constant throughout the region, resulting in the use of a uniform value across the entire model domain. In Table 2, the values of the aforementioned aquifer parameters, as determined through the model calibration/validation procedure, are presented. As already mentioned, especially regarding the transport parameters, i.e., effective porosity and dispersivities, the same values are also used in the case of the nitrate fate and transport model developed in the current study;
v. The aquifer is primarily replenished by rainfall, irrigation return flows and losses from water supply and wastewater networks, with additional recharge taking place at the southern and northern boundaries; particularly concerning the first group of aquifer recharge sources, the study area is divided into several recharge zones (Figure 3b), taking into account factors such as local hydrological conditions, land use types and administrative boundaries within the region;
vi. The groundwater is abstracted from numerous wells to meet irrigation, domestic and livestock needs (Figure 4a); the pumping rates of the wells vary across municipal districts and specific water uses, under the assumption that each district's domestic and livestock needs are met by the same wells;
vii. The numerical model's spatial discretization involves the construction of a square grid in the horizontal plane, where cells are uniformly sized (100-m side) across the model domain;

viii. The numerical model's temporal discretization includes: (a) month-long stress periods accounting for the aquifer recharge and withdrawal regimes, (b) a pumping (1 May–30 September, 153 days) and a non-pumping (1 October–30 April, 212 days) period within each year in an effort to incorporate diverse temporal patterns regarding both groundwater abstraction for irrigation and irrigation return flows and (c) a 13-year simulation period (2001–2014) for calibration and validation purposes;

ix. The calibration of the flow model was performed using 13 observation wells monitored during November 2002 and 12 observation wells monitored during April 2003 [46] (Mean Error = −0.176 m, Mean Absolute Error = 1.502 m, Root Mean Square Error = 1.735 m and Mean Relative Error = 1.02%), and the validation was performed using 12 observation wells monitored during November 2010 [48] (Mean Error = −0.478 m, Mean Absolute Error = 1.822 m, Root Mean Square Error = 2.059 m and Mean Relative Error = 1.36%). The transport model was calibrated from 2011 to 2014 using chloride concentration measurements from 6 observation wells [48] (Mean Error = −0.10 mg/L, Mean Absolute Error = 6.93 mg/L, Root Mean Square Error = 9.29 mg/L and Mean Relative Error = 3.36%); and,

x. Low hydraulic head values (in relation to the mean sea level) are observed especially in the central part of the Nea Moudania aquifer due to the extensive usage of groundwater within that area (approximately 75% of the total water abstractions originating from this location). As a result, reversal of the natural groundwater flow occurs, leading to the influx of seawater along the coastline (southern boundary) and towards the interior of the aquifer (Figure 4b).

Table 2. Values of the various aquifer parameters as determined during the calibration/validation of the existing groundwater model (adapted from Siarkos and Latinopoulos, 2016 [49]).

Zone	Hydraulic Conductivity (m/day)	Effective Porosity/Specific Yield	Specific Storage (m^{-1})	Longitudinal Dispersivity (m)	Transverse Dispersivity (m)
1	0.575	0.150	0.00004	75	7.5
2	0.514	0.125	0.00012		
3	0.325	0.145	0.00019		
4	0.154	0.070	0.00015		
5	0.114	0.050	0.00006		
6	0.344	0.140	0.00020		

2.3.2. Nitrate Fate and Transport Model Development

In order to build the nitrate fate and transport model, apart from obtaining the velocity field from the groundwater flow model previously presented, the following have to be properly defined (Figure 2, Step 1): (a) the model boundary conditions, that is, specifying the type of boundary conditions for nitrate transport along the aquifer limits, (b) the model initial conditions, that is, determining the spatial distribution of nitrate concentrations at the beginning of the simulation, (c) the amount of nitrate entering the aquifer system, that is, quantifying nitrate leaching from the soil to groundwater, which is directly linked to the on-ground nitrogen loadings deriving from various nitrogen pollution sources, and (d) the physical or chemical processes occurring during the transport of nitrates in groundwater.

At first, the boundary and initial conditions used for the development of the nitrate fate and transport model are illustrated in Figure 5a. Regarding the boundary conditions, the eastern and western boundaries were set as zero-dispersive boundaries, since no flow is observed through these parts of the aquifer (Figure 3a), while the southern and northern boundaries were represented by constant concentration boundaries (CCB). In the case of southern boundary, a zero-nitrate concentration was assigned to it accounting for the inflow of seawater into the aquifer through this boundary. On the other hand, the northern boundary was split into two segments, each designated with a different nitrate concentration (2 mg/L for the left segment and 8 mg/L for the right segment), determined on the basis of the background concentrations observed near the respective segments. In both segments, nitrate concentrations remain constant over time for the entire simulation. With respect to the model initial conditions, i.e., initial nitrate distribution, they were derived by applying the Kriging method to nitrate concentrations measured during November 2001 across 16 monitoring wells [46].

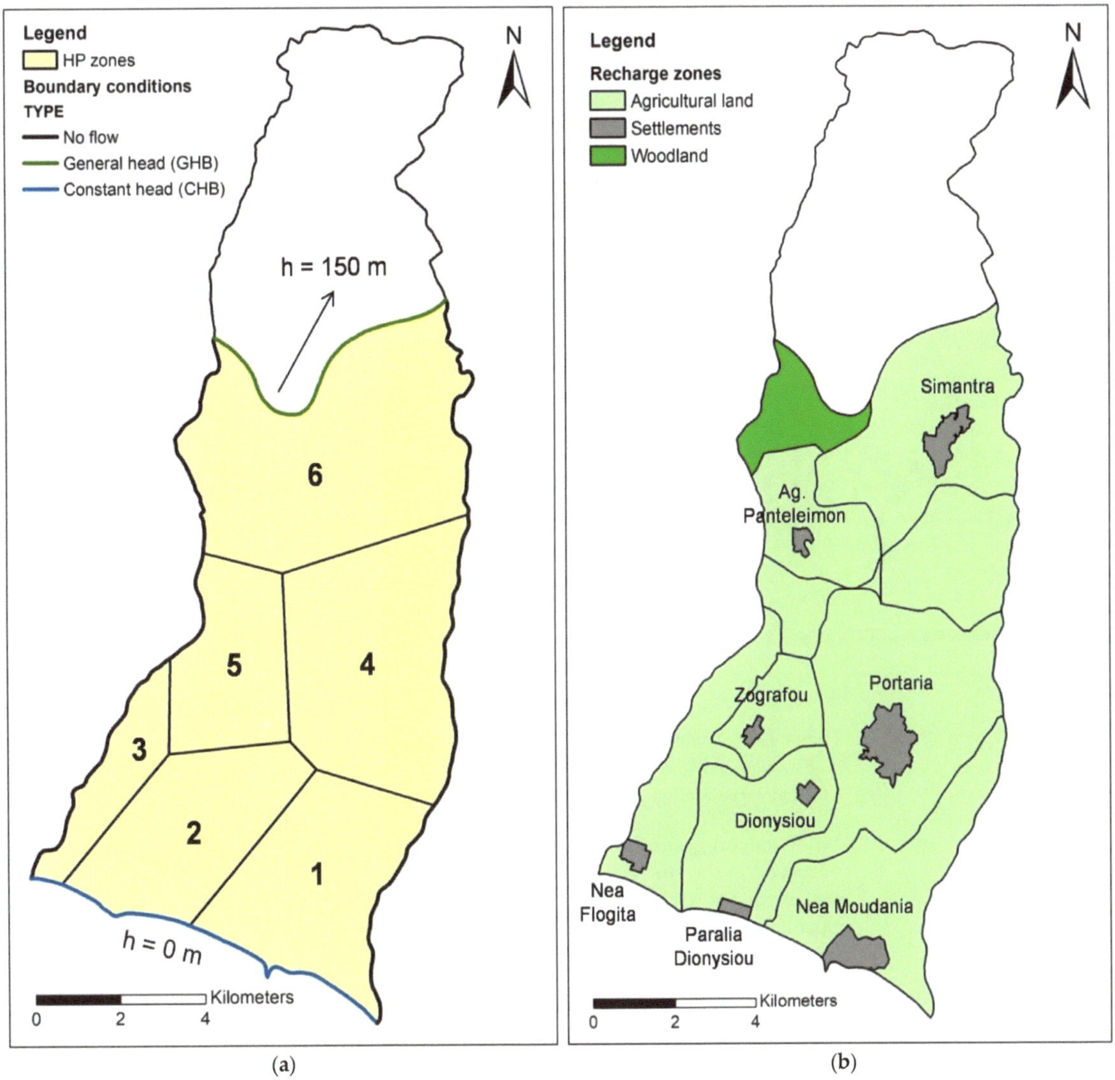

Figure 3. (**a**) Flow boundary conditions, along with the six distinct hydro-geological zones (HP zones) of the study area (defined based on a number of pumping tests in individual wells), and (**b**) the recharge zones in the study area (delineated based on several factors such as hydrological conditions, land use types and administrative boundaries).

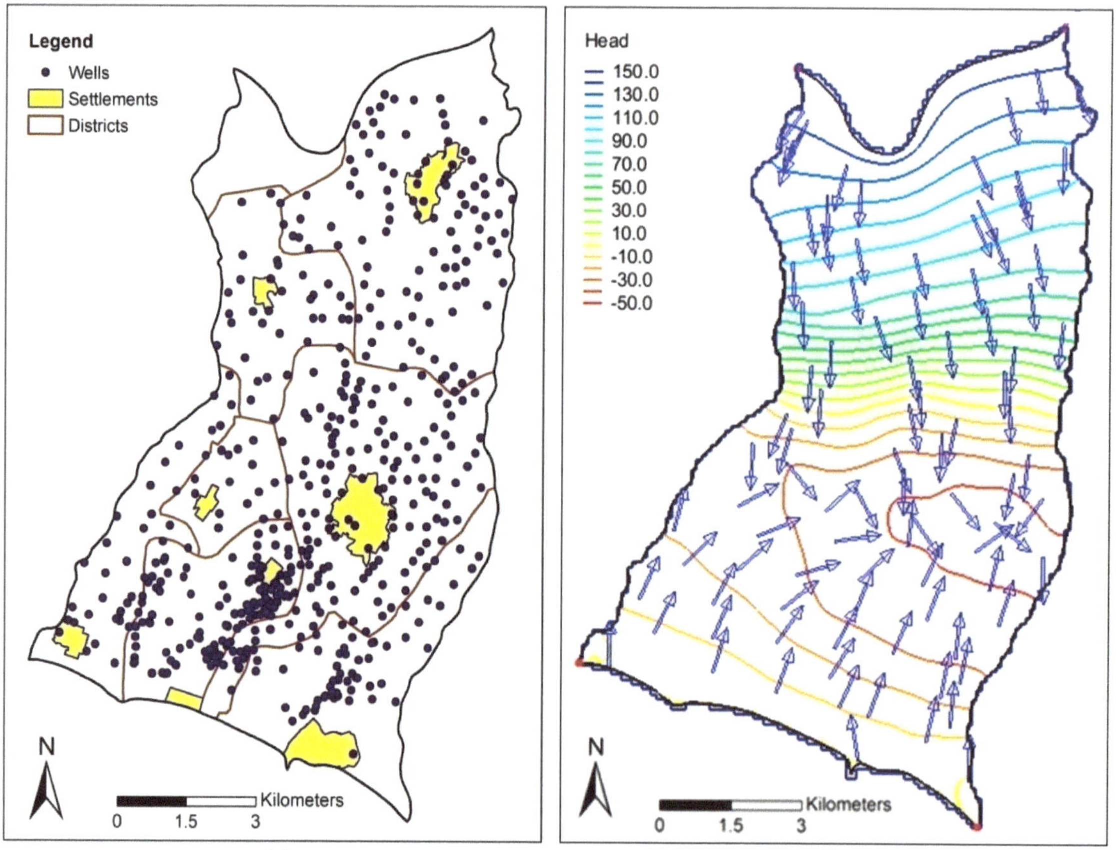

(a) (b)

Figure 4. (**a**) The abstraction wells (518 in total) operating in the region, and (**b**) hydraulic head distribution, along with groundwater flow direction in the Nea Moudania aquifer at the end of the calibration/validation period (2014).

Next, on-ground nitrogen loadings originating from the potential nitrogen sources observed in the study area (see Section 2.1) have to be estimated in order to determine the amount of nitrate leaching entering the groundwater system through infiltration. The calculation of nitrogen loading for each of these sources was performed according to the type of land use it is associated with, while taking into consideration the different municipal districts and settlements within the study area, thus accounting for the spatial distribution of on-ground nitrogen sources and corresponding loadings. Using this approach, also applied to the calculation of groundwater recharge, nitrogen loading was computed at the resolution of individual recharge zone (Figure 3b), thus directly linking the nitrate inputs to the net recharge of each individual zone within the study area [19,36]. Under these conditions, in order to capture, on the one hand, the temporal patterns of groundwater recharge (i.e., monthly time-step) and, on the other hand, the temporal variability in certain nitrogen source applications, e.g., fertilizer application, irrigation, deposition, which generally vary within the year [54], nitrate loading for all nitrogen sources was computed on a monthly basis.

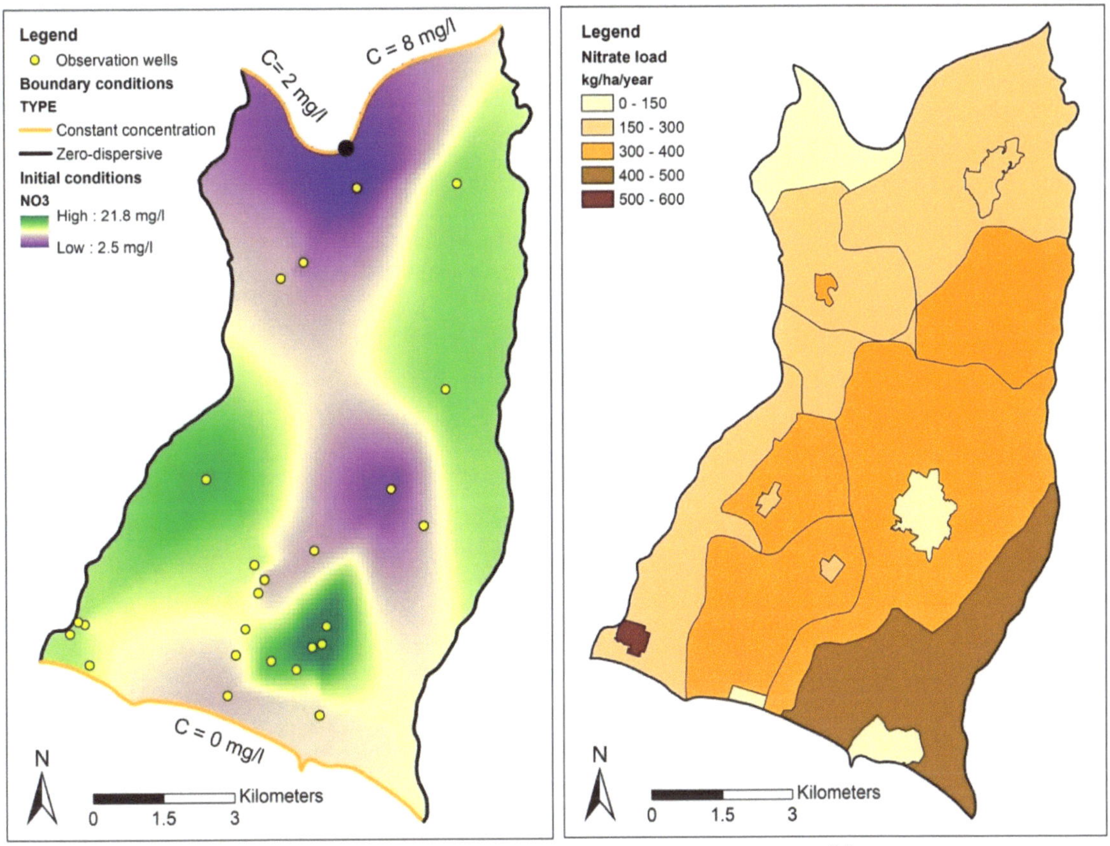

(a) (b)

Figure 5. (**a**) Boundary and initial conditions for the nitrate fate and transport model, and (**b**) distribution of annual nitrogen loading in the study area at a recharge zone level.

Based on the preceding analysis, nitrogen loading per month at a recharge zone level was determined taking into account the types of nitrogen sources in each zone, in addition to the following information regarding the estimation of the corresponding loading from each source [10,54–56]:

- Fertilizers: For the different types of crops spotted in the study area, nitrogen loading from agricultural fertilizers was calculated by multiplying the suggested fertilizer application rate for each crop (kg/ha), as listed in Table 3, with the actual fertilized area, while taking into consideration the fertilizer application timing for each crop, i.e., the specific months in which fertilizers are generally applied.

Table 3. Suggested amount of nitrogen per crop type (kg/ha) [57].

Trees			Grains		Vegetables		Vineyards
Olives	Apricots	Other	Wheat	Other	Tomatoes	Other	
750	150	200	150	100	350	150	150

- Irrigation water: Nitrogen loading from irrigation water was calculated by multiplying the mean concentration of nitrates (mg/L) in groundwater, as obtained by previous research conducted in the study area [46], with the amount of water used for irrigation, after converting nitrate (NO_3) concentrations into nitrate–nitrogen (NO_3-N) concentrations.

- Septic systems: Nitrogen loading from septic systems was calculated by multiplying the corresponding nitrogen production per capita (0.012 kg/d) [54] with the actual population (permanent and seasonal).
- Wastewater network: Nitrogen loading from wastewater networks was calculated by multiplying the corresponding nitrogen concentration in wastewater (35 mg/L) [58] with the estimated water leakage in the network.
- Livestock: For the different types of dairy animals detected in the study area, nitrogen loading from livestock was calculated by multiplying the corresponding nitrogen production per head for each animal type (kg/d), as presented in Table 4, with the actual number of animals.

Table 4. Nitrogen production per head for each animal type (kg/d) [59].

Cattle	Sheep	Goats	Pigs	Poultry
0.2250	0.0125	0.0135	0.0520	0.0011

- Atmospheric deposition (wet): Nitrogen loading from precipitation was calculated by multiplying the mean concentration of nitrogen in precipitation (3.0 mg/L) [60] with the amount of water from precipitation (after subtracting the runoff volumes).

In an effort to provide a comprehensive overview of the spatial and temporal distribution of on-ground nitrogen loading in the region, several key figures based on the aforementioned calculations are provided. More specifically, in Figures 5b and 6, the spatial distribution of annual nitrogen loading at the recharge zone level and the percentage breakdown of nitrogen loadings from the different sources present in the study area are depicted, respectively, while in Figure 7, the monthly variation in nitrogen loading is shown. As is apparent, and in accordance with the findings of previous studies, agricultural fertilizers constitute the predominant nitrogen source in the Nea Moudania aquifer, accounting for approximately 90% of the total nitrogen generated in the area. This clearly suggests that management strategies should be tailored to address this specific source. However, the highest nitrogen loading per hectare (590 kg/ha/year) is observed in an urban area, and more specifically in the settlement of Nea Flogita, due to the presence of septic systems within the settlement (Table 1) and the substantial increase in population during the tourist season. As far as the agricultural areas are concerned, the Nea Moudania district exhibits the highest nitrogen loading per hectare (404 kg/ha/year), primarily due to the notably higher percentage of olive trees (53% of the cultivated land) within the district (nitrogen requirements for olive trees significantly exceed those of other crops, as shown in Table 3). With respect to the temporal variations in nitrogen loading, it is observed that the peak amounts coincide with months mainly dedicated to nitrogen fertilizer application (i.e., February, March, June, August and November). Notably, the highest levels are recorded in February (1.18×10^6 kg), aligning with the primary fertilization period for olive trees.

After calculating nitrogen loading, nitrate loading reaching the groundwater system was determined using the same spatial (per recharge zone) and temporal (per month) resolution. For this task, an export coefficient model was applied, according to which the nitrate loading is calculated as a fraction of the total on-ground nitrogen loading, accounting for the several biochemical transformations of nitrogen occurring in the soil, such as mineralization, immobilization, nitrification, denitrification and plant uptake [17,25,56]. In particular, nitrate loading is calculated by multiplying nitrogen loading with predefined export coefficients depending on the types of nitrogen sources [14,40,54,55,61,62]. The export coefficients adopted in the current study for the different types of nitrogen sources within the area of interest were obtained from relevant literature sources [50,54,61], and are presented in Table 5. It should be mentioned that, particularly for agricultural fertilizers, since the export coefficient does not receive a fixed value like other pollution sources but falls within a specific range, its exact value was determined during the process of calibrating the nitrate fate and transport model. The initial value assigned to the export coefficient, which was accordingly adjusted, was equal to 0.40. As a final step, the export coefficient model results, i.e., nitrate loadings, were converted into nitrate concentrations in soil leachate by dividing them by groundwater recharge volumes, while maintaining the previous spatial (per recharge zone) and temporal (per month) discretization. The resulting monthly values for nitrate leaching in each recharge zone were introduced into the transport model using the recharge concentration boundary condition.

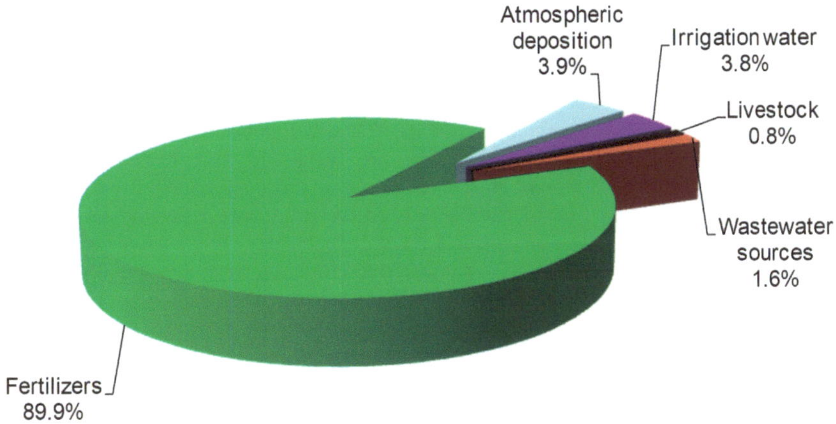

Figure 6. Percentages of on-ground nitrogen loadings from the different sources present in the study area.

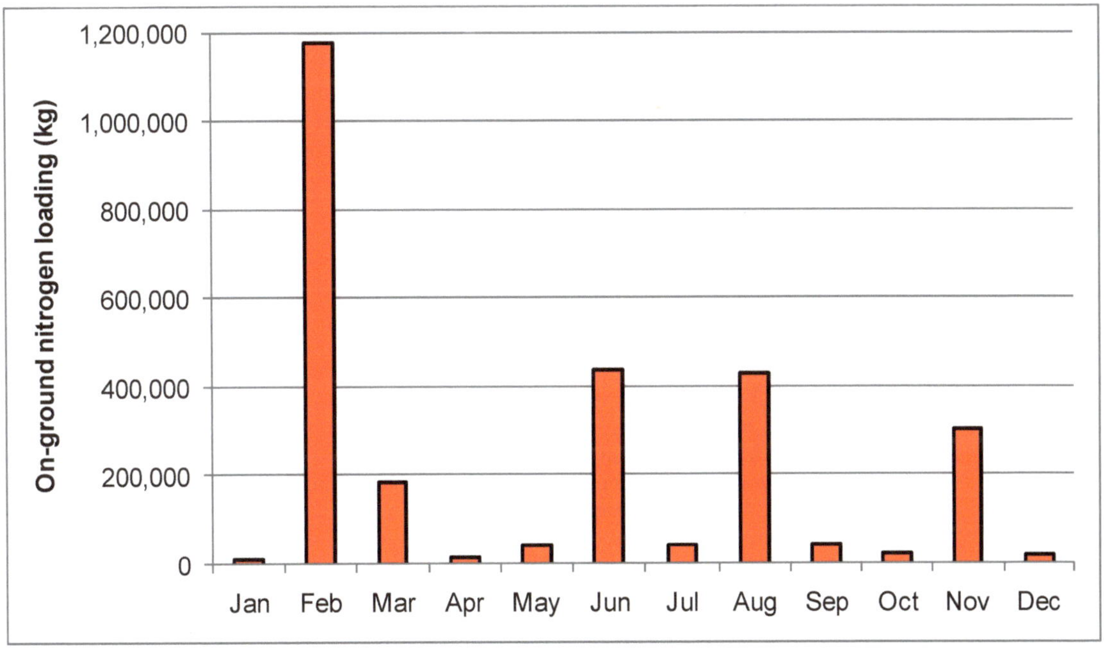

Figure 7. Monthly on-ground nitrogen loading distribution for the study area.

Table 5. Export coefficients for the nitrogen sources present in the study area [50,54,61].

Fertilizers	Irrigation Water	Wastewater Sources	Livestock	Atmospheric Deposition
0.30–0.50	0.25	0.68	0.30	0.25

Finally, regarding the physical or chemical processes taking place during nitrate transport in groundwater, denitrification is generally recognized as the dominant chemical reaction affecting nitrate concentration in groundwater [17,39,63,64]. On the other hand, sorption is commonly neglected given nitrate's high mobility in soil water and groundwater systems [5,17,65], due to the fact that it is highly soluble and negatively charged [6,66]. However, in this study, denitrification was also considered to be negligible. This assumption is based primarily on the chemical analyses from various studies conducted in the region (e.g., [46,48,67]), which indicate the absence of nitrites (NO_2^-) in local groundwater. In general, nitrites are used as indicators of denitrification, given their role as intermediate products within the overall process [5,68,69]. In addition, it should be noted that, while denitrification mainly occurs at the pore scale where local conditions may differ, at a regional scale, nitrates generally behave as conservative contaminants, exhibiting no significant retardation compared to groundwater movement [35,70]. In conclusion, in the present study, nitrate transport was simulated by considering only the advective–dispersive mechanism, since molecular diffusion was also neglected given its generally minor role [27,35].

2.3.3. Nitrate Fate and Transport Model Calibration

The nitrate fate and transport model was calibrated via a trial-and-error procedure by altering the value of the export coefficient for agricultural fertilizers—which actually translates into modifying nitrate leaching—until the simulated nitrate concentrations at the calibration targets closely fit the concentrations observed in the field. Due to the absence of long-term nitrate concentration data, model calibration relied on a selected set of scattered observed values obtained from nitrate concentration measurements (22 in total) in 12 water-supply wells during April 2011, November 2011 and November 2013 [48]. In order to test the overall performance of the calibration procedure, error-based measures, such as Mean Error (ME), Mean Absolute Error (MEA), Root Mean Squared Error (RMSE) and Mean Relative Error (MRE), in conjunction with a correlation-based measure, specifically the coefficient of determination (R^2), were used.

2.3.4. Simulation of Fertilizer Application Scenarios

To assess the effectiveness of fertilization restrictions on nitrate concentrations and select the most suitable option regarding the reduction in fertilizer application rates, five alternative scenarios, applying the calibrated nitrate fate and transport model over a 20-year time period (2014–2034), were formed and simulated. In these scenarios, the fertilizer application rates on croplands were reduced by 10% (Scenario 1—S1), 20% (Scenario 2—S2), 30% (Scenario 3—S3), 40% (Scenario 4—S4) and 50% (Scenario 5—S5) compared to the current rates, which actually resulted in a corresponding decrease in nitrate loading within the agricultural areas and a subsequent decrease in nitrate leaching into the aquifer. The rationale of this choice lies on the fact that, in the present study, it is aimed not to develop specific spatial allocations of fertilizer application for the reference area, but rather to examine the general impact of fertilizer usage regulations on nitrate concentrations. Furthermore, it is important to note that, when performing regional decision analysis concerning nitrate contamination, management options are generally applied to the entire area of interest. Limiting the implementation of the management options to certain areas might be challenging, primarily due to political constraints [6]. At this point, it should be mentioned that a do-nothing scenario (Baseline scenario—S0) corresponding to current fertilization patterns (existing practices) was also simulated to act as a benchmark for comparing the various fertilization restriction scenarios.

2.4. Economic Analysis

In order to quantify the economic consequences of fertilizer restrictions, expected crop-yield losses due to reduced fertilizer usage need to be estimated as the initial step. In the available literature, several approaches establishing the relationship between nutrient application and agricultural product yield can be found [71]. In the present study, the analytical solution developed by Murrell (2009) [72] was adopted. According to this solution, the relationship between crop-yield losses and the amount of nutrient applied (specifically nitrogen in this case) can be expressed as a second-degree polynomial (quadratic) function, defined as follows:

$$y_i = a + b \times x_i + c \times x_i^2 \tag{1}$$

where y_i is crop yield and x_i is nutrient rate for the ith scenario, a is the yield without added nutrient, b is the component of linear slope of the quadratic function and c is the coefficient of curvature. Figure 8 depicts the graphical representation of the above Equation (1), indicating that the crop yield reaches its minimum with no nutrient applied, at a value denoted as a. According to the same figure, as nutrient application increases, the yield also increases until it reaches the maximum point (y_{eor}), which signifies the economically optimal expected yield attained by applying the recommended nutrient rate (x_{eor}). Further application beyond this point leads to decreased production, indicating diminishing returns or negative effects on yield despite additional nutrient application [72].

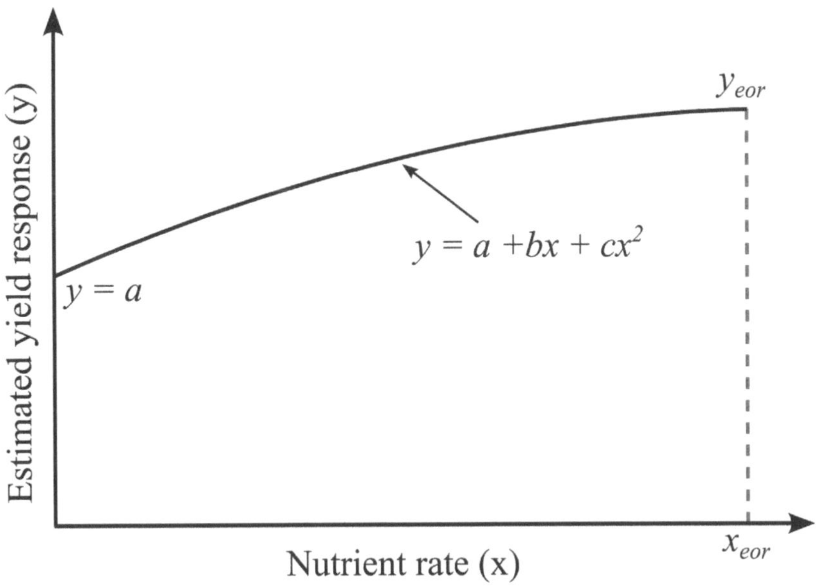

Figure 8. A conceptual model of a quadratic response to a recommended nutrient rate [72].

To determine the coefficients a, b and c in Equation (1), the values of the following parameters must be known: crop yield with no fertilizer applied, crop yield with the recommended amount of fertilizer applied, the recommended rate and the cost per unit of fertilizer, as well as the price of a unit of harvested crop. In particular, determining the expected crop yield with no fertilizer applied (minimum crop yield) constitutes a challenging task due to the scarcity of available data. Its estimation can be quite difficult or even impossible in real conditions due to the variability in soil characteristics within crop fields. Hence, obtaining a single value corresponding to the entire study area is challenging. For the study's purposes, it was assumed that the minimum crop yield equals 50% of the expected yield when fertilizer is applied.

After estimating crop-yield losses, the agricultural income for the do-nothing and fertilization reduction scenarios was calculated in order to ascertain the economic consequences of implementing fertilization restrictions. For this task, various agro-economic data, including crop yield and crop selling price, as well as labor costs, seed expenses, fertilizer and pesticide costs, were used, and incorporated into an economic model that considers both farmers' revenue and crop production costs. As a final step, the potential economic losses for every fertilization reduction scenario were determined as the difference in agricultural income between the baseline scenario and the reduction scenarios.

2.5. Decision Analysis

An important step in decision-making analysis for selecting the best option from a set of alternatives is to identify and define the decision criteria, which generally plays a critical role in determining the final outcome [6,29]. As already mentioned, the proposed methodological framework integrates two broad sets of criteria, namely environmental and economic, in order to assess the desirability of the applied management options, that is, of the various fertilization reduction scenarios (S1–S5). In the present analysis, for each set, two distinct criteria were defined (four in total), based on the specific objectives outlined in the study. In particular, these direct objectives involve: (a) decreasing nitrate occurrences in groundwater, (b) reducing health risks for the exposed population and (c) minimizing income losses for farmers. In this context, the following four decision criteria were incorporated into the analysis: (a) nitrate buildup in groundwater (kg), (b) population exposed to poor-quality water (NO_3 > 25 mg/L), (c) net cost incurred by farmers (EUR) due to production losses and (d) cost per concentration reduction (EUR/mg per L) (CPCR). Each fertilization reduction scenario was appraised for these decision criteria employing the nitrate fate and transport model and the economic model previously described. More analytically, each of the aforementioned criteria was determined as follows:

- Nitrate mass: The nitrate buildup in the aquifer under study is directly derived from the numerical model, as it is a key output of the model.
- Population: The population exposed to poor-quality water refers to the total number of residents in the region supplied with water containing nitrate concentrations exceeding 25 mg/L. To determine this population for each settlement in the study area, the nitrate concentration in the water provided from the corresponding water-supply wells (see Table 1) was found. For this task, the nitrate concentration at each corresponding well (critical receptor) in every settlement was obtained from the numerical model, and then the mean concentration value (considering blended water) was computed, given that all wells in each settlement have the same pumping rate.
- Cost: The net cost is calculated through the economic analysis as the difference in agricultural income between the do-nothing scenario and any fertilization reduction scenarios.
- CPCR: The CPCR criterion is computed by applying the following equation:

$$CPCR = \frac{COST_i}{AC_o - AC_i} \quad (2)$$

where $COST_i$ is the net cost incurred from the ith scenario as previously defined, and AC_o and AC_i are the average nitrate concentrations at the critical receptors (water-supply wells) corresponding to the do-nothing and the ith scenario, respectively.

To evaluate the effectiveness of the different fertilization reduction scenarios according to the predefined decision criteria and select the most suitable option, the Analytic Hierarchy Process (AHP) was applied. AHP is a powerful multi-criteria decision analysis technique, frequently used for dealing with complex problems [6,73–75]. In this method, a series of pair-wise comparisons is set up in order to estimate the weights of the predefined decision criteria, by ranking the importance of the criteria on a scale from 1/9 (absolute unimportance) to 9 (absolute importance). A consistency ratio (CR) that measures the coherence of judgments made by decision makers is also computed. For CR ≤ 0.1, the evaluation procedure is considered acceptable; otherwise, the process should be repeated and the PCMs should be re-established [6,75–77]. After estimating the weights, the overall priority of each scenario (S1–S5) is calculated by summing up the results obtained by multiplying the weight of each criterion with its corresponding (normalized) value. Once the overall priorities are determined, the scenarios can be ranked, and the most appropriate one can be selected based on the highest overall priority value.

3. Results and Discussion

3.1. Numerical Modeling

In this section, the results of the nitrate fate and transport model calibration procedure are presented and discussed, as well as those referring to model application, particularly those involving the spatio-temporal variation of nitrate concentrations in the study area, along with the concentrations in water-supply wells under the various fertilizer application scenarios (S0–S5).

3.1.1. Model Calibration Results

In Figure 9, a scatterplot of the simulated versus observed nitrate concentrations for the entire set of measurements considered during the model calibration procedure is depicted, together with the values of the several evaluation measures (ME, MAE, MRSE, MRE, and R^2) and the locations of the observation wells (12 wells). By considering the observed variability in nitrate concentrations, it can be inferred that the calculated errors are reasonably acceptable, indicating an overall successful calibration. Hence, the calibrated nitrate fate and transport model can be employed in order to project nitrate concentrations in response to future fertilizer application scenarios. Finally, it should be mentioned that, during the model calibration and after a series of model runs, the export coefficient for agricultural fertilizers was found to be equal to 0.42.

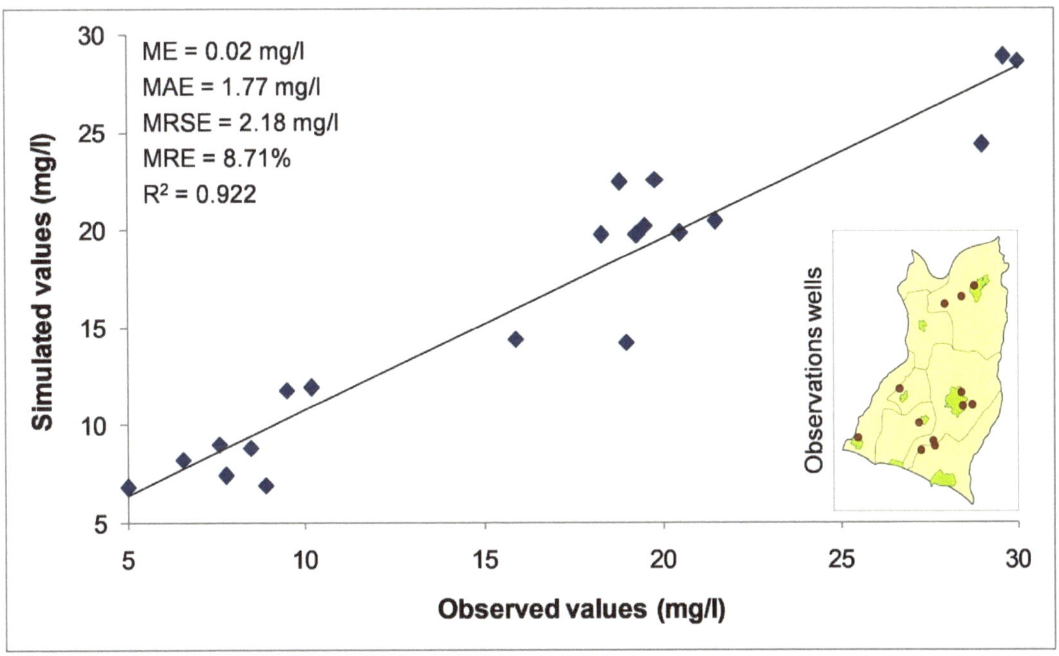

Figure 9. Scatterplot of the simulated versus observed nitrate concentrations, along with the values of the several evaluation measures employed in the study and the locations of the observation wells (12 in total) used for nitrate concentration measurements.

3.1.2. Model Application Results

First, the spatial distribution of nitrate concentrations for the baseline (S0) and different fertilization reduction scenarios (S1–S5) by the end of the simulation period (2034) is depicted in Figure 10. According to the modeling results, elevated nitrate concentrations exceeding the indicative threshold of 25 mg/L are mainly expected within the broader central part of the region in all considered scenarios. In particular, in Scenarios S0 and S1, nitrate levels are projected to locally surpass the upper limit of 50 mg/L. Several factors contribute to the presence of elevated concentrations within this specific area, including: (a) the combined effect of high initial nitrate concentrations in certain locations (Figure 5a) and nitrogen loading (Figure 5b) from fertilizers; (b) the direction of groundwater flow, which is towards the central part of the region (Figure 4b); and (c) the values of aquifer parameters associated with mass transport, such as the effective porosity (Table 2). Particularly regarding the latter, the lower values of effective porosity in the two central HP Zones (Figure 3a), as they were determined during the calibration/validation of the existing groundwater model (see Section 2.3.1), result in higher groundwater velocity values in that area and, therefore, in higher nitrate concentrations. In addition, elevated concentrations are anticipated in the settlement of Nea Flogita, situated in the south-western part of the region (Figure 1), due to the relatively high nitrate loading observed within it (Figure 5b) and originating from wastewater sources. On the other hand, lower nitrate concentrations are generally expected in the north-western part of the region, where both initial nitrate concentrations and nitrate loading are low (Figure 5a,b), as well as along the southern boundary of the aquifer (coastline), which was defined as a constant concentration boundary of 0 mg/L (Figure 5a) due to the intrusion of seawater from this part of the aquifer.

Furthermore, upon comparing the outcomes of the diverse fertilizer application scenarios (S0–S5), it is evident that nitrate concentrations in the region generally decrease under the fertilization reduction scenarios (S1–S5) in relation to the baseline scenario (S0). This decrease, as expected, becomes more pronounced when moving from Scenario S1 to Scenario S5, that is, to higher rates of reduction in fertilizer application (from 10% to 50%). The same conclusion can be drawn from Table 6, which presents the minimum, maximum, and mean nitrate concentrations in the region for all concerned scenarios. As is clear, the higher the reduction in fertilizer application, the lower both the maximum and mean nitrate concentrations. However, even though the mean nitrate concentration remains below the threshold of 25 mg/L in all the applied scenarios, the maximum nitrate level is above the aforementioned "guidance value" in every case; particularly in Scenarios 0 and 1, where, as previously noted, it exceeds the uppermost limit of 50 mg/L. Only when the fertilizer application rate is reduced by 20% (Scenario S2) does the maximum nitrate concentration fall below 50 mg/L.

Table 6. Minimum, maximum and mean nitrate concentrations within the study area both at the beginning (2014) and end (2034) of the projection period under the various fertilizer application scenarios (S0–S5).

Metrics	2014	2034					
		S0	S1	S2	S3	S4	S5
Min	1.2	0.1	0.1	0.1	0.1	0.1	0.1
Max	31.8	54.6	51.9	49.2	46.5	44.0	43.0
Mean	16.5	24.5	23.5	22.6	21.6	20.7	19.7

Next, in order to distinctly localize the areas exhibiting relatively high nitrate concentrations, i.e., 25 mg/L, Figure 11 shows the spatial distribution of concentrations exceeding this threshold under various fertilizer application scenarios (S0–S5). Once again, it is apparent that, in all the scenarios considered, nitrate concentrations exceeding the "guidance value" of 25 mg/L are mainly detected within the broader central region of the study area as a result of the combined effect of initial nitrate concentrations, nitrogen loading and groundwater flow conditions. However, as the fertilizer application rate decreases (i.e., when moving from Scenario 1 to Scenario 5), the areas with concentrations above 25 mg/L tend to shrink, yet not fully diminish. This last statement implies that, despite a 50% reduction in fertilizer application, it is anticipated that nitrate concentrations above 25 mg/L will persist in a considerable part of the region, particularly concentrated in its central part.

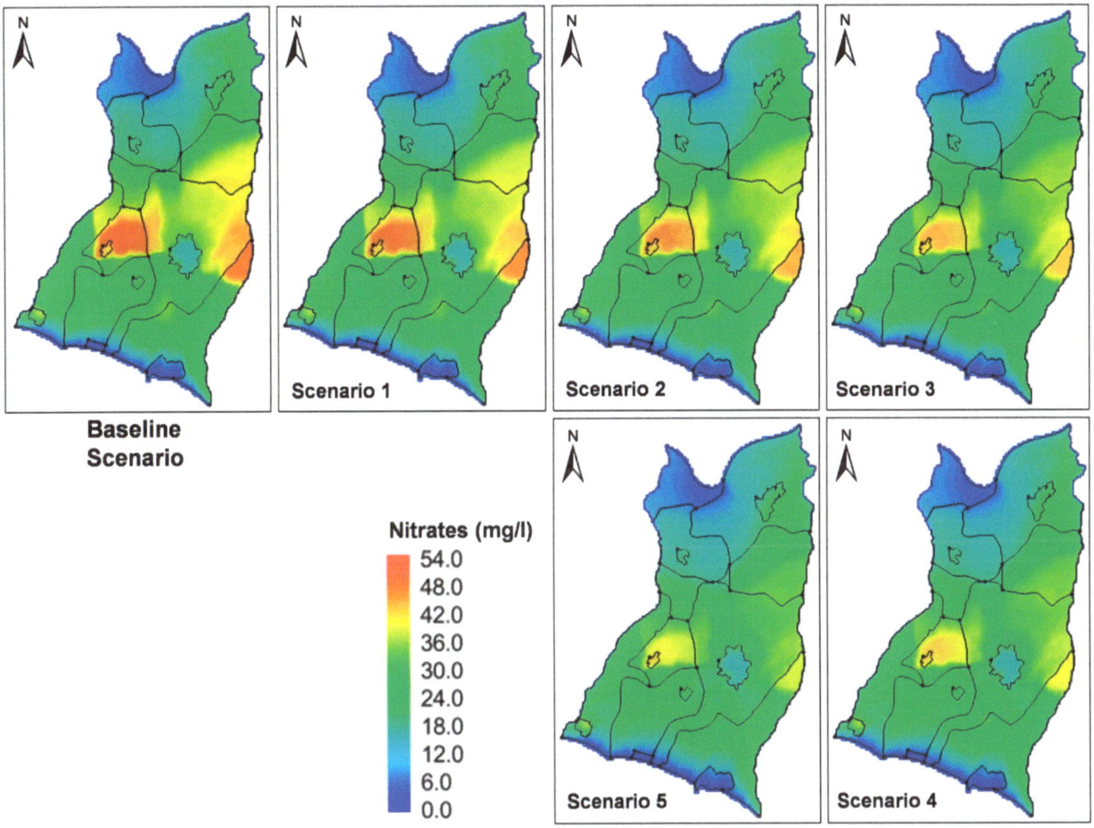

Figure 10. Nitrate concentration (in mg/L) distribution under both the baseline (S0) and different fertilization reduction scenarios (S1–S5) by the end of the simulation period (2034).

Regarding the temporal variation in nitrates, Figure 12 displays a time-series of nitrate mass in the aquifer throughout the projection period (2014–2034) for all fertilizer application scenarios. As is clear, an increase in nitrate mass over time is observed across all scenarios. However, this increase is comparatively lower in scenarios with higher reductions in fertilizer application. Table 6 supports the same conclusion, showing an increase in both maximum and mean nitrate concentrations within the study area from 2014 to 2034 under all scenarios (nitrate concentrations are identical for all scenarios in 2014). Thus, it is clear that, regardless of the varying levels of nitrogen loading in each applied scenario, a consistent upward trend in nitrate mass over time prevails in the aquifer under study. The predominant cause of this trend generally lies in the limited nitrate discharge from the groundwater system due to the existing boundary conditions and the absence of nitrate attenuation processes, like denitrification. In addition to the overall variation in nitrate mass throughout the entire projection period, Figure 12 also shows the annual nitrate mass variation during an indicative time period (October 2023–September 2024) for Scenario 0. In this case, it is worth highlighting the substantial increase in nitrate mass during February, aligning precisely with the monthly variation in nitrogen loading in the region (Figure 7), according to which the highest levels are observed during this particular month.

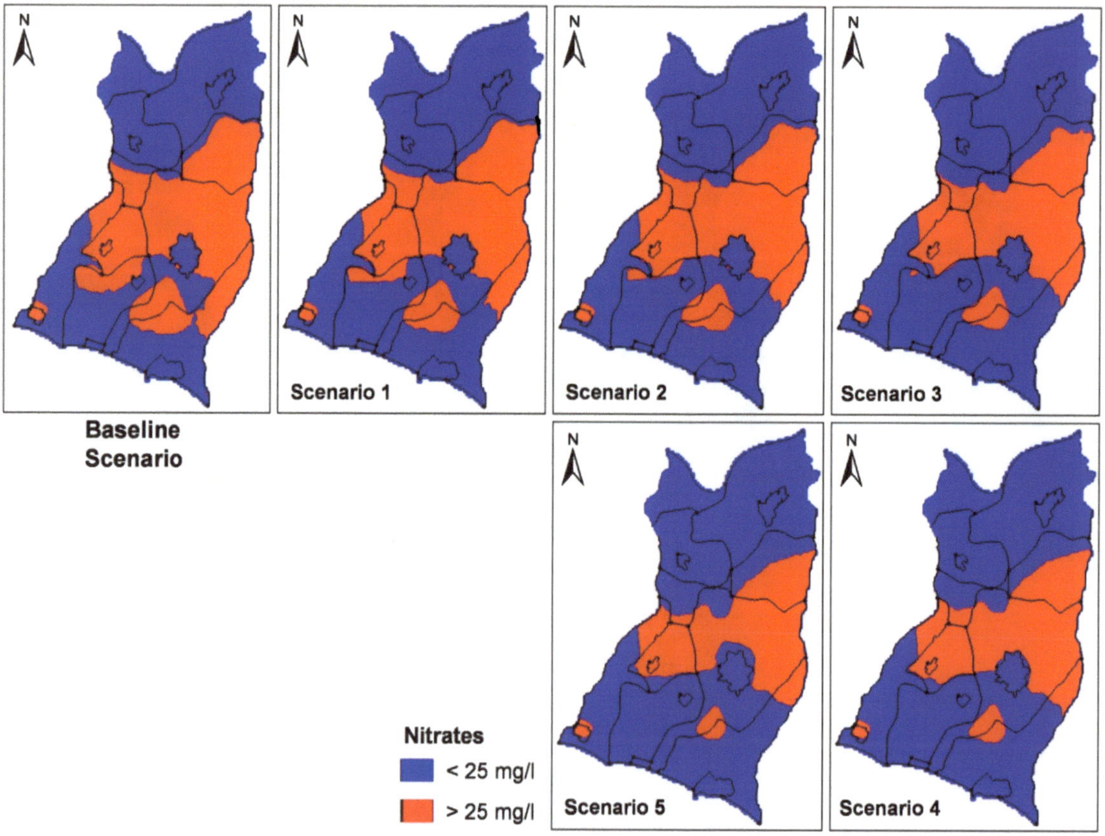

Figure 11. Spatial distribution of nitrate concentration exceeding 25 mg/L under both the baseline (S0) and different fertilization reduction scenarios (S1–S5) by the end of the simulation period (2034).

In the final part of the modeling results, Figure 13 depicts the nitrate concentrations in water-supply wells for both the baseline (S0) and different fertilization reduction scenarios (S1–S5) by the end of the simulation period (2034). These findings are of great importance as they reveal the actual magnitude of nitrate contamination and the associated level of risk. They may yield valuable insights into the quality of the extracted groundwater, particularly its suitability for meeting drinking water standards. In this context, the following conclusions can be drawn: (a) none of the water-supply wells in the study area are expected to exceed the maximum allowable limit of 50 mg/L during the implementation of the fertilization reduction scenarios (S1–S5), except for one well that marginally exceeds this threshold (NO_3 = 50.7 mg/L) during the baseline scenario (S0); (b) the number of water-supply wells exceeding the indicative threshold of 25 mg/L varies among the fertilization reduction scenarios (S1–S5), and specifically stands at 17 for Scenario 1, 16 for Scenario 2, 13 for Scenario 3, 11 for Scenario 4 and 9 for Scenario 5, thus rendering Scenario 5 the most effective in terms of reducing nitrate concentrations in water-supply wells; and (c) the implementation of fertilization reduction scenarios has no effect at all on the nitrate concentrations in the three water-supply wells situated near the settlement of Nea Flogita in the south-western part of the region, meaning that, in order to reduce nitrate contamination in this certain settlement, measures related to the proper management of wastewater should be considered.

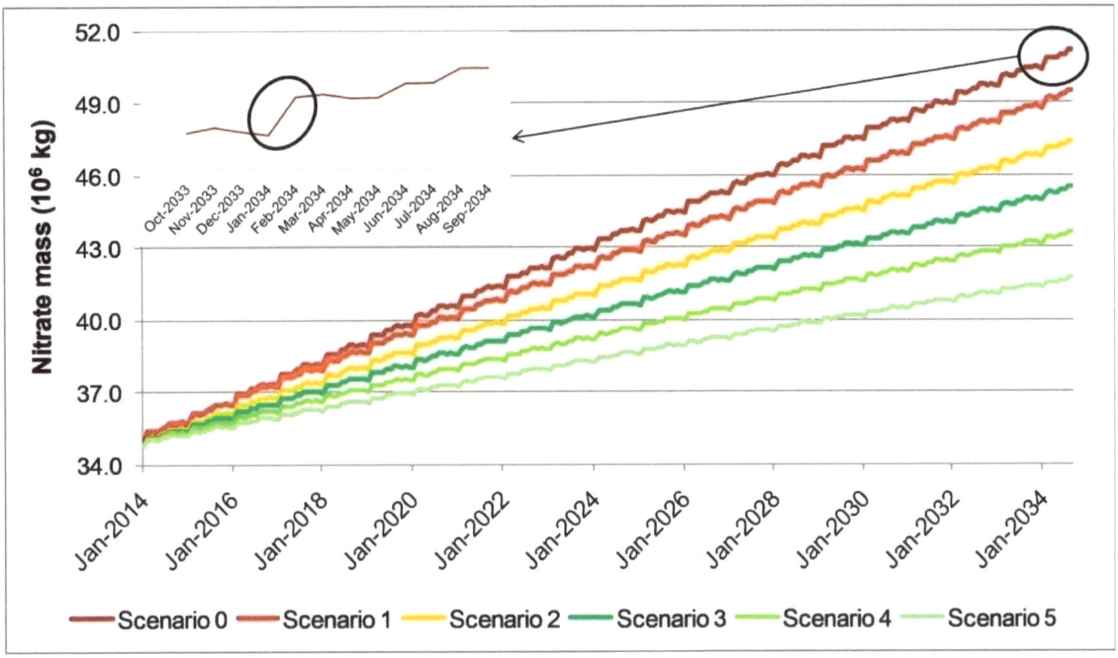

Figure 12. Time-series of nitrate mass (in 10^6 kg) in the aquifer under study for both the baseline (S0) and different fertilization reduction scenarios (S1–S5) throughout the projection period (2014–2034).

However, due to the fact that a different number of water-supply wells serves each settlement (Table 1) and the water provided to each settlement is actually blended water coming from all the corresponding water-supply wells, it is important to know the nitrate concentrations in this blended water. In Table 7, the nitrate concentrations in the blended water provided by the corresponding water-supply wells to individual settlements for the different fertilization reduction scenarios (S1–S5) are presented. According to these results, it becomes evident that, as the fertilizer application decreases, there is a related reduction in the nitrate concentrations observed in the blended water, as expected. The most important conclusion, however, is associated with the number of settlements receiving blended water with nitrate concentrations exceeding the "guidance level" of 25 mg/L. This metric serves as a crucial indicator, indicating a population potentially exposed to water of inadequate quality. As it turns out, the number of settlements exceeding the 25 mg/L nitrate concentration threshold stands at three for Scenario 1, two for Scenario 2 and one for Scenarios 3, 4, and 5. It is, therefore, apparent that, even with a 50% reduction in fertilizer application, the nitrate concentrations in blended water provided to a certain settlement (the settlement of Zografou, Figure 1) are generally expected to stay above the 25 mg/L threshold.

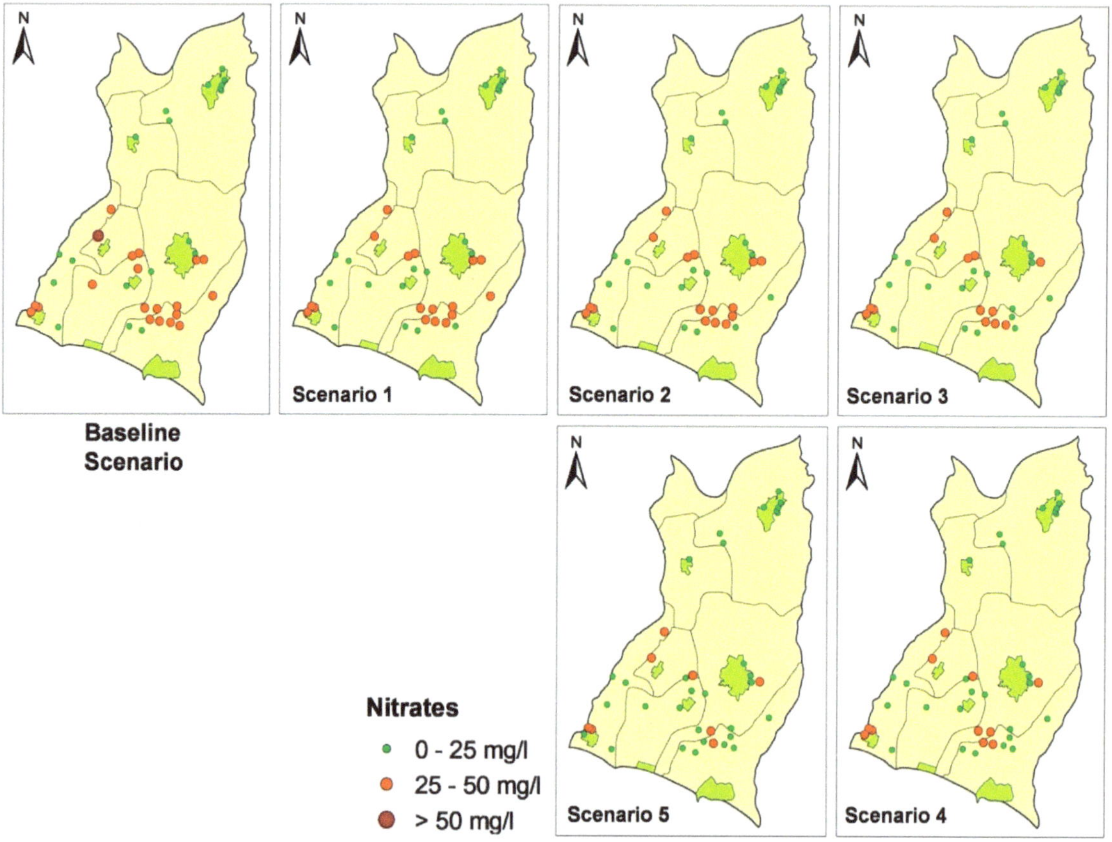

Figure 13. Nitrate concentrations (in mg/L) in water-supply wells (critical receptors) under both the baseline (S0) and different fertilization reduction scenarios (S1–S5) by the end of the simulation period (2034).

Table 7. Nitrate concentration (in mg/L) in the blended water provided by the corresponding water-supply wells to each settlement under the different fertilization reduction scenarios (S1–S5).

Settlements	S1	S2	S3	S4	S5
Nea Moudania	26.6	25.7	24.8	23.9	23.0
Nea Flogita	24.2	23.8	23.4	23.1	22.7
Dionysiou Paralia Dionysiou	25.4	24.8	23.8	22.8	21.8
Portaria	25.6	24.5	23.5	22.4	21.3
Zografou	41.4	39.5	37.5	35.5	33.6
Ag. Panteleimon	13.6	13.1	12.6	12.1	11.5
Simantra	20.3	20.0	19.6	19.3	19.0

3.2. Economic Analysis

In this section, the outcomes of the agro-economic analysis in terms of crop-yield losses in the various fertilization reduction scenarios, along with the estimated farmers' income losses, are presented and discussed. It is important, however, to mention that only the three principal crops—wheat, olives, and apricots—detected in the study area were included in the analysis, since these crops, as already mentioned, comprise 96% of the total cultivated land in the region. Therefore, the decrease in the farmers' income resulting from the reduced production of the aforementioned crops quite accurately reflects the total income loss in the region associated with agricultural activities.

First, in Table 8, the agro-economic data related to the three crops considered in the analysis, and required in order to apply Equation (1) in estimating the crop-yield losses resulting from the reduced fertilizer usage (see Section 3.2), are given. Next, Table 9 presents the estimated crop-yield losses for each crop and fertilization reduction scenario (S1–S5) in both net absolute values and percentages (in parentheses). The results indicate that fertilizer reduction has, in general, a greater impact on wheat (considering the percentage reduction in crop-yield), which is mainly attributable to the smaller variations between the expected crop-yield (3500 kg/ha) and the yield with no nutrient applied (1750 kg/ha) as compared to the other two crops.

Table 8. The agro-economic data related to the three crops analyzed in the present study and required for estimating crop-yield losses (all necessary data were acquired from the Directorate-General for Agricultural Economics and Veterinary of the Region of Central Macedonia, Greece).

Agro-Economic Data	Wheat	Olives	Apricots
Suggested rate of nitrogen (kg/ha)	150	750	150
Cost of unit of nitrogen (EUR/kg)	1.375	1.375	1.375
Price of unit of crop (EUR/kg)	0.36	1.20	1.10
Expected yield (kg/ha)	3500	15,000	30,000
Expected yield with no nitrogen (kg/ha)	1750	7500	15,000

Table 9. Crop-yield losses (in kg/ha) for each crop and fertilization reduction scenario (S1–S5) (the percentage reduction in crop yield is given in parentheses).

Scenarios	Wheat	Olives	Apricots
S1	69 (2.0%)	152 (1.0%)	167 (0.6%)
S2	162 (4.6%)	438 (2.9%)	630 (2.1%)
S3	278 (7.9%)	855 (5.7%)	1389 (4.6%)
S4	418 (11.9%)	1406 (9.4%)	2445 (8.2%)
S5	581 (16.6%)	2090 (13.9%)	3797 (12.7%)

Subsequently, the results of the economic analysis in terms of farmers' income losses due to fertilizer restrictions are presented. However, to obtain these results, the farmers' income for the baseline (S0) and fertilization reduction scenarios (S1–S5) were initially calculated. For this task, the revenue generated by farmers and the costs associated with crop production were considered, and the latter was subtracted from the former. Farmers' revenues were calculated by multiplying the total crop production by the respective selling price. Alternatively, production costs were calculated by summing the expenses incurred for labor, involving the costs associated with both human and machinery hours, together with material expenses, including seeds, fertilizers, and pesticides. In Table 10, the results derived from the aforementioned procedure for each type of crop considered in the analysis, namely wheat, olives and apricots, in the case of the baseline scenario (S0), are provided. From these findings, it is clear that olives play a substantial role in the total farmers' income in the region, making up roughly 66% of the overall income, followed by apricots, with a contribution of about 31%.

Table 10. Estimation of farmers' annual income (EUR) for each type of crop considered in the economic analysis in the case of the baseline scenario (S0) (all necessary data were acquired from the Directorate-General for Agricultural Economics and Veterinary of the Region of Central Macedonia, Greece).

	Wheat	Olives	Apricots
a. Area (ha)	4190.7	2502.0	595.6
b. Crop yield (kg/ha)	3500	15,000	30,000
c. Total crop yield: (a*b) (kg)	14,667,450	37,530,000	17,868,000
d. Selling price (EUR/kg)	0.36	1.20	1.10
A. REVENUE: (c*d) (EUR)	5,280,282	45,036,000	19,654,800
I. TOTAL REVENUE: (A) (EUR)	5,280,282	45,036,000	19,654,800
e. Labor costs (EUR/ha)	300	1300	1600
f. Seed expenses (EUR/ha)	175	0.0	0.0
g. Nitrogen-rich fertilizers costs (EUR/ha)	206	1031	206
h. Pesticides costs (EUR/ha)	70	480	1600
i. Materials expenses: (f+g+h) (EUR/ha)	451	1511	1806
B. TOTAL LABOR COSTS: (e*a) (EUR)	1,257,210	3,252,600	952,960
C. TOTAL MATERIALS EXPENSES: (i*a) (EUR)	1,891,053	3,781,148	1,075,803
D. OTHER COSTS: 5%*(A) (EUR)	264,014	2,251,800	982,740
II. TOTAL COSTS: (B+C+D) (EUR)	3,412,277	9,285,548	3,011,503
III. INCOME: (I–II) (EUR)	1,868,005	35,750,453	16,643,298

The same process as above was applied consistently across all fertilization reduction scenarios (S1–S5), with adjustments solely focused on the crop yield (referred to as "b"), as indicated by the values presented in Table 9, and fertilizer costs (referred to as "g"). In Table 11, the estimated farmers' income losses for each crop, together with the total income losses resulting from the implementation of different fertilization reduction scenarios (S1–S5), are presented. It is evident that the higher the reduction in fertilizer application, the higher the economic losses for each crop. Specifically, olive productions suffer the greatest loss of annual income in net absolute terms, ranging from 381,943 EUR in Scenario 1 (10% reduction in fertilizer application) to 5,703,246 EUR in Scenario 5 (50% reduction in fertilizer application). In the case of the most economically adverse scenario, i.e., Scenario 5, wheat indicates a loss of 400,535 EUR, while apricots show a loss of 2,301,839 EUR, culminating in a total income reduction of 8,405,620 EUR across all three crops. To provide a clearer insight into the economic impact of fertilization restrictions on individual crops, Figure 14 shows the percentage decrease in yearly earnings for each crop under various fertilization reduction scenarios. Among all three crops, wheat exhibits the greatest percentage income reduction, notably a 21.4% decrease with a 50% reduction in fertilizer usage (Scenario 5). Olives and apricots, on the other hand, demonstrate income reductions of 16.0% and 13.8%, respectively, under the same scenario.

Table 11. Estimated farmers' annual income losses (EUR) for each crop, together with the total income losses resulting from the implementation of fertilization reduction scenarios (S1–S5).

Scenarios	Wheat	Olives	Apricots	Total
S1	12,459	381,943	91,657	486,059
S2	59,315	1,146,091	367,545	1,572,951
S3	139,135	2,283,888	827,664	3,250,687
S4	253,353	3,803,891	1,472,636	5,529,880
S5	400,535	5,703,246	2,301,839	8,405,620

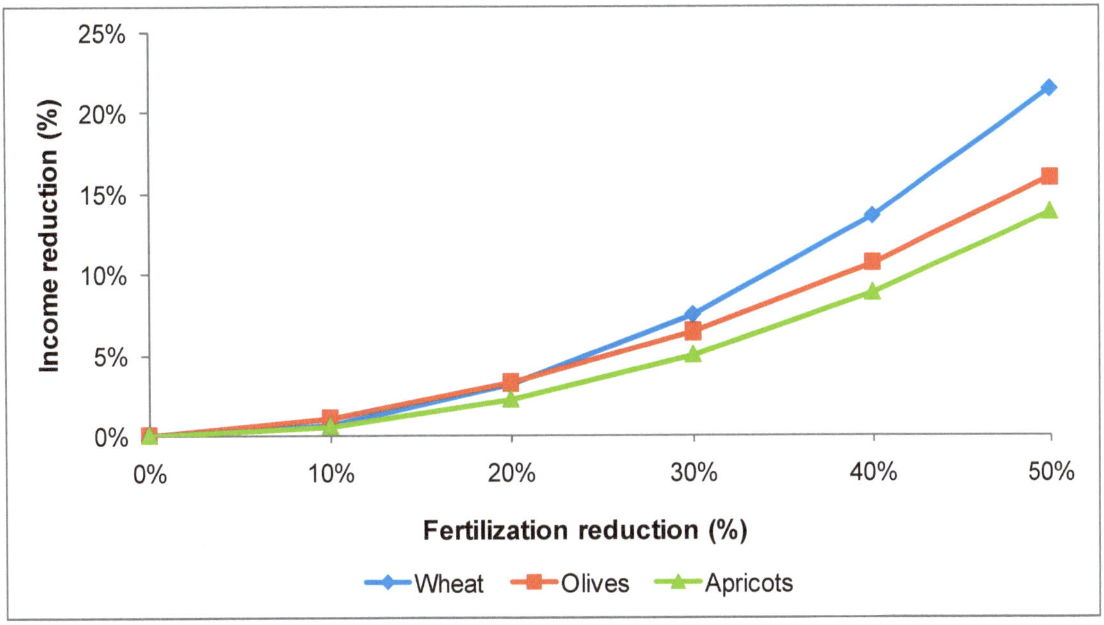

Figure 14. Percentage income reduction for each crop under various fertilization reduction scenarios (S1–S5).

3.3. Decision Analysis

In this section, the results of the decision analysis procedure regarding the estimation of decision criteria weights, along with scenarios ranking, are presented and discussed. First, in Table 12, a pair-wise comparison matrix for the four decision criteria considered in the analysis, as determined based on the importance order of criteria, is provided, along with the estimated criteria weights. It is worthwhile to mention that, among the selected criteria, the highest preference (greatest importance) was placed on the criterion referring to the population at risk, considering that the primary objective of the management measures, i.e., the fertilization reduction scenarios, is safeguarding public health. Regarding the remaining criteria, the order is as follows: CPCR, cost and nitrate mass.

Table 12. Pair-wise comparison matrix for the predefined decision criteria (CR = 0.0 < 0.1), along with the results of criteria weights estimation.

Criteria	Nitrate Mass	Population	Cost	CPCR	Weights
Nitrate mass	1.000	0.250	0.750	0.500	0.12
Population	4.000	1.000	3.000	2.000	0.48
Cost	1.333	0.333	1.000	0.667	0.16
CPCR	2.000	0.500	1.500	1.000	0.24

Next, Table 13 displays the values calculated for all decision criteria in each fertilization reduction scenario (S1–S5) as determined through the nitrate fate and transport model and the economic analysis, in addition to the corresponding scenario rankings. At this point, it should be mentioned that the ranking scores, as shown at the last column of Table 13, were computed following the normalization of the values of the selected criteria. According to the results, it can be evidently concluded that Scenario 3, entailing a 30% reduction in fertilizer application, is superior to all other scenarios, thus standing out as the most favorable option for the area under study, considering both the environmental and economic perspectives. The performance of this alternative scenario, particularly when compared to Scenarios 4 and 5 which encompass higher reductions in fertilizer application, is primarily attributed to the consistency across all the aforementioned scenarios in terms of the population exposed to water with nitrate concentrations exceeding the "guidance value" of 25 mg/L. Under these conditions, and given the fact that both net cost and CPCR increase in the case of Scenarios 4 and 5, the ranking score of Scenario 3 turns out to be higher than that of the other scenarios. However, Scenario 4 claimed the second position following Scenario 3, while Scenario 5 achieved the third place in the ranking scheme. As far as the remaining scenarios are concerned, despite a larger population being exposed to poor-quality water in Scenario 1, it outperforms Scenario 2 due to the lower net cost and CPCR value.

Table 13. The decision criteria's values calculated for each fertilization reduction scenario (S1–S5), along with the corresponding scenario rankings (S3, which involves a 30% reduction in nitrogen-rich fertilizers was identified as the most suitable option for the area under study).

Scenarios	Nitrate ($\times 10^6$ kg)	Population	Cost (EUR)	CPCR (EUR/mg/L)	Ranking
S1	49.5	42,080	486,059	567,554	0.511
S2	47.4	25,945	1,572,951	995,862	0.308
S3	45.5	903	3,250,687	1,379,508	0.713
S4	43.6	903	5,529,880	1,763,412	0.686
S5	41.7	903	8,405,620	2,152,457	0.673

At this point, it is worthwhile to mention that these findings, and particularly the designation of Scenario3 as the most recommendable option for the study area, seem to offer a rather "stable" solution, with minimal susceptibility to uncertainties linked to certain problem parameters, such as the "export coefficient" used for calculating nitrate leaching. This is mainly attributable to both the assumptions made during the formulation of the numerical model (involving conservative, advective-dispersive transport for the nitrates) and the model outcomes concerning nitrate concentrations in the blended water provided to the various settlements in the study area (see Table 7).

In conclusion, it can be inferred that, considering both environmental and economic aspects, a moderate reduction in fertilizer application (Scenario 3) was found to be the most suitable solution for the area under study. Even though higher reductions in fertilizers lead to a greater decrease in nitrate concentrations in the local groundwater resources, it appears that these reductions do not actually have a direct impact on the population exposed to poor-quality water. Conversely, higher reductions result in greater economic ramifications for farmers without actually offering a more effective solution regarding the quality of water provided to the residents of a certain settlement in the study area, since nitrate concentrations in drinking water remain above 25 mg/L. Under these conditions, with a portion of the local population in the study area still at risk, it becomes evident that, despite the implementation of fertilization restrictions, they alone do not offer an absolute solution in an effort to comply with drinking water standards. In this context, it is important to explore a more holistic management approach that integrates fertilizer regulations with more thorough strategies (i.e., abandoning existing wells with nitrate concentrations above 25 mg/L and constructing new ones within the potentially non-contaminated part of the aquifer). However, the solution identified through the proposed framework can serve as a foundation and be combined with additional measures with the ultimate goal of ensuring complete public health safety in the study area.

4. Conclusions

In the present study, an integrated framework aimed at properly managing groundwater nitrate contamination resulting from the excessive use of fertilizers in rural areas was developed. Specifically, the framework attempts to investigate and evaluate the implementation of fertilization restriction policies by considering both environmental and economic aspects in an effort to identify and propose the most appropriate option. To achieve this, numerical modeling was applied to study the spatio-temporal variation of nitrate concentrations, while an agro-economic analysis was performed to assess the economic impact of fertilization restrictions. As a final step, a multi-criteria decision analysis was carried out, incorporating criteria associated with both the environmental and economic assessment of fertilizer regulations, while placing specific emphasis on ensuring the quality of water provided to local population (i.e., reducing health risks). To test the methodology, five different scenarios, in which fertilizer application was reduced by certain rates (10%, 20%, 30%, 40% and 50%), were formed, evaluated and ranked.

The case study implementation revealed that, even though a 50% reduction in fertilizers resulted in lower nitrate concentrations in local groundwater resources, this option did not emerge as the most favorable one when considering all the criteria examined. Instead, a 30% reduction was identified as the most efficient solution specifically designated for the area being studied (site-specific finding), as it not only carries a similar level of risk to local residents compared to scenarios with higher reduction rates, but also leads to reduced economic losses for farmers. It then becomes clear that implementing such an approach is crucial when faced with selecting the most appropriate option among alternatives with conflicting objectives. Within this context, the proposed framework could serve as a useful and practical set of guidelines in designing efficient fertilizer regulation policies that consider both environmental and economic perspectives, considerably contributing to the proper management of nitrate contamination caused by agricultural fertilizers in rural areas. It might also provide valuable insights for conducting a more comprehensive management of nitrate contamination, particularly in cases where fertilizer restrictions fall short of fully guaranteeing water quality.

However, in order to expand the applicability of the proposed framework, future research could extend its analysis beyond fertilizer restrictions, encompassing additional policies which could implemented to control groundwater nitrate contamination. Such policies could either include different mechanisms for reducing fertilizer application, such as pricing policies (e.g., taxes) and changes in land use (e.g., from croplands to grassland or forest), or focus on addressing other potential sources of nitrate contamination, such as manure application, animal feedlot operations and urban wastewater sources (e.g., septic systems). In this context, additional criteria associated with the societal costs resulting from reductions in crop productivity could be considered in the decision-making process, thus leading to a more holistic evaluation of the impacts of fertilizer regulations. Additional work might also involve an all-inclusive analysis of the modeling system on the basis of the "Origin (land surface)—Pathway (vadose zone)—Target (aquifer)" model, thus allowing for a more detailed calculation of on-ground nitrogen loading (e.g., on a farm level) to be conducted, while also modeling nitrogen dynamics in the unsaturated zone. Finally, an update to the nitrate fate and transport model presented in the current study should be considered, specifically involving a more rigorous model validation, using new datasets of nitrate measurements from the area being studied.

Author Contributions: Conceptualization, I.S.; methodology, I.S. and Z.M.; software, I.S.; validation, I.S.; formal analysis, I.S.; investigation, I.S.; resources, I.S.; writing—original draft preparation, I.S. and Z.M.; writing—review and editing, I.S., Z.M. and P.L.; visualization, I.S.; supervision, P.L. All authors have read and agreed to the published version of the manuscript.

Funding: This research received no external funding.

Data Availability Statement: The data used in this study are contained within the article. Additional data are available upon request from the corresponding author.

Acknowledgments: Part of the research was conducted by Ilias Siarkos as a PHD candidate at the School of Civil Engineering, Aristotle University of Thessaloniki, Thessaloniki, Greece. Moreover, the authors would particularly like to thank Stefanos Sevastas for his assistance in acquiring the necessary agro-economic data used in the economic analysis.

Conflicts of Interest: The authors declare no conflict of interest.

References

1. De Filippis, G.; Ercoli, L.; Rossetto, R. A spatially distributed, physically-based modeling approach for estimating agricultural nitrate leaching to groundwater. *Hydrology* **2021**, *8*, 8. [CrossRef]
2. Mirzaee, M.; Safavi, H.R.; Taheriyoun, M.; Rezaei, F. Multi-objective optimization for optimal extraction of groundwater from a nitrate-contaminated aquifer considering economic-environmental issues: A case study. *J. Contam. Hydrol.* **2021**, *241*, 103806. [CrossRef]
3. Shultz, C.D.; Gates, T.K.; Bailey, R.T. Evaluating best management practices to lower selenium and nitrate in groundwater and streams in an irrigated river valley using a calibrated fate and reactive transport model. *J. Hydrol.* **2018**, *566*, 299–312. [CrossRef]
4. Sishu, F.K.; Tilahun, S.A.; Schmitter, P.; Steenhuis, T.S. Investigating nitrate with other constituents in groundwater in two contrasting tropical highland watersheds. *Hydrology* **2023**, *10*, 82. [CrossRef]
5. Wei, X.; Bailey, R.T.; Records, R.M.; Wible, T.G.; Arabi, M. Comprehensive simulation of nitrate transport in coupled surface-subsurface hydrologic systems using the linked SWAT-MODFLOW-RT3D model. *Environ. Modell. Softw.* **2019**, *122*, 104242. [CrossRef]
6. Almasri, M.N. Nitrate contamination of groundwater: A conceptual management framework. *Environ. Impact Asses.* **2007**, *27*, 220–242. [CrossRef]
7. Ji, W.; Xiao, J.; Toor, G.S.; Li, Z. Nitrate-nitrogen transport in streamwater and groundwater in a loess covered region: Sources, drivers, and spatiotemporal variation. *Sci. Total Environ.* **2021**, *761*, 143278. [CrossRef]
8. Latinopoulos, P. Nitrate contamination of groundwater: Modeling as a tool for risk assessment, management and control. In *Groundwater Pollution Control*; Katsifarakis, K.L., Ed.; WIT Press: Southampton, UK, 1999; pp. 1–48.
9. Sullivan, T.P.; Gao, Y.; Reimann, T. Nitrate transport in a karst aquifer: Numerical model development and source evaluation. *J. Hydrol.* **2019**, *573*, 432–448. [CrossRef]
10. Viers, J.H.; Liptzin, D.; Rosenstock, T.S.; Jensen, V.B.; Hollander, A.D.; McNally, A.; King, A.M.; Kourakos, G.; Lopez, E.M.; De La Mora, N.; et al. Nitrogen sources and loading to groundwater. In *Addressing Nitrate in California's Drinking Water with a Focus on Tulare Lake Basin and Salinas Valley Groundwater*; Technical Report 2; University of California: Davis, CA, USA, 2012.
11. Buskulic, P.; Parlov, J.; Kovac, Z.; Nakic, Z. Estimation of nitrate background value in groundwater under the long-term human impact. *Hydrology* **2023**, *10*, 63. [CrossRef]
12. Sidiropoulos, P.; Mylopoulos, N.; Vasiliades, L.; Loukas, A. Stochastic nitrate simulation under hydraulic conductivity uncertainty of an agricultural basin aquifer at Eastern Thessaly, Greece. *Environ. Sci. Pollut. R.* **2021**, *28*, 65700–65715. [CrossRef]
13. Wick, K.; Heumesser, C.; Schmid, E. Groundwater nitrate contamination: Factors and indicators. *J. Environ. Manag.* **2012**, *111*, 178–186. [CrossRef]
14. Zhang, H.; Yang, R.; Wang, Y.; Ye, R. The evaluation and prediction of agriculture-related nitrate contamination in groundwater in Chengdu Plain, southwestern China. *Hydrogeol. J.* **2019**, *27*, 785–799. [CrossRef]
15. Ameur, M.; Aouiti, S.; Hamzaoui-Azaza, F.; Cheikha, L.B.; Gueddari, M. Vulnerability assessment, transport modeling and simulation of nitrate in groundwater using SI method and modflow-MT3DMS software: Case of Sminja aquifer, Tunisia. *Environ. Earth Sci.* **2021**, *80*, 220. [CrossRef]
16. Correa-Gonzalez, A.; Hernandez-Bedolla, J.; Martinez-Cinco, M.A.; Sanchez-Quispe, S.T.; Hernandez-Hernandez, M.A. Assessment of nitrate in groundwater from diffuse sources considering spatiotemporal patterns of hydrological systems using a coupled SWAT/MODFLOW/MT3DMS model. *Hydrology* **2023**, *10*, 209. [CrossRef]
17. Almasri, M.N.; Kaluarachchi, J.J. Modeling nitrate contamination of groundwater in agricultural watersheds. *J. Hydrol.* **2007**, *343*, 211–229. [CrossRef]
18. Bryan, N.S.; Van Grinsven, H. The role of nitrate in human health. *Adv. Agron.* **2013**, *119*, 153–182.
19. Sidiropoulos, P.; Tziatzios, G.; Vasiliades, L.; Mylopoulos, N.; Loukas, A. Groundwater nitrate contamination integrated modeling for climate and water resources scenarios: The case of Lake Karla over-exploited aquifer. *Water* **2019**, *11*, 1201. [CrossRef]
20. Wolfe, A.H.; Patz, J.A. Reactive nitrogen and human health: Acute and long-term implications. *Ambio* **2002**, *31*, 120–125. [CrossRef]
21. Powlson, D.S.; Addiscott, T.M.; Benjamin, N.; Cassman, K.G.; De Kok, T.M.; Van Grinsven, H.; L'hirondel, J.; Avery, A.A.; Van Kessel, C. When does nitrate become a risk for humans? *J. Environ. Qual.* **2008**, *37*, 291–295. [CrossRef]
22. Zhang, F.-X.; Miao, Y.; Ruan, J.-G.; Meng, S.-P.; Dong, J.-D.; Yin, H.; Huang, Y.; Chen, F.-R.; Wang, Z.-C.; Lai, Y.-F. Association between nitrite and nitrate intake and risk of gastric cancer: A systematic review and meta-analysis. *Med. Sci. Monit.* **2019**, *25*, 1788–1799. [CrossRef]
23. Xin, J.; Wang, Y.; Shen, Z.; Liu, Y.; Wang, H.; Zheng, X. Critical review of measures and decision support tools for groundwater nitrate management: A surface-to-groundwater profile perspective. *J. Hydrol.* **2021**, *598*, 126386. [CrossRef]
24. Bastani, M.; Harter, T. Source area management practices as remediation tool to address groundwater nitrate pollution in drinking supply wells. *J. Contam. Hydrol.* **2019**, *226*, 103521. [CrossRef]
25. Pena-Haro, S.; Pulido-Velazquez, M.; Sahuquillo, A. A hydro-economic modelling framework for optimal management of groundwater nitrate pollution from agriculture. *J. Hydrol.* **2009**, *373*, 193–203. [CrossRef]

26. Kumar, P.; Thakur, P.K.; Bansod, B.K.S.; Debnath, S.K. Groundwater: A regional resource and a regional governance. *Environ. Dev. Sustain.* **2018**, *20*, 1133–1151. [CrossRef]
27. Siarkos, I.; Sevastas, S.; Latinopoulos, D. Using a hydroeconomic model to evaluate alternative methods applied for the delineation of protection zones. *Desalin. Water. Treat.* **2018**, *133*, 315–326. [CrossRef]
28. Siarkos, I.; Arfaoui, M.; Tzoraki, O.; Zammouri, M.; Hamzaoui-Azaza, F. Implementation and evaluation of different techniques to modify DRASTIC method for groundwater vulnerability assessment: A case study from Bouficha aquifer, Tunisia. *Environ. Sci. Pollut. Res.* **2023**, *30*, 89459–89478. [CrossRef]
29. Almasri, M.N.; Kaluarachchi, J.J. Multi-criteria decision analysis for the optimal management of nitrate contamination of aquifers. *J. Environ. Manag.* **2005**, *74*, 365–381. [CrossRef]
30. Liu, R.; Zhang, P.; Wang, X.; Chen, Y.; Shen, Z. Assessment of effects of best management practices on agricultural non-point source pollution in Xiangxi River watershed. *Agric. Water Manag.* **2013**, *117*, 9–18. [CrossRef]
31. Samadi-Darafshani, M.; Safavi, H.R.; Golmohammadi, M.H.; Rezaei, F. Assessment of the management scenarios for groundwater quality remediation of a nitrate-contaminated aquifer. *Environ. Monit. Assess.* **2021**, *193*, 190. [CrossRef]
32. Yang, L.; Zheng, C.; Andrews, C.B.; Wang, C. Applying a regional transport modeling framework to manage nitrate contamination of groundwater. *Groundwater* **2021**, *59*, 292–307. [CrossRef]
33. Almasri, M.N.; Kaluarachchi, J.J. Assessment and management of long-term nitrate pollution of ground water in agriculture-dominated watersheds. *J. Hydrol.* **2004**, *295*, 225–245. [CrossRef]
34. Reed, E.M.; Wang, D.; Duranceau, S.J. Evaluating nitrate management in the Volusia Blue Springshed. *J. Environ. Eng.* **2018**, *144*, 05018001. [CrossRef]
35. Karlovic, I.; Posavec, K.; Larva, O.; Markovic, T. Numerical groundwater flow and nitrate transport assessment in alluvial aquifer of Varazdin region, NW Croatia. *J. Hydrol. Reg. Stud.* **2022**, *41*, 101084. [CrossRef]
36. Musacchio, A.; Mas-Pla, J.; Soana, E.; Re, V.; Sacchi, E. Governance and groundwater modelling: Hints to boost the implementation of the EU Nitrate Directive. The Lombardy Plain case, N Italy. *Sci. Total Environ.* **2021**, *782*, 146800. [CrossRef]
37. Rawat, M.; Sen, R.; Onyekwelu, I.; Wiederstein, T.; Sharda, V. Modeling of groundwater nitrate contamination due to agricultural activities—A systematic review. *Water* **2022**, *14*, 4008. [CrossRef]
38. Zhang, H.; Yang, R.; Guo, S.; Li, Q. Modeling fertilization impacts on nitrate leaching and groundwater contamination with HYDRUS-1D and MT3DMS. *Paddy Water Environ.* **2020**, *18*, 481–498. [CrossRef]
39. Hou, C.; Chu, M.L.; Botero-Acosta, A.; Guzman, J.A. Modeling field scale nitrogen non-point source pollution (NPS) fate and transport: Influences from land management practices and climate. *Sci. Total Environ.* **2021**, *759*, 143502. [CrossRef]
40. Zhang, H.; Hiscock, K.M. Modelling the effect of forest cover in mitigating nitrate contamination of groundwater: A case study of the Sherwood Sandstone aquifer in the East Midlands, UK. *J. Hydrol.* **2011**, *399*, 212–225. [CrossRef]
41. Gomann, H.; Kreins, P.; Kunkel, R.; Wendland, F. Model based impact analysis of policy options aiming at reducing diffuse pollution by agriculture—A case study for the river Ems and a sub-catchment of the Rhine. *Environ. Modell. Softw.* **2005**, *20*, 261–271. [CrossRef]
42. Graveline, N.; Rinaudo, J.D.; Segger, V.; Lambrecht, H.; Casper, M.; Elsass, P.; Grimm-Strele, J.; Gudera, T.; Koller, R.; Van Dijk, P. Integrating economic and groundwater models for developing long-term nitrate concentration scenarios in a large aquifer. In *Aquifer Systems Management: Darcy's Legacy in a World of Impending Water Shortage*; Chery, L., Marsily, G., Eds.; Taylor & Francis: New York, NY, USA, 2007; pp. 483–495.
43. Pena-Haro, S.; Llopis-Albert, C.; Pulido-Velazquez, M.; Pulido-Velazquez, D. Fertilizer standards for controlling groundwater nitrate pollution from agriculture: El Salobral-Los Llanos case study, Spain. *J. Hydrol.* **2010**, *392*, 174–187. [CrossRef]
44. Liu, G.; Chen, L.; Wei, G.; Shen, Z. New framework for optimizing best management practices at multiple scales. *J. Hydrol.* **2019**, *578*, 124133. [CrossRef]
45. Dotoli, M.; Epicoco, N.; Falagario, M. Multi-criteria decision making techniques for the management of public procurement tenders: A case study. *Appl. Soft. Comput.* **2020**, *88*, 106064. [CrossRef]
46. Latinopoulos, P. Development of water resources management plan for water supply and irrigation. In *Final Report Prepared for: Municipality of Nea Moudania*; Aristotle University of Thessaloniki: Thessaloniki, Greece, 2003. (In Greek)
47. Syridis, G. Lithostromatographical, Biostromatographical and Paleostromatographical Study of Neogene-Quaternary Formation of Chalkidiki Peninsula. Ph.D. Thesis, Aristotle University of Thessaloniki, Thessaloniki, Greece, 1990. (In Greek)
48. Siarkos, I. Developing a Methodological Framework Using Mathematical Simulation Models in Order to Investigate the Operation of Coastal Aquifer Systems: Application in the Aquifer of Nea Moudania. Ph.D. Thesis, Aristotle University of Thessaloniki, Thessaloniki, Greece, 2015. (In Greek)
49. Siarkos, I.; Latinopoulos, P. Modeling seawater intrusion in overexploited aquifers in the absence of sufficient data: Application to the aquifer of Nea Moudania, northern Greece. *Hydrogeol. J.* **2016**, *24*, 2123–2141. [CrossRef]
50. Shamrukh, M.; Corapcioglu, M.Y.; Hassona, F.A.A. Modeling the effect of chemical fertilizers on ground water quality in the Nile Valley aquifer, Egypt. *Groundwater* **2001**, *39*, 59–67. [CrossRef]
51. Siarkos, I.; Latinopoulos, D.; Mallios, Z.; Latinopoulos, P. A methodological framework to assess the environmental and economic effects of injection barriers against seawater intrusion. *J. Environ. Manag.* **2017**, *193*, 532–540. [CrossRef] [PubMed]
52. Guo, W.; Langevin, C.D. User's guide to SEAWAT: A computer program for simulation of three-dimensional variable-density groundwater flow. In *Techniques of Water-Resources Investigations, Book 6-A7*; USGS: Reston, VA, USA, 2002.

53. Zheng, C.; Wang, P.P. *MT3DMS: A Modular Three-Dimensional Multispecies Transport Model for Simulation of Advection, Dispersion, and Chemical Reactions of Contaminants in Groundwater Systems—Documentation and User's Guide*; Contract Report SERD-99-1; US Army Corps of Engineers: Washington, DC, USA, 1999.
54. Almasri, M.N.; Kaluarachchi, J.J. Implications of on-ground nitrogen loading and soil transformations on ground water quality management. *J. Am. Water Resour. Assoc.* **2004**, *40*, 165–186. [CrossRef]
55. Hajhamad, L.; Almasri, M.N. Assessment of nitrate contamination of groundwater using lumped-parameter models. *Environ. Modell. Softw.* **2009**, *24*, 1073–1087. [CrossRef]
56. Kaluarachchi, J.; Almasri, M. *Conceptual Model of Fate and Transport of Nitrate in the Extended Sumas-Blaine Aquifer, Whatcom County, Washington*; Project Report; Utah State University: Logan, UT, USA, 2002.
57. Ministry of Rural Development and Food. *Data on the Integrated Management System of Cultivated Areas*; Ministry of Rural Development and Food: Athens, Greece, 2007.
58. Zhang, J.; Jorgensen, S.E. Modelling of point and non-point nutrient loadings from a watershed. *Environ. Modell. Softw.* **2005**, *20*, 561–574. [CrossRef]
59. ASAE. *Manure Production and Characteristics*; D384.2; American Society of Agricultural Engineers: St. Joseph, MI, USA, 2005.
60. Meisinger, J.J.; Randall, G.W. Estimating nitrogen budgets for soil-crop systems. In *Managing Nitrogen for Groundwater Quality and Farm Profitability*; Follet, R.F., Keeney, D.R., Cruse, R.M., Eds.; Soil Science Society of America: Madison, WI, USA, 1991; pp. 85–124.
61. Cox, S.E.; Kahle, S.C. *Hydrogeology, Ground-Water Quality, and Sources of Nitrate in Lowland Glacial Aquifers of Whatcom County, Washington, and British Columbia, Canada*; Water-Resources Investigations Report 98-4195; U.S. Geological Survey: Reston, VA, USA, 1999.
62. Zhang, H.; Hiscock, K.M. Modelling response of groundwater nitrate concentration in public supply wells to land-use change. *Q. J. Eng. Geol. Hydrog.* **2016**, *49*, 170–182. [CrossRef]
63. Peterson, E.W.; Hayden, K.M. Transport and fate of nitrate in the streambed of a low-gradient stream. *Hydrology* **2018**, *5*, 55. [CrossRef]
64. Puig, R.; Soler, A.; Widory, D.; Mas-Pla, J.; Domenech, C.; Otero, N. Characterizing sources and natural attenuation of nitrate contamination in the Baix Ter aquifer system (NE Spain) using a multi-isotope approach. *Sci. Total Environ.* **2017**, *580*, 518–532. [CrossRef] [PubMed]
65. Eryigit, M.; Engel, B. Spatiotemporal modelling of groundwater flow and nitrate contamination in an agriculture-dominated watershed. *J. Environ. Inform.* **2022**, *39*, 125–135. [CrossRef]
66. Kourgialas, N.N.; Karatzas, G.P.; Koubouris, G.C. A GIS policy approach for assessing the effect of fertilizers on the quality of drinking and irrigation water and wellhead protection zones (Crete, Greece). *J. Environ. Manag.* **2017**, *189*, 150–159. [CrossRef] [PubMed]
67. Xefteris, A.; Anastasiadis, P.; Latinopoulos, P. Groundwater chemical characteristics in Kalamaria Plain, Halkidiki Peninsula, Greece. *Fresen. Environ. Bull.* **2004**, *13*, 1158–1167.
68. Rivett, M.O.; Buss, S.R.; Morgan, P.; Smith, J.W.N.; Bemment, C.D. Nitrate attenuation in groundwater: A review of biogeochemical controlling processes. *Water Res.* **2008**, *42*, 4215–4232. [CrossRef] [PubMed]
69. Tesoriero, A.; Liecscher, S.; Cox, S. Mechanism and rate of denitrification in an agricultural watershed: Electron and mass balance along ground water flow paths. *Water Resour. Res.* **2000**, *36*, 1545–1559. [CrossRef]
70. Psarropoulou, E.T.; Karatzas, G.P. Pollution of nitrates—Contaminant transport in heterogeneous porous media: A case study of the coastal aquifer of Corinth, Greece. *Global Nest J.* **2014**, *16*, 9–23.
71. Dhakal, C.; Lange, K. Crop yield response functions in nutrient application: A review. *Agron. J.* **2021**, *113*, 5222–5234. [CrossRef]
72. Murrell, T.S. *A Procedure for Estimating Yield Loss from Nutrient Rate Reductions*; Report, International Plant Nutrition Institute: Norcross, GA, USA, 2009.
73. Nasri, G.; Hajji, S.; Aydi, W.; Bougariou, E.; Allouche, N.; Bouri, S. Water vulnerability of coastal aquifers using AHP and parametric models: Methodological overview and a case study assessment. *Arab. J. Geosci.* **2021**, *14*, 59. [CrossRef]
74. Nkeki, F.N.; Bello, E.I.; Agbaje, I.G. Flood risk mapping and urban infrastructural susceptibility assessment using a GIS and analytic hierarchical raster fusion approach in the Ona River Basin, Nigeria. *Int. J. Disast. Risk Re.* **2022**, *77*, 103097. [CrossRef]
75. Yang, J.; Tang, Z.; Jiao, T.; Muhammad, A.M. Combining AHP and genetic algorithms approaches to modify DRASTIC model to assess groundwater vulnerability: A case study from Jianghan Plain, China. *Environ. Earth Sci.* **2017**, *76*, 426. [CrossRef]
76. Barzegar, R.; Moghaddam, A.A.; Adamowski, J.; Nazemi, A.H. Delimitation of groundwater zones under contamination risk using a bagged ensemble of optimized DRASTIC frameworks. *Environ. Sci. Pollut. Res.* **2019**, *26*, 8325–8339. [CrossRef] [PubMed]
77. Wu, H.; Chen, J.; Qian, H. A modified DRASTIC model for assessing contamination risk of groundwater in the northern suburb of Yinchuan, China. *Environ. Earth Sci.* **2016**, *75*, 483. [CrossRef]

hydrology

MDPI

Article

An Integrated Remote Sensing and GIS-Based Technique for Mapping Groundwater Recharge Zones: A Case Study of SW Riyadh, Central Saudi Arabia

Eman Mohamed M. EL-Bana [1], Haya M. Alogayell [1], Mariam Hassan Sheta [2] and Mohamed Abdelfattah [3,4,*]

[1] Department of Geography and Environmental Sustainability, College of Humanities and Social Sciences, Princess Nourah bint Abdulrahman University, P.O. Box 84428, Riyadh 11671, Saudi Arabia
[2] Environmental Sciences Department, Faculty of Science, Port Said University, Port Said 42522, Egypt; mariam_sheta@sci.psu.edu.eg
[3] Geology Department, Faculty of Science, Port Said University, Port Said 42522, Egypt
[4] EDYTEM, University Savoie Mont-Blanc, 73370 Le Bourget du Lac, France
* Correspondence: mohamed_abdelfattah@sci.psu.edu.eg or mohamed.abdelfattah@univ-smb.fr

Abstract: It might be difficult to find possible groundwater reservoir zones, especially in arid or hilly regions. In the twenty-first century, remotely sensed satellite imagery may present a new opportunity to locate surface and subsurface water resources more quickly and affordably. In order to identify groundwater potential zones, the current study was conducted in Central Saudi Arabia, southwest of Riyadh. The present analysis employed a multi-criteria approach that relies on remote sensing and geographic information systems. The variables employed in this technique include geology, rainfall, elevation, slope, aspect, hillshade, drainage density, lineaments density, and Land Use/Land Cover (LULC). The Analytical Hierarchical Process (AHP) was used for assigning weights to the parameters, and the corresponding significance of each parameter's several classes for groundwater potentiality. Different groundwater potential zones were identified by the study: very high (16.8%), high (30%), medium (26.7%), low (18.6%), and very low (7.9%). Only two of the observation wells were located in the "medium" potential zone, but the other ten wells were observed in the "very high and high" potential zones, according to the validation survey. Consequently, the results may demonstrate that the current approach, which combines improved conceptualization with AHP to define and map groundwater potential zones, has a greater chance of producing accurate results and can be used to reduce the threat of drought in broader arid regions.

Keywords: remote sensing; GIS; multicriteria; groundwater potential zones; SW Riyadh; Saudi Arabia

Citation: EL-Bana, E.M.M.; Alogayell, H.M.; Sheta, M.H.; Abdelfattah, M. An Integrated Remote Sensing and GIS-Based Technique for Mapping Groundwater Recharge Zones: A Case Study of SW Riyadh, Central Saudi Arabia. *Hydrology* **2024**, *11*, 38. https://doi.org/10.3390/hydrology11030038

Academic Editors: Augustina Clara Alexander and Pantelis Sidiropoulos

Received: 24 January 2024
Revised: 27 February 2024
Accepted: 29 February 2024
Published: 3 March 2024

1. Introduction

Water access is essential for socioeconomic growth. Freshwater availability has historically served as a crucial indication of human prosperity and is important for achieving sustainable development objectives. However, in recent years, there has been a widespread worsening of water availability due to rapid population growth, high water requirements for construction, industrialization, and irrigation. Furthermore, freshwater availability has declined due to global climate change in many areas, including developing nations, especially in arid and semi-arid regions [1,2].

The primary supply of freshwater for a variety of applications is groundwater [3,4]. Prior research found that 26% of the world's sustainable freshwater sources come from groundwater [5]. Approximately 50% of the world's water requirements for residential use come from groundwater, which serves as the main source of water for industry, households, and agriculture, according to other studies [6,7]. In the majority of African communities, particularly those in dry and semi-arid regions, groundwater is one of the most valuable natural resources. Roughly 20% of the water utilized for agriculture, 40% for industry, and

50% for potable consumption comes from groundwater supplies [8]. Consequently, it is necessary to investigate these priceless resources for those who frequently have no other possibilities. Thus, it is critical to have a thorough grasp of groundwater systems over location and time. Groundwater reserves are subject to significant fluctuations in availability within short timeframes. This also holds true when research shows that groundwater resources are still not sufficiently investigated to satisfy the community's high demands [9]. Groundwater fills the pore spaces, cavities, and joints in the soil that are found inside rocks and geologic strata. The occurrence and infiltration of groundwater in rocks is determined by the hydraulic conductivity of the materials due to their lithologic properties such as porosity, permeability, and fluid transport via geologic structures [10–12]. Consequently, the two standard approaches for locating groundwater are drilling and geoelectric observation, both of which are quite costly and time-consuming.

As numerous studies have demonstrated, aquifer recharge zones are currently identified globally using the integration of remote sensing (RS) and geographic information system (GIS) techniques [13–16]. By considering similarities with the research region, these studies successfully identify the most advantageous locations for aquifer recharge [17–28]. A variety of characteristics have been used to define global potential zones, including rainfall, topography, lithology, geomorphology, slope, drainage pattern, and lineaments. Multi-criteria evaluation (MCE) is a widely utilized technique that is frequently used to assess various criteria together, and the results of this methodology offer persuasive proof of the aquifer's state [29–34].

GIS and remote sensing (RS) have the potential to be employed for groundwater investigation, as well as for regional predictions [35,36]. Big geographical data may be aggregated and analyzed to forecast and find new water sources by using GIS-based technology and/or data-driven methodologies [37,38]. Several research projects have demonstrated the practicality of combining GIS and RS to locate possible groundwater resource locations [35,39–41]. Several researchers have adopted and used a variety of methods for identifying groundwater potential zones (GWPZ), including the logistic regression model [42], random forest model [43], and frequency ratio [28,44]. The GIS and RS-based analytical hierarchy process (AHP) is considered a simple, effective, reliable, and cost-effective approach in this case [45–47]. Saaty (1977) gave the first explanation of AHP [48]. The intended outcomes are obtained by the AHP technique, which reduces complex results to a set of paired data [49]. It is commonly acknowledged that AHP is a very powerful method for assessing output reliability, reducing bias, and applying it in an innovative setting [50]. In order to evaluate aquifer potential and identify recharge zones, it is imperative that prior parameters, including rainfall, geology, geomorphology, slope, lineament density, and drainage density, be integrated into the weight overlay model. This comprehensive method uses the AHP to compute weightage, allowing for a more accurate evaluation of the aquifer's prospects as well as the identification of areas with notable recharging capability. Numerous investigations have been conducted worldwide to determine the groundwater potential zones using GIS and RS for the evaluation of groundwater resources in the Maknassy basin, Tunisia [51], Ghana [52], Burdur, Turkey [53], Egypt [54,55], Saudi Arabia [56], Kurdistan region, Iran [57], West Bengal, India [58], Eritrea [59], Uganda [60], Kenya [61], Huay Sai area, Thailand [62], and China [63].

The arid environment of the Kingdom of Saudi Arabia (KSA) makes renewable water resources scarce [64]. These limited water supplies and growing unpredictability brought on by climate change present significant problems for water management in the KSA. Another negative impact of using underground water from aquifers is the depletion of resources which have taken decades to recover again because the current annual rainfall has no direct effect [65,66]. The nation receives an estimated 158.47 billion m^3 of rainwater yearly [67]. According to the previous studies, the largest single alluvial reservoir in the KSA has an overall capacity of 84 billion m^3 [68–70]. Over the past two decades, about 254.5 billion m^3 of groundwater has been utilized from Saudi Arabia's aquifers to satisfy the needs of the country's expanding agricultural industry [67]. However, over the past

20 years, only 41.04 billion m^3 of water has been recharged into the deep aquifers [67]. The KSA's main supply of drinkable and agricultural water comes from groundwater [71,72]. Of the country's water requirement, 80% is met by the 17 billion m^3/year of groundwater extracted in the KSA over the previous thirty years [71–73]. The KSA developed a plan to reduce the amount of water needed for agriculture by using modern irrigation techniques in order to adhere to the water preservation strategy [74]. The amount of water utilized for agriculture has decreased by 2.5% on average each year as a result of this strategy. It is essential to assess groundwater potential zones for agricultural expansion due to the limitations on the supply of water and the potential for growing the cultivated area.

Therefore, this study used a wide range of modern RS and GIS tools for broad investigation with the goal of identifying the groundwater promising zones and the most suitable spots for groundwater recharge. In the current study, the most effective groundwater recharge potentiality zones were predicted by processing and combining data from numerous satellite images, involving Shuttle Radar Topography Mission (STRM) and Landsat imagery. Utilizing GIS spatial analysis, several layers including topography, geology, soil, slope drainage density, LULC, and lineament maps were combined. The optimal groundwater recharge spots are successfully determined by this method. Water management in the research area is greatly aided by the findings from potential zone maps' classification into five zones, ranging from extremely high to very low groundwater recharge potential zones, using the GIS reclassification tool. The accuracy of the resultant map, which displays several neglected groundwater-capable places, is verified using drill wells. The groundwater potential zone (GWPZ) map can be used by developers and other stakeholders to direct future development. It is a crucial input for the use of sustainable agriculture in the interim. This study aims to map the distribution of shallow aquifers that facilitate industrial development, agriculture, and human settlement. This is accomplished by identifying locations in the Saudi Arabian province of SW Riyadh where aquifer recharge takes place, thoroughly evaluating the parameters that influence aquifers, and, in the end, creating a GWPZ map using MCE methodologies.

2. The Study Area

The study area was located in the geographical region of Al-Riyadh in central Saudi Arabia, roughly 150 km southwest of the country's capital, Riyadh (Figure 1a). Figure 1b shows the location of the wells. It has a surface area of 2340 km^2, and its elevation varies in altitude, starting at 645 m above sea level downstream of the region (the eastern part) to about 1277 m at the highest point of the watershed (the western section) (Figure 1c).

Saudi Arabia's terrain is mostly made up of harsh desert regions with hot temperatures. The research region exhibits a very warm climate, especially from May to September when monthly mean temperatures can reach above 30 °C (Figure 2a). In the KSA, the annual average precipitation ranges from 50 to 100 mm [75]. Riyadh and the Red Sea coast have the highest rates of rainfall in January, March, and April (Figure 2b). At Riyadh, the daily evaporation maximum in the summer may exceed 18.5 mm/day, with an average yearly evaporation of 2900 mm [75].

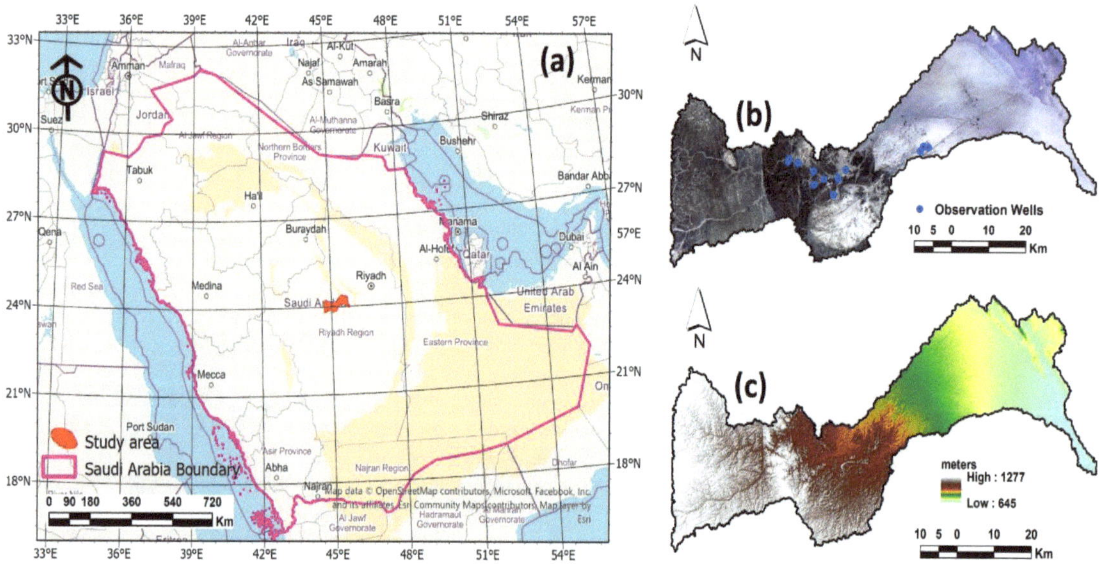

Figure 1. (**a**) Map showing the location of the study area in Saudi Arabia's boundary (**b**) Recent landsat-8 image showing the locations of wells (**c**) Digital Elevation Model of the study area.

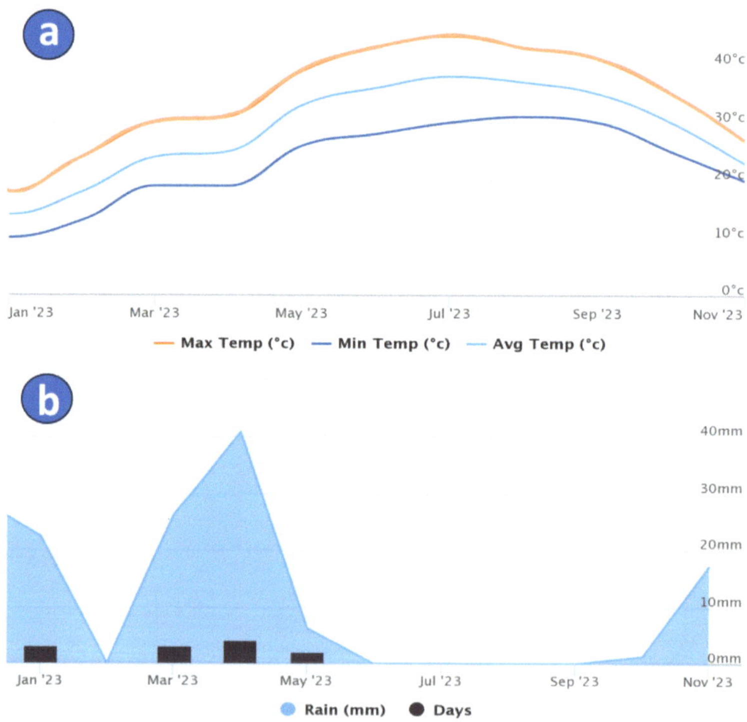

Figure 2. (**a**) The monthly temperature and (**b**) The monthly rainfall of the study area during 2023 [76].

Geological and Hydrogeological Setting

Tectonically, the Arabian Plate is where the KSA is located. Rocks that have undergone Late Precambrian metamorphism, both volcanic and volcano-sedimentary, comprise its composition [75]. The regional split of the plate is composed of two geological regimes: the Arabian Platform and the Arabian Shield. The Arabian Shield basements' tilt is the reason for this strata's eastern dip towards the Arabian Gulf [75]. The following characteristics make up the subsurface geology of the study area (Figure 3) [75]:

1. Rocks that are colluvial, alluvial, aeolian, and superficial. In the western section, quaternary sediments comprise an unconfined aquifer, while in the eastern region, they constitute a confined aquifer.
2. The Ash Shiqqah of the Permian Khuff Formation, which is made of shale and large conglomerates. There are fossilized marl, dolomite, sandstone, and gypsum in the upper part. A less significant discontinuous aquifer is the Khuff Formation.
3. Saq Sandstone, which is Cambrian to Ordovician in age. It depicts a regional aquifer system. The Saq Sandstone is above and frequently dipping east.
4. The Precambrian basement's Arabian shield complex.

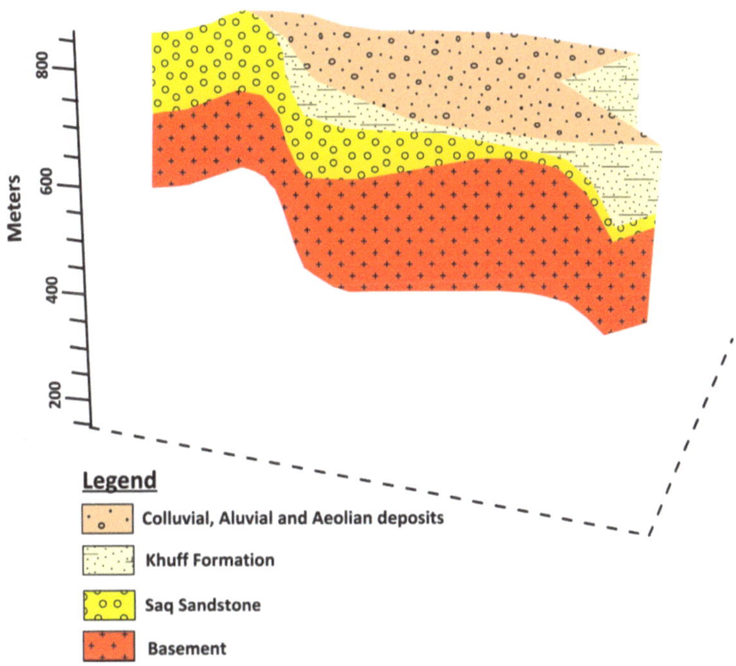

Figure 3. 3D subsurface stratigraphic units around the study area [75].

In addition to the subsurface geology, Figure 4 shows the surface geological map of the study area. Hydrogeologically, the Saq Sandstone is the main sedimentary aquifer system in the research area. It is a large aquifer that covers a large area in the region and holds a substantial quantity of groundwater (approximately 280 billion m^3) [75]. Groundwater yields vary widely, typically from 50 to 100 L/s [75]. Except in cases when significant drops in water levels have caused a reversal of the groundwater hydraulic gradient, the natural flow direction of groundwater is northeast [75]. Precipitation, catchments downstream, and vertical infiltration from surface water systems replenish the Saq Sandstone. However, aquifer discharge occurs locally due to throughflow and irrigation bores in local production.

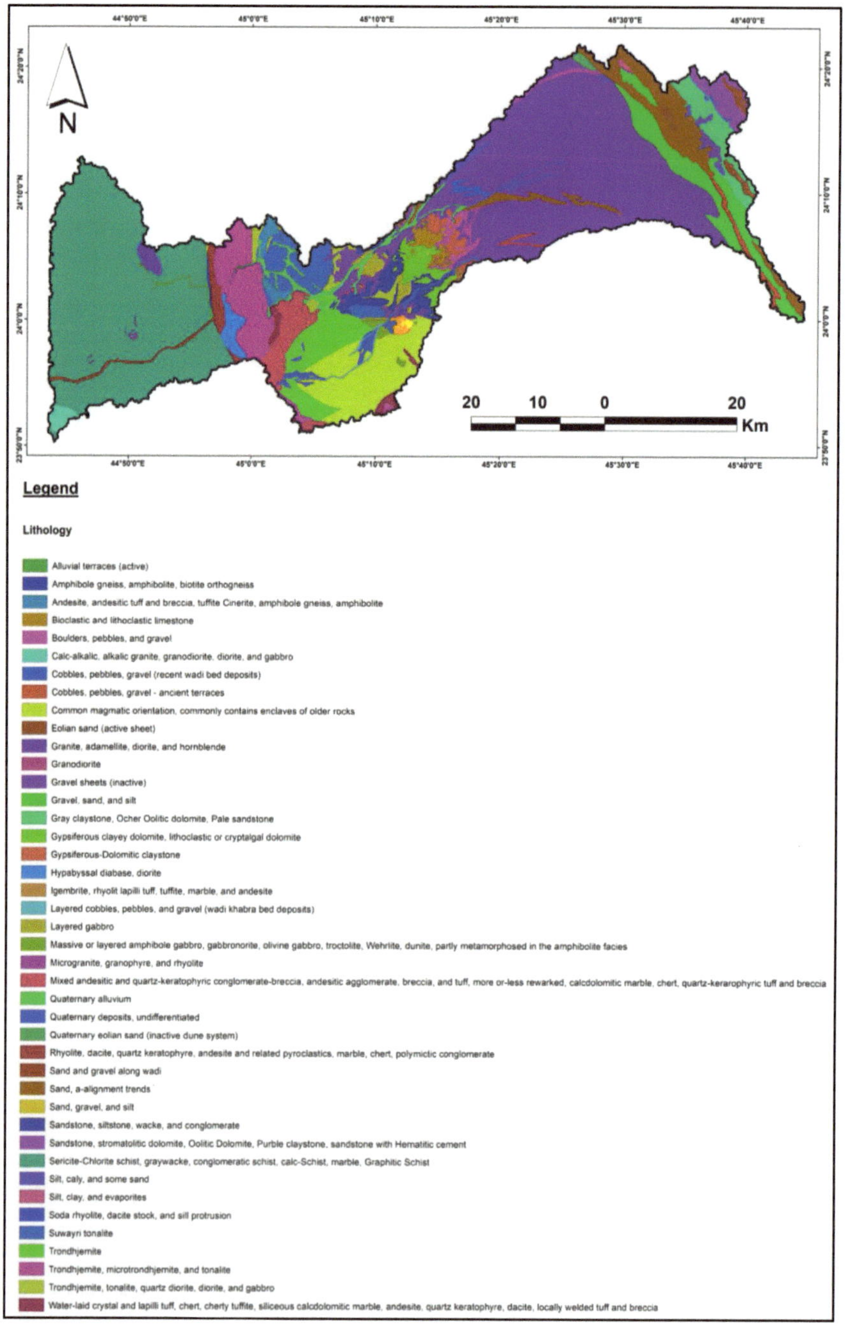

Figure 4. Geological map of the study area [77].

3. Materials and Methods

The present research used remote sensing imagery in conjunction with existing geological, hydrological, and other data to define possible recharge areas for water resources. Using a GIS-based technique, eleven significant thematic maps such as geology, elevation, slope, aspect, hill shade, flow direction, stream order, drainage density, lineaments,

lineaments density, and NDVI were combined (Figure 5). Remote sensing images were processed using ENVI v5.3 and ArcGIS v10.8 software. During the field investigation, information about groundwater was acquired, including groundwater well locations, the abundance and distribution of irrigated crops, and flow discharge.

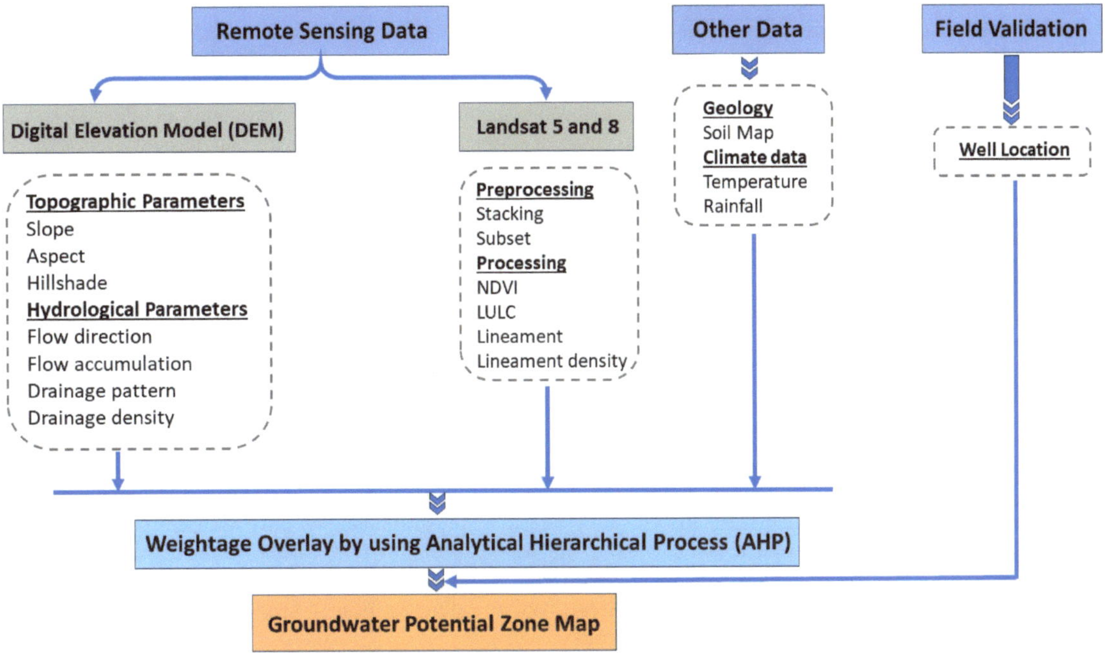

Figure 5. The flow chart illustrates the used methodology in the research.

In order to assess the topographical features, one of the Digital Elevation Models (DEMs) employed in this study was the Shuttle Radar Topography Mission (SRTM) data (30 m cell size). NASA provided SRTM DEM data (1 arc second). The Deterministic eight-neighbor (8D) technique was used to automatically generate a drainage system [78]. Then, using a GIS tool, this was transformed into a drainage density map. Furthermore, the fill-difference method was used to figure out the depressions to identify areas of water accumulation [79].

The operational land imager (OLI) and thermal infrared sensor (TIRS) are carried by Landsat 8, which was launched on 11 February 2013. The wavelength ranges considered in the study are shortwave infrared and visible and near-infrared. Image processing and enhancement techniques were used for a Landsat-8 OLI scene (path 166/row 44) that was acquired on 5 February 2023. Bands 1, 2, 3, 4, 5, and 7 were merged with a spatial resolution of 30 m and projected to WGS-84, UTM Zone 38 N. The vegetation cover that can be identified utilizing groundwater resources was shown by using data from Landsat 5 and 8. Consequently, we evaluated the vegetation areas by using the visible infrared bands and computed the normalized difference vegetation index (NDVI) by using the following equation for displaying the vegetation areas [80].

$$\text{NDVI} = (\text{NIR band} - \text{Red band})/(\text{NIR band 5} + \text{Red band})$$

For the purpose of this study's NDVI change detection, we employed not only Landsat OLI but also a Landsat-5 scene (path 166/row 43), which was obtained on 25 August 2000 (bands 7, 4, and 2 in R, G, and B, respectively). A LULC map was also produced from the landsat-8 OLI scene using a supervised classification method to identify the surface features within the study area. Using the maximum likelihood classifier (MLC) for parametric input data and field investigation, the resulting map is categorized into four classes based on supervised classification.

By integrating RS and GIS data applying the weighted overlay approach, the potential of groundwater resources was figured out. Proof of groundwater conditions was provided for every theme layer, including DEM, rainfall, geology, drainage density, slope, topography, soil, lineaments, flow accumulation, and flow direction maps. The rank of each layer—which is obtained by dividing the whole number of the layer rankings by the rank of each map—was converted to generate the GIS layer. The process of dividing each class's rank by the combined values of all the layer classes yields the capability values (CVi). The groundwater potential zone map is created by multiplying these capability values by the relevant layer weight in each theme map. Following that, the potential occurrences are combined in a linear combination equation with the aid of the Raster Calculator tool. The steps provided are followed to complete the integration procedure.

$$\text{GWP} = \sum \text{CVi} \times \text{Wi} \quad (1)$$

where GWP = groundwater potential, CVi = capability value (weight of inter-map class), Wi = map weight.

Research on the integration of geographic information systems (GIS), remote sensing (RS), and GPS has grown popular in the fields of groundwater hydrology and environmental monitoring. Recent developments in RS, GIS, and GPS, and higher-level processing will make it easier to produce and handle a variety of data in a fast and economical manner. The use of RS and GIS, the necessity of integration models, hydrological modeling, and the outcome are all covered in this flowchart (Figure 5). We will be more capable of managing large data volumes and transferring knowledge about rapidly depleting societal resources if these issues have been resolved on a philosophical and technical level.

4. Results and Discussion

4.1. Topographical Parameters

Slope, aspect, and hillshade are examples of topographical factors that were extracted to achieve the target of the study.

4.1.1. Slope

Slope analysis is crucial for determining the best places for water infiltration since sites with low slopes have limited surface runoff and high percolating rates, whereas higher slopes encourage runoff and quick drainage of water [81]. According to the slope map, places with high slopes were found in the upstream section, particularly at the level of the basement in the center of the study area, whereas areas with low slopes were found in the downstream on the eastern side (Figure 6a). Runoff flows from the western side to the eastern side, and therefore the eastern side is generally more susceptible to groundwater recharge.

4.1.2. Aspect

The physical characteristics of soil structure, curved materials, and possible groundwater zones are all controlled by aspect, which reflects the vegetation, moisture retention, and inclination of rock bedding [82]. The present study reveals that the groundwater potential is positively weighted in the central and eastern regions of the study area, while the western portion of the study exhibits a negatively weighted groundwater potential zone map due to its more severe slope and gradient values (Figure 6b).

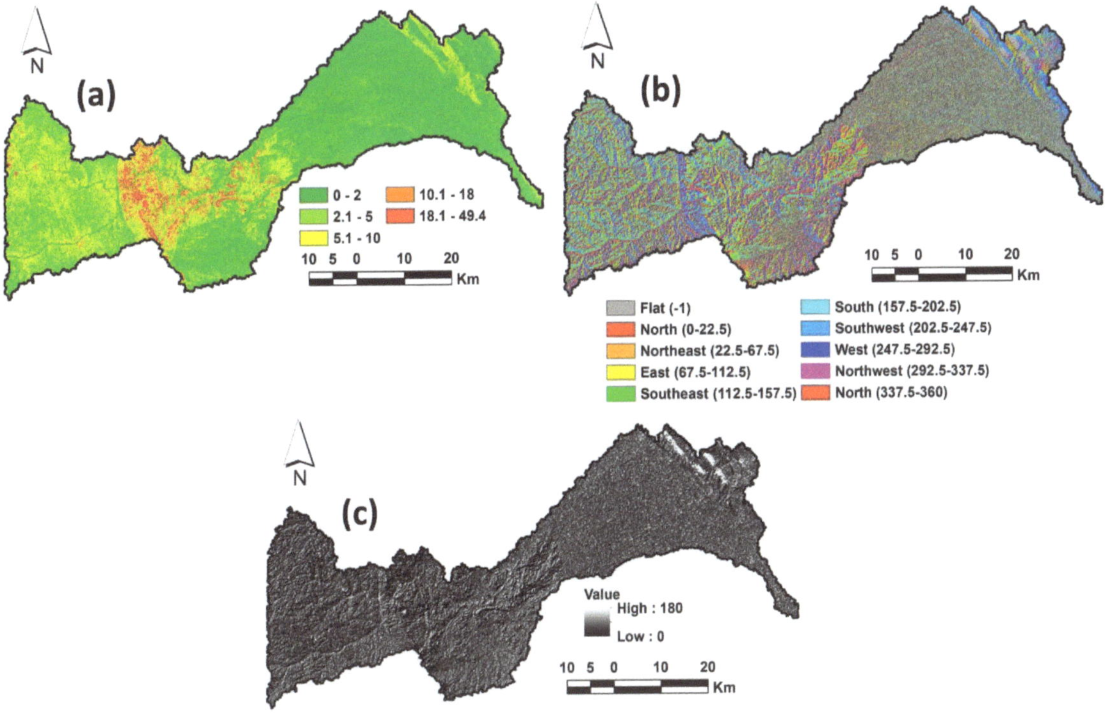

Figure 6. Topographical maps including: (**a**) Slope, (**b**) Aspect and (**c**) Hillshade.

4.1.3. Hillshade

Hillshade refers to a grayscale representation of the surface that uses the sun's relative position to determine how the relevant surface is colored. This feature uses azimuth and altitude data to define the sun's path. ArcGIS software 10.8's spatial analysis tool was used to extract the hillshade map. In the study area, the hillshade value varies from 0 to 180 (Figure 6c). Interpreting the entire basin area's steep terrain and surface topography depends heavily on this map.

4.2. Hydrological Parameters

4.2.1. Flow Direction

The network of streams was identified using a flow direction map, which depicts the movement of water along a surface that continually flows in the steepest downward direction. The river flows from the central to the eastern and northeastern regions, as shown by the flow direction map (Figure 7a). The relative proximity of the channel structure is indicated by the flow direction. Permeability factor and stream/flow direction are the opposite. More runoff in the direction of greatest flow indicates less water penetration below the surface (Table 1), and conversely [82].

Table 1. Layer map weights and their rank for groundwater recharge potential zones.

Factor	Weight	Category	Rank
Lineament Density (km/km^2)			
0–0.01		Very Low	1
0.11–0.3		Low	2
0.31–0.5	27.6	Moderate	3
0.51–0.7		High	4
0.71–1.1		Very High	5
Slope			
0–2		Very High	5
2.1–5		High	4
5.1–10	12.3	Moderate	3
10.1–18		Low	2
18.1–49.4		Very Low	1
Topography			
645–733		Very High	5
733–815		High	4
815–899	8.7	Moderate	3
899–960		Low	2
960–1277		Very Low	1
Soil			
Sandy soil and gravel		Very High	4
Alluvial	8.3	High	3
Sandstone with limestone		Moderate	2
Hard rock		Low	1
Flow Direction			
<2	5.2	Very High	5
2–8.		High	4
8–32		Moderate	3
32–64		Low	2
>64		Very Low	1
Flow Accumulation			
<2	5.7	Very High	5
2–3		High	4
3–5		Moderate	3
5–6		Low	2
>6		Very Low	1
LULC			
Vegetation	9.6	Moderate	2
Hard rocks		Low	1
Wadi deposits		Very High	4
Sedimentary rocks		High	3
Drainage Density (km/km^2)			
7.8–51.8	8.9	Very High	5
51.81–95.93		High	4
95.93–140.06		Moderate	3
140.06–184.18		Low	2
184.18–228.31		Very Low	1
Rainfall (mm/Year)			
<20	13.6	Very Low	1
20–60		Low	2
60–90		Moderate	3
90–130		High	4
>130		Very High	5

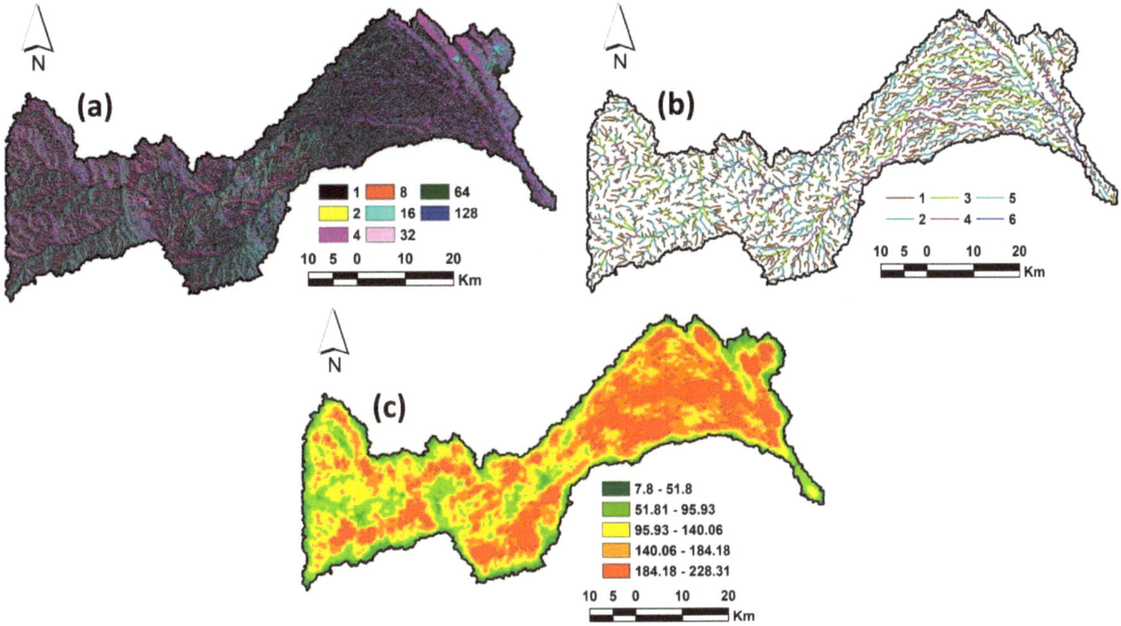

Figure 7. Hydrological maps including: (**a**) Flow direction, (**b**) stream orders and (**c**) Stream density.

4.2.2. Drainage Density

Furthermore, one of the main factors influencing the existence of groundwater in a particular basin is the drainage density [83]. It is calculated by dividing a basin's total area by the sum of all of its drainages [84]. Therefore, higher surface water runoff is accelerated by increased drainage density, but slower drainage density permits water to migrate more slowly and finally percolate into the ground surface [85,86]. Thus, the automatically generated stream networks in the study region were extracted, and the stream order was defined into six classes as a result (Figure 7b).

The uneven distribution of high to extremely high drainage densities in the northern, eastern, and southern regions of the study area suggests a limited groundwater potential. The medium to extremely low drainage densities in the remaining areas essentially encourage groundwater accumulation. The estimated drainage density in this investigation indicated a well-connected drainage system with extremely low to very high density, with values that varied from 7.8 to 228.3 lengths per unit area (Figure 7c). The Saaty model was utilized to rank and reclassify the drainage density categories based on their impact on groundwater supply. In comparison to low density drainage zones, which were rated higher with an average of 50%, high drainage density areas received a rating relatively lower, of 3%. According to investigations conducted in the coastal aquifer of eastern India, it was decided to impose a lower groundwater impact on an area with higher drainage density and larger effects on the lower drainage density [87].

4.2.3. Lineament

Lineaments on the surface of the Earth represent subsurface geological structures, including faults, fractures, and cracks with no visible fracture movement [88]. Because of their ability to allow groundwater distribution, they enhance secondary porosity, which makes them useful for locating possible groundwater regions [89]. A high-lineament density area may have significant groundwater potential (Table 1). Essential information on surface and subsurface fracture systems is provided by lineaments, which could influence groundwater flow and accumulation [88]. This has been used to distribute groundwater resources and find potential sites for runoff water accumulation [79]. Promising areas for groundwater potential are indicated by higher lineament density, which also represent permeability zones [88]. ArcGIS's Spatial Analyst was used to digitize the lineaments (Figure 8a) of the basin from Landsat-8 and SRTM data and create a lineament density map. Five classes which represent 54.3, 13.8, 13.3, 12.3, and 6.16% of the lineament density map, respectively, are identified on the map (Figure 8b).

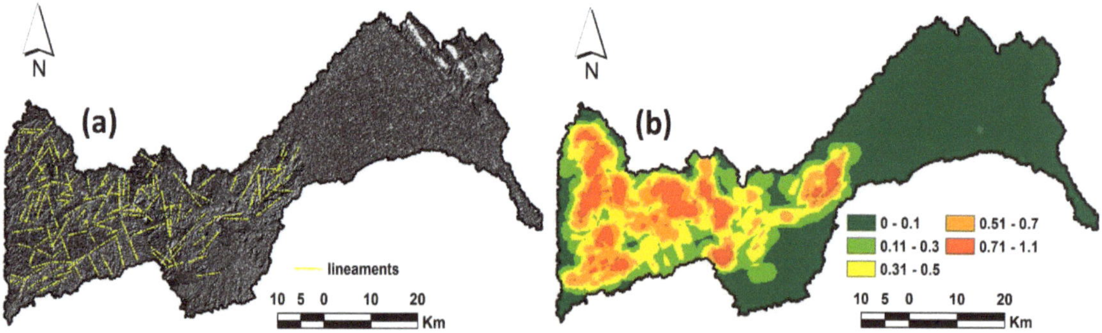

Figure 8. (**a**) The extracted lineament map (**b**) The lineament density map.

4.3. NDVI

One commonly utilized variable to assess groundwater potential zones is the NDVI [90] that represents the coverage and density of the vegetation. Landsat images (Figure 9a,b) were used to create the NDVI maps. The NDVI values range from −1 to 1. Higher NDVI values indicate extensive vegetation. To figure out how the vegetation cover has changed between 2000 and 2023, the research area's NDVI result maps were extracted. The findings demonstrate that the vegetation cover expanded from 6 km^2 to 30 km^2 during a period of 23 years (Figure 9c,d). Generally, groundwater resources in the study area were used as a water source for vegetation in this period.

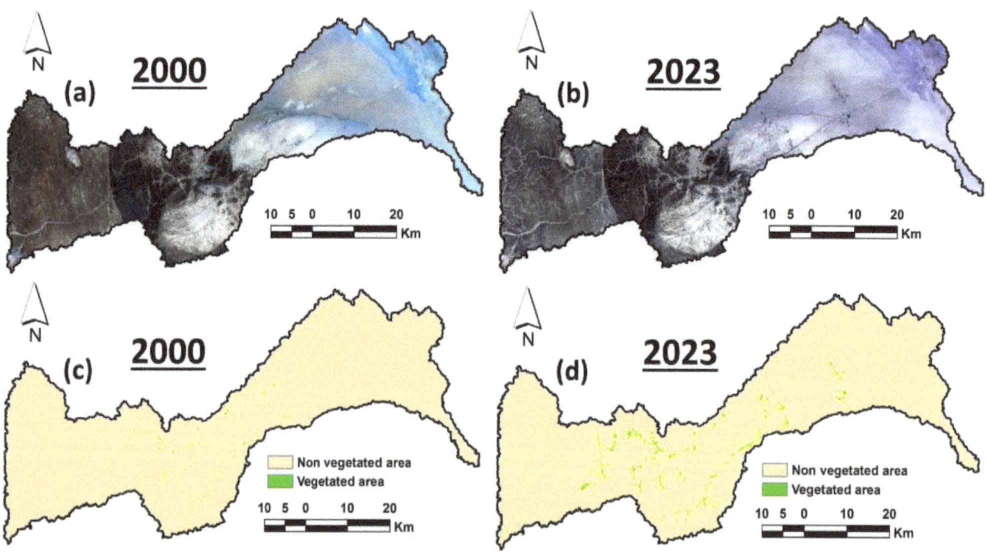

Figure 9. The temporal satellite images including: (**a**) Landsat-5 image for 2000 and (**b**) Landsat-8 image for 2023, (**c**) NDVI map for 2000, and (**d**) NDVI map for 2023.

4.4. LULC

The data were combined with a Landsat-8 OLI image for 2023, which helped classify land use and land cover (LULC). During field investigations, sample collections were gathered to generate reflectance values, or spectral signatures, for the supervised classification process, which determined the contents of the clusters, such as hard rocks, vegetation, wadi deposits, and sedimentary rocks. According to Figure 10, there are four primary classes in the LULC map. For roughly 40% of the overall area, there exist both hard rocks and sedimentary rocks. However, wadi deposits and vegetation cover make up 1.3% and 18.4% of the entire region, respectively. Using a confusion matrix with the ARCGIS program 10.8, the output map displays a 94% accuracy assessment rate.

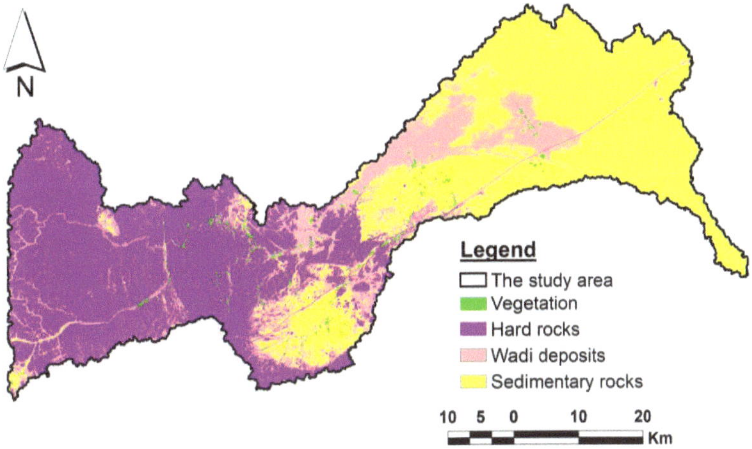

Figure 10. LULC map of the study area.

4.5. Groundwater Potential Zones

Prior to the integration of the thematic layers via the weighted overlay approach, consistency ratios were established for every thematic level and its subclasses. The results demonstrated the validity and reasonable consistency of the judgment matrices used in this investigation. As a result, the reclassified theme layers were integrated by utilizing the weighted overlay strategy, which depends on their weight, to create the groundwater potential map. The study region was classified, using the natural breaks/Jenks technique in the GIS platform, into five groups (i.e., very low, low, moderate, high, and very high groundwater potential zones) based on the values of the pixels (Figure 11). The majority filter in the GIS platform was used for additional processing of the map in order to reduce the pixel speckling.

The delineated map revealed discernible variations in the groundwater potential of the studied area. While areas of the district with subsurface hard rock are extremely vulnerable to a groundwater problem, approximately 7.9% of the region's distribution has very low groundwater potential. Regarding the possibility for groundwater, the map further designates 18.6% of the zone as semi-critical. However, it was discovered that there were still moderate, high, and very high groundwater potential zones in around 26.7%, 30%, and 16.8% of the study area, respectively. Because of the greater availability of runoff water and good groundwater recharge, a substantial potential for groundwater was found in several areas of the province along the flooded areas of the main rivers. As depicted in Figure 11, the southern and eastern regions of the examined basin have the highest concentration of streams, lineaments, and geological fractures, making them the most favorable locations for future drilling. These outcomes agree with the previous research [91,92].

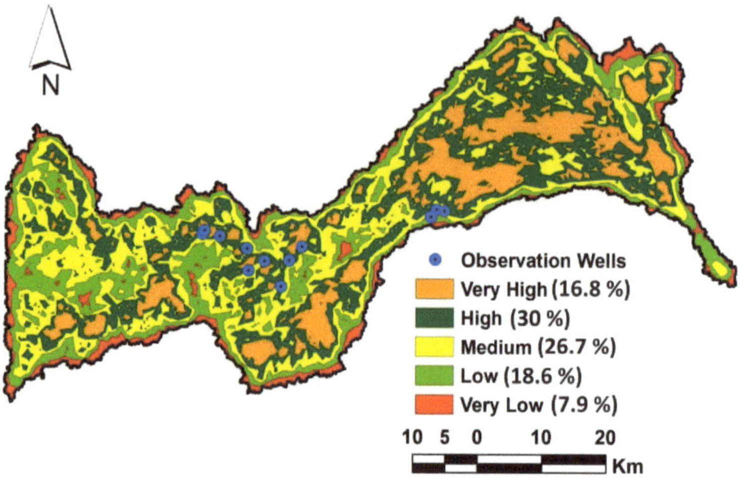

Figure 11. Groundwater potential recharge zones using AHP model and field data.

The research also included measurable cross-validation maps of groundwater prospect zones based on receiver operating characteristics (ROC). For a number of the variable's cut-off themes, the ROC display a true positive rate against an untrue positive rate; each concept on the path represents a sensitivity pair in relation to a threshold. Conversely, the area under the curve (AUC) determines the degree to which a factor can distinguish between the two sets of data. AUC values between 0.5 and 0.6 indicate poor forecast accuracy. Additionally, according to previous research [36,93], the 0.6–0.7, 0.7–0.8, 0.8–0.9, and 0.9–1 levels suggest very good, average, excellent, and good forecast accuracy, respectively. The AUC of the potential area map was 0.85, indicating that the results of the AHP approach were confirmed for good predictions. Thus, the current work reached a significantly higher accuracy rate with the largest number of appropriate theme maps by utilizing the AHP approach [94–96].

The identified potential zones can offer crucial methods for choosing new well regions in order to remove vigorous water resource management from the watershed zones. The application of a groundwater harvesting method may benefit from this result. Seeing the work's boundaries, creative potential, and limitations is fascinating. Although this study chose eleven theme layers, more accurate findings may have been obtained by including other data, such as hydro-geological and geomorphology maps (geomorphology and water level). The use of groundwater, water management, limited abstraction, sophisticated water framework, water awareness, alternative cropping techniques, and the conservation of home and industrial water are the main focuses of sustainable strategies. Long-term water management would benefit from the extraction of non-polluted groundwater. Because of the chosen methodology, the analysis has certain theoretical restrictions. AHP is additionally capable of suppressing errors. Groundwater zone rules are often significantly impacted by the volume of supply for agriculture and the amount of drinking water consumed; these issues were not included in this study. Nevertheless, the findings are precise and reasonable given the factors in the analysis method and prediction. The river basin's groundwater management can be improved with the use of the AHP data.

4.6. Field Validation

Field observations were used to validate the groundwater recharge potential map of the research area. To verify the combined results of RS and GIS, we collected data on the current wells (Table 2). The primary objectives of the field trip were to locate the wells and measure the depth and salinity of the water. Other information included soil type, topography, and surface geology (Figure 12a,b). These wells were drilled into the extremely high and high potential zones in the central and eastern parts of the region. The wells were between 43 and 150 m deep (Figure 12c,d). The bottoms of these wells laid close to silt and clay beds, which act as barriers to deeper water infiltration. The salinity of groundwater varies from 1970 to 6600 mg/L. The groundwater's chemical makeup and suitability for irrigation were determined by a thorough investigation that we carried out before in previous work. Groundwater samples are appropriate for irrigation, according to the results [97]. The primary constituents of the surface soil sequence are gravel, sand, and silty sand. Consequently, the central and Eastern regions of the studied area (Figure 11) have been developed by local farmers, who irrigate many crops (Figure 12e,f) with good-quality groundwater. In addition, it was observed that groundwater is used for farming, and the majority of the wells in the research area have groundwater levels between 40 and 60 m deep. This demonstrates that rainfall, particularly during periods of precipitation, clearly replenishes the groundwater reservoir. This development aligns with the high potential area shown in the GW recharge potential map (Figure 11).

Figure 12. Field images showing: (**a**,**b**) The different geological and geomorphological units, (**c**,**d**) Water level observation from wells and (**e**,**f**) An example of farms in the study area.

Table 2. Wells dataset during the field validation.

X	Y	Elevation (m)	TDS (mg/L)	W.L (m)	Well Age (Years)
534,413.00 m E	2,665,349.00 m N	770	4405	110	14
533,274.00 m E	2,665,531.00 m N	777	4510	100	13
532,506.00 m E	2,664,425.00 m N	782	3720	150	12–15
514,308.00 m E	2,659,802.00 m N	822.5	3620	60	3
512,494.00 m E	2,657,740.00 m N	832.7	3290	60	3
511,331.00 m E	2,653,835.00 m N	871.8	2510	60	40
508,984.00 m E	2,657,588.00 m N	835.2	2520	60	7
506,704.00 m E	2,656,193.00 m N	862.1	5610	--	--
506,528.00 m E	2,659,460.00 m N	850.6	6180	57	more than 20
502,644.00 m E	2,661,355.00 m N	860.7	3370	43	more than 35
500,481.00 m E	2,662,234.00 m N	877.2	1970	50	1
500,206.00 m E	2,661,665.00 m N	878.3	6600	45	25–30

5. Conclusions

For sustainable development in arid and hyper-arid regions, groundwater is especially crucial as a water source. RS and GIS techniques are efficiently used to reveal, assess, and monitor exploratory data for water resources in different climatic conditions. RS and GIS techniques were used to study the SW Riyadh area in central Saudi Arabia. The primary objective was to locate possible recharge and groundwater potential zones. The geology, geomorphic, climatic, and hydrological characteristics are represented by twelve GIS conceptual layers that we combined in this instance. The lithology, elevation, slope, aspect, hillshade, drainage pattern, drainage density, NDVI, and lineament density are these different layers. Using an AHP overlay model based on geographic information systems, they were processed, normalized, and integrated.

The findings show the areas with potential for groundwater, which are divided into five categories: very high, high, moderate, low, and very low. These together represent 16.8, 30, 26.7, 18.6, and 7.9% of the total area. The model demonstrated good performance, as tested by field observation and the AUC (0.85). The locations that show promise are those that use artificial abstraction to supply groundwater. Overall, the GWPZ that is produced utilizing RS and GIS is very beneficial to decision makers and sustainability considerations. Additionally, this strategy can be used regionally and in locations with similar circumstances.

Author Contributions: Conceptualization, E.M.M.E.-B., H.M.A., M.H.S. and M.A.; methodology, E.M.M.E.-B., H.M.A., M.H.S. and M.A.; funding acquisition, E.M.M.E.-B. and H.M.A.; validation, E.M.M.E.-B. and H.M.A.; statical analysis, M.A.; writing—original draft preparation M.A., writing—review and editing, E.M.M.E.-B., H.M.A., M.H.S. and M.A. All authors have read and agreed to the published version of the manuscript.

Funding: This research received no external funding.

Data Availability Statement: Data are contained within the article.

Acknowledgments: Princess Nourah bint Abdulrahman University Researchers Supporting Project number (PNURSP2024R243), Princess Nourah bint Abdulrahman University, Riyadh, Saudi Arabia.

Conflicts of Interest: The authors declare no conflicts of interest.

References

1. Faramarzi, M.; Abbaspour, K.C.; Vaghefi, S.A.; Farzaneh, M.R.; Zehnder, A.J.B.; Srinivasan, R.; Yang, H. Modeling Impacts of Climate Change on Freshwater Availability in Africa. *J. Hydrol.* **2013**, *480*, 85–101. [CrossRef]
2. Abdelfattah, M.; Abdel-Aziz Abu-Bakr, H.; Aretouyap, Z.; Sheta, M.H.; Hassan, T.M.; Geriesh, M.H.; Shaheen, S.E.-D.; Alogayell, H.M.; M EL-Bana, E.M.; Gaber, A. Mapping the Impacts of the Anthropogenic Activities and Seawater Intrusion on the Shallow Coastal Aquifer of Port Said, Egypt. *Front. Earth Sci.* **2023**, *11*, 1204742. [CrossRef]

3. Shemsanga, C.; Muzuka, A.N.N.; Martz, L.; Komakech, H.; Mcharo, E. Indigenous Knowledge on Development and Management of Shallow Dug Wells of Dodoma Municipality in Tanzania. *Appl. Water Sci.* **2018**, *8*, 59. [CrossRef]
4. Gamal, G.; Hassan, T.M.; Gaber, A.; Abdelfattah, M. Groundwater Quality Assessment along the West of New Damietta Coastal City of Egypt Using an Integrated Geophysical and Hydrochemical Approaches. *Environ. Earth Sci.* **2023**, *82*, 107. [CrossRef]
5. Elbeih, S.F. An Overview of Integrated Remote Sensing and GIS for Groundwater Mapping in Egypt. *Ain Shams Eng. J.* **2015**, *6*, 1–15.
6. Kemper, K.E. Groundwater—From Development to Management. *Hydrogeol. J.* **2004**, *12*, 3–5. [CrossRef]
7. Xiao, R.; He, X.; Zhang, Y.; Ferreira, V.G.; Chang, L. Monitoring Groundwater Variations from Satellite Gravimetry and Hydrological Models: A Comparison with in-Situ Measurements in the Mid-Atlantic Region of the United States. *Remote Sens.* **2015**, *7*, 686–703. [CrossRef]
8. Molden, D. Water Responses to Urbanization. *Paddy Water Environ.* **2007**, *5*, 207–209. [CrossRef]
9. Altchenko, Y.; Villholth, K.G. Transboundary Aquifer Mapping and Management in Africa: A Harmonised Approach. *Hydrogeol. J.* **2013**, *21*, 1497–1517. [CrossRef]
10. Owolabi, S.T.; Madi, K.; Kalumba, A.M.; Alemaw, B.F. Assessment of Recession Flow Variability and the Surficial Lithology Impact: A Case Study of Buffalo River Catchment, Eastern Cape, South Africa. *Environ. Earth Sci.* **2020**, *79*, 187. [CrossRef]
11. Barlow, P.M.; Leake, S.A. *Streamflow Depletion by Wells: Understanding and Managing the Effects of Groundwater Pumping on Streamflow*; US Geological Survey: Reston, VA, USA, 2012; Volume 1376.
12. Barker, J.A. A Generalized Radial Flow Model for Hydraulic Tests in Fractured Rock. *Water Resour. Res.* **1988**, *24*, 1796–1804.
13. Chowdhury, A.; Jha, M.K.; Chowdary, V.M. Delineation of Groundwater Recharge Zones and Identification of Artificial Recharge Sites in West Medinipur District, West Bengal, Using RS, GIS and MCDM Techniques. *Environ. Earth Sci.* **2010**, *59*, 1209–1222. [CrossRef]
14. Meshram, K.S.; Tripathi, M.P.; Mukharjee, A.P. Effect of Artificial Recharge Structures on Ground Water Availability in Semi-Critical Area in Chhattisgarh. Ph.D. Thesis, Indira Gandhi KrishiVishwavidyalaya, Raipur, India, 2010.
15. Samson, S.; Elangovan, K. Delineation of Groundwater Recharge Potential Zones in Namakkal District, Tamilnadu, India Using Remote Sensing and GIS. *J. Indian Soc. Remote Sens.* **2015**, *43*, 769–778. [CrossRef]
16. Taheri, A.; Zare, M. Groundwater Artificial Recharge Assessment in Kangavar Basin, a Semi-Arid Region in the Western Part of Iran. *Afr. J. Agric. Res.* **2011**, *6*, 4370–4384.
17. Murthy, K.S.R. Ground Water Potential in a Semi-Arid Region of Andhra Pradesh-a Geographical Information System Approach. *Int. J. Remote Sens.* **2000**, *21*, 1867–1884. [CrossRef]
18. Srinivasa Rao, Y.; Reddy, T.; Nayudu, P. Groundwater Targeting in a Hard-Rock Terrain Using Fracture-Pattern Modeling, Niva River Basin, Andhra Pradesh, India. *Hydrogeol. J.* **2000**, *8*, 494–502.
19. Jaiswal, R.K.; Mukherjee, S.; Krishnamurthy, J.; Saxena, R. Role of Remote Sensing and GIS Techniques for Generation of Groundwater Prospect Zones towards Rural Development—An Approach. *Int. J. Remote Sens.* **2003**, *24*, 993–1008. [CrossRef]
20. Srivastava, P.K.; Bhattacharya, A.K. Groundwater Assessment through an Integrated Approach Using Remote Sensing, GIS and Resistivity Techniques: A Case Study from a Hard Rock Terrain. *Int. J. Remote Sens.* **2006**, *27*, 4599–4620. [CrossRef]
21. Dinesh Kumar, P.K.; Gopinath, G.; Seralathan, P. Application of Remote Sensing and GIS for the Demarcation of Groundwater Potential Zones of a River Basin in Kerala, Southwest Coast of India. *Int. J. Remote Sens.* **2007**, *28*, 5583–5601. [CrossRef]
22. Mondal, N.C.; Das, S.N.; Singh, V.S. Integrated Approach for Identification of Potential Groundwater Zones in Seethanagaram Mandal of Vizianagaram District, Andhra Pradesh, India. *J. Earth Syst. Sci.* **2008**, *117*, 133–144. [CrossRef]
23. Chowdhury, A.; Jha, M.K.; Chowdary, V.M.; Mal, B.C. Integrated Remote Sensing and GIS-based Approach for Assessing Groundwater Potential in West Medinipur District, West Bengal, India. *Int. J. Remote Sens.* **2009**, *30*, 231–250. [CrossRef]
24. Dhakate, R.; Chowdhary, D.K.; Gurunadha Rao, V.V.S.; Tiwary, R.K.; Sinha, A. Geophysical and Geomorphological Approach for Locating Groundwater Potential Zones in Sukinda Chromite Mining Area. *Environ. Earth Sci.* **2012**, *66*, 2311–2325. [CrossRef]
25. Nag, S.K.; Ghosh, P. Delineation of Groundwater Potential Zone in Chhatna Block, Bankura District, West Bengal, India Using Remote Sensing and GIS Techniques. *Environ. Earth Sci.* **2013**, *70*, 2115–2127. [CrossRef]
26. Balakrishna; Balaji Shrikant Maury, S.; Narshimulu, G. Ground Water Potential in Fractured Aquifers of Ophiolite Formations, Port Blair, South Andaman Islands Using Electrical Resistivity Tomography (ERT) and Vertical Electrical Sounding (VES). *J. Geol. Soc. India* **2014**, *83*, 393–402.
27. Sonkamble, S.; Satishkumar, V.; Amarender, B.; Sethurama, S. Combined Ground-Penetrating Radar (GPR) and Electrical Resistivity Applications Exploring Groundwater Potential Zones in Granitic Terrain. *Arab. J. Geosci.* **2014**, *7*, 3109–3117. [CrossRef]
28. Al-Abadi, A.M. Groundwater Potential Mapping at Northeastern Wasit and Missan Governorates, Iraq Using a Data-Driven Weights of Evidence Technique in Framework of GIS. *Environ. Earth Sci.* **2015**, *74*, 1109–1124. [CrossRef]
29. Mezni, I.; Chihi, H.; Hammami, M.A.; Gabtni, H.; Baba Sy, M. Regionalization of Natural Recharge Zones Using Analytical Hierarchy Process in an Arid Hydrologic Basin: A Contribution for Managed Aquifer Recharge. *Nat. Resour. Res.* **2022**, *31*, 867–895. [CrossRef]
30. Khan, M.Y.A.; ElKashouty, M.; Tian, F. Mapping Groundwater Potential Zones Using Analytical Hierarchical Process and Multicriteria Evaluation in the Central Eastern Desert, Egypt. *Water* **2022**, *14*, 1041. [CrossRef]
31. Khan, M.Y.A.; ElKashouty, M.; Subyani, A.M.; Tian, F.; Gusti, W. GIS and RS Intelligence in Delineating the Groundwater Potential Zones in Arid Regions: A Case Study of Southern Aseer, Southwestern Saudi Arabia. *Appl. Water Sci.* **2022**, *12*, 3. [CrossRef]

32. Pradhan, R.M.; Guru, B.; Pradhan, B.; Biswal, T.K. Integrated Multi-Criteria Analysis for Groundwater Potential Mapping in Precambrian Hard Rock Terranes (North Gujarat), India. *Hydrol. Sci. J.* **2021**, *66*, 961–978. [CrossRef]
33. Diriba, D.; Karuppannan, S.; Takele, T.; Husein, M. Delineation of Groundwater Potential Zonation Using Geoinformatics and AHP Techniques with Remote Sensing Data. *Heliyon* **2024**, *10*, e25532. [CrossRef] [PubMed]
34. Feujio, D.H.A.; Aretouyap, Z.; Tchato, S.C.; Legrand, C.N.I.I.; Djomdi, E.; Madadjeu, N.N.; Nguimgo, C.N.; Kpoumie, A.N. Application of Analytical Hierarchy Process to Assess Groundwater Potential for a Sustainable Management in the Menoua Division. *Heliyon* **2024**, *10*, e24310. [CrossRef] [PubMed]
35. Arulbalaji, P.; Padmalal, D.; Sreelash, K. GIS and AHP Techniques Based Delineation of Groundwater Potential Zones: A Case Study from Southern Western Ghats, India. *Sci. Rep.* **2019**, *9*, 2082. [CrossRef] [PubMed]
36. Ajay Kumar, V.; Mondal, N.C.; Ahmed, S. Identification of Groundwater Potential Zones Using RS, GIS and AHP Techniques: A Case Study in a Part of Deccan Volcanic Province (DVP), Maharashtra, India. *J. Indian Soc. Remote Sens.* **2020**, *48*, 497–511. [CrossRef]
37. Yariyan, P.; Avand, M.; Omidvar, E.; Pham, Q.B.; Linh, N.T.T.; Tiefenbacher, J.P. Optimization of Statistical and Machine Learning Hybrid Models for Groundwater Potential Mapping. *Geocarto Int.* **2022**, *37*, 3877–3911. [CrossRef]
38. Sun, T.; Cheng, W.; Abdelkareem, M.; Al-Arifi, N. Mapping Prospective Areas of Water Resources and Monitoring Land Use/Land Cover Changes in an Arid Region Using Remote Sensing and GIS Techniques. *Water* **2022**, *14*, 2435. [CrossRef]
39. Abdelkareem, M.; El-Baz, F.; Askalany, M.; Akawy, A.; Ghoneim, E. Groundwater Prospect Map of Egypt's Qena Valley Using Data Fusion. *Int. J. Image Data Fusion* **2012**, *3*, 169–189. [CrossRef]
40. Melese, T.; Belay, T. Groundwater Potential Zone Mapping Using Analytical Hierarchy Process and GIS in Muga Watershed, Abay Basin, Ethiopia. *Glob. Challenges* **2022**, *6*, 2100068. [CrossRef]
41. Priya, U.; Iqbal, M.A.; Salam, M.A.; Nur-E-Alam, M.; Uddin, M.F.; Islam, A.R.M.T.; Sarkar, S.K.; Imran, S.I.; Rak, A.E. Sustainable Groundwater Potential Zoning with Integrating GIS, Remote Sensing, and AHP Model: A Case from North-Central Bangladesh. *Sustainability* **2022**, *14*, 5640. [CrossRef]
42. Pourtaghi, Z.S.; Pourghasemi, H.R. GIS-Based Groundwater Spring Potential Assessment and Mapping in the Birjand Township, Southern Khorasan Province, Iran. *Hydrogeol. J.* **2014**, *22*, 643–662. [CrossRef]
43. Naghibi, S.A.; Pourghasemi, H.R.; Dixon, B. GIS-Based Groundwater Potential Mapping Using Boosted Regression Tree, Classification and Regression Tree, and Random Forest Machine Learning Models in Iran. *Environ. Monit. Assess.* **2016**, *188*, 1–27. [CrossRef] [PubMed]
44. Guru, B.; Seshan, K.; Bera, S. Frequency Ratio Model for Groundwater Potential Mapping and Its Sustainable Management in Cold Desert, India. *J. King Saud Univ.* **2017**, *29*, 333–347. [CrossRef]
45. Machiwal, D.; Jha, M.K.; Mal, B.C. Assessment of Groundwater Potential in a Semi-Arid Region of India Using Remote Sensing, GIS and MCDM Techniques. *Water Resour. Manag.* **2011**, *25*, 1359–1386. [CrossRef]
46. Ishizaka, A.; Labib, A. Review of the Main Developments in the Analytic Hierarchy Process. *Expert Syst. Appl.* **2011**, *38*, 14336–14345. [CrossRef]
47. Maity, D.K.; Mandal, S. Identification of Groundwater Potential Zones of the Kumari River Basin, India: An RS & GIS Based Semi-Quantitative Approach. *Environ. Dev. Sustain.* **2019**, *21*, 1013–1034.
48. Saaty, T.L. A Scaling Method for Priorities in Hierarchical Structures. *J. Math. Psychol.* **1977**, *15*, 234–281. [CrossRef]
49. Saaty, T.L. *The Analytic Hierarchy Process: Planning, Priority Setting, Resource Allocation*; McGraw-Hill International: New York, NY, USA, 1980.
50. Baghel, S.; Tripathi, M.P.; Khalkho, D.; Al-Ansari, N.; Kumar, A.; Elbeltagi, A. Delineation of Suitable Sites for Groundwater Recharge Based on Groundwater Potential with RS, GIS, and AHP Approach for Mand Catchment of Mahanadi Basin. *Sci. Rep.* **2023**, *13*, 9860. [CrossRef]
51. Chenini, I.; Ben, M.A.; El May, M. Groundwater Recharge Zone Mapping Using GIS-Based Multi-Criteria Analysis: A Case Study in Central Tunisia (Maknassy Basin). *Water Resour. Manag.* **2010**, *24*, 921–939. [CrossRef]
52. Gumma, M.K.; Pavelic, P. Mapping of Groundwater Potential Zones across Ghana Using Remote Sensing, Geographic Information Systems, and Spatial Modeling. *Environ. Monit. Assess.* **2013**, *185*, 3561–3579. [CrossRef]
53. Sener, E.; Davraz, A.; Ozcelik, M. An Integration of GIS and Remote Sensing in Groundwater Investigations: A Case Study in Burdur, Turkey. *Hydrogeol. J.* **2005**, *13*, 826–834. [CrossRef]
54. Abdalla, F. Mapping of Groundwater Prospective Zones Using Remote Sensing and GIS Techniques: A Case Study from the Central Eastern Desert, Egypt. *J. Afr. Earth Sci.* **2012**, *70*, 8–17. [CrossRef]
55. Gaber, A.; Mohamed, A.K.; ElGalladi, A.; Abdelkareem, M.; Beshr, A.M.; Koch, M. Mapping the Groundwater Potentiality of West Qena Area, Egypt, Using Integrated Remote Sensing and Hydro-Geophysical Techniques. *Remote Sens.* **2020**, *12*, 1559. [CrossRef]
56. Abdekareem, M.; Abdalla, F.; Al-Arifi, N.; Bamousa, A.O.; El-Baz, F. Using Remote Sensing and GIS-Based Frequency Ratio Technique for Revealing Groundwater Prospective Areas at Wadi Al Hamdh Watershed, Saudi Arabia. *Water* **2023**, *15*, 1154. [CrossRef]
57. Rahmati, O.; Nazari Samani, A.; Mahdavi, M.; Pourghasemi, H.R.; Zeinivand, H. Groundwater Potential Mapping at Kurdistan Region of Iran Using Analytic Hierarchy Process and GIS. *Arab. J. Geosci.* **2015**, *8*, 7059–7071. [CrossRef]

58. Roy, S.; Hazra, S.; Chanda, A.; Das, S. Assessment of Groundwater Potential Zones Using Multi-Criteria Decision-Making Technique: A Micro-Level Case Study from Red and Lateritic Zone (RLZ) of West Bengal, India. *Sustain. Water Resour. Manag.* **2020**, *6*, 1–14. [CrossRef]
59. Solomon, S.; Ghebreab, W. Hard-Rock Hydrotectonics Using Geographic Information Systems in the Central Highlands of Eritrea: Implications for Groundwater Exploration. *J. Hydrol.* **2008**, *349*, 147–155. [CrossRef]
60. Lubang Nobert, J.B.; Mtalo, E.G. Integrated Remote Sensing and GIS Techniques for Groundwater Exploration in Semi-Arid Region: A Case of Karamoja Region-Uganda. *Tanzan. J. Eng. Technol.* **2008**, *2*, 31–42.
61. Kuria, D.N.; Gachari, M.K.; Macharia, M.W.; Mungai, E. Mapping Groundwater Potential in Kitui District, Kenya Using Geospatial Technologies. *Int. J. Water Resour. Environ. Eng.* **2012**, *4*, 15–22.
62. Konkul, J.; Rojborwornwittaya, W.; Chotpantarat, S. Hydrogeologic Characteristics and Groundwater Potentiality Mapping Using Potential Surface Analysis in the Huay Sai Area, Phetchaburi Province, Thailand. *Geosci. J.* **2014**, *18*, 89–103. [CrossRef]
63. Li, Y.; Abdelkareem, M.; Al-Arifi, N. Mapping Potential Water Resource Areas Using GIS-Based Frequency Ratio and Evidential Belief Function. *Water* **2023**, *15*, 480. [CrossRef]
64. Al-Omran, A.M.; Aly, A.A.; Al-Wabel, M.I.; Sallam, A.S.; Al-Shayaa, M.S. Hydrochemical Characterization of Groundwater under Agricultural Land in Arid Environment: A Case Study of Al-Kharj, Saudi Arabia. *Arab. J. Geosci.* **2016**, *9*, 68. [CrossRef]
65. Mahmoud, S.H.; Alazba, A.A. The Potential of in Situ Rainwater Harvesting in Arid Regions: Developing a Methodology to Identify Suitable Areas Using GIS-Based Decision Support System. *Arab. J. Geosci.* **2015**, *8*, 5167–5179. [CrossRef]
66. Mahmoud, S.H.; Alazba, A.A. Identification of Potential Sites for Groundwater Recharge Using a GIS-Based Decision Support System in Jazan Region-Saudi Arabia. *Water Resour. Manag.* **2014**, *28*, 3319–3340. [CrossRef]
67. Al-Rashed, M.F.; Sherif, M.M. Water Resources in the GCC Countries: An Overview. *Water Resour. Manag.* **2000**, *14*, 59–75. [CrossRef]
68. Khouri, J.; Agha, W.; Al-Deroubi, A. Water Resources in the Arab World and Future Perspectives. In Proceedings of the Symposium on Water Resources and Uses in the Arab World, Kuwait, 1986.
69. Ukayli, M.A.; Husain, T. Comparative Evaluation of Surface Water Availability, Wastewater Reuse and Desalination in Saudi Arabia. *Water Int.* **1988**, *13*, 218–225. [CrossRef]
70. Abdulrazzak, M.J. Water Supplies versus Demand in Countries of Arabian Peninsula. *J. Water Resour. Plan. Manag.* **1995**, *121*, 227–234. [CrossRef]
71. Alghamdi, A.G.; Aly, A.A.; Aldhumri, S.A.; Al-Barakaha, F.N. Hydrochemical and Quality Assessment of Groundwater Resources in Al-Madinah City, Western Saudi Arabia. *Sustainability* **2020**, *12*, 3106. [CrossRef]
72. Saud, A.G.; Abdullah, S.A. Water Resources and Reuse in Al-Madinah. In Proceedings of the International Conference on Water Conservation in Arid Regions (ICWCAR'09), Water Research Center-King Abdulaziz University, Jeddah, Saudi Arabia, 12–14 October 2009; pp. 12–14.
73. Ali, I.; Hasan, M.A.; Alharbi, O.M.L. Toxic Metal Ions Contamination in the Groundwater, Kingdom of Saudi Arabia. *J. Taibah Univ. Sci.* **2020**, *14*, 1571–1579. [CrossRef]
74. MOEP (The Ministry of Economy and Planning). *The Ninth Development Plan (2010–2014)*; MOEP (The Ministry of Economy and Planning): Riyadh, Saudi Arabia, 2014.
75. Yihdego, Y.; Drury, L. Mine Water Supply Assessment and Evaluation of the System Response to the Designed Demand in a Desert Region, Central Saudi Arabia. *Environ. Monit. Assess.* **2016**, *188*, 619. [CrossRef]
76. Available online: https://www.worldweatheronline.com/riyadh-weather-averages/ar-riyad/sa.aspx (accessed on 22 November 2023).
77. Available online: https://ngp.sgs.gov.sa/ (accessed on 3 December 2023).
78. O'Callaghan, J.F.; Mark, D.M. The Extraction of Drainage Networks from Digital Elevation Data. *Comput. Vis. Graph. Image Process.* **1984**, *28*, 323–344. [CrossRef]
79. Zhu, Q.; Abdelkareem, M. Mapping Groundwater Potential Zones Using a Knowledge-Driven Approach and GIS Analysis. *Water* **2021**, *13*, 579. [CrossRef]
80. Abdelkareem, M.; Abdalla, F. Revealing Potential Areas of Water Resources Using Integrated Remote-Sensing Data and GIS-Based Analytical Hierarchy Process. *Geocarto Int.* **2021**, *37*, 8672–8696. [CrossRef]
81. Mogaji, K.A.; Lim, H.S.; Abdullah, K. Regional Prediction of Groundwater Potential Mapping in a Multifaceted Geology Terrain Using GIS-Based Dempster–Shafer Model. *Arab. J. Geosci.* **2015**, *8*, 3235–3258. [CrossRef]
82. Pande, C.B.; Moharir, K.N.; Panneerselvam, B.; Singh, S.K.; Elbeltagi, A.; Pham, Q.B.; Varade, A.M.; Rajesh, J. Delineation of Groundwater Potential Zones for Sustainable Development and Planning Using Analytical Hierarchy Process (AHP), and MIF Techniques. *Appl. Water Sci.* **2021**, *11*, 186. [CrossRef]
83. Megahed, H.A.; Farrag, A.E.-H.A.; Mohamed, A.A.; D'Antonio, P.; Scopa, A.; AbdelRahman, M.A.E. Groundwater Recharge Potentiality Mapping in Wadi Qena, Eastern Desert Basins of Egypt for Sustainable Agriculture Base Using Geomatics Approaches. *Hydrology* **2023**, *10*, 237. [CrossRef]
84. Tolche, A.D. Groundwater Potential Mapping Using Geospatial Techniques: A Case Study of Dhungeta-Ramis Sub-Basin, Ethiopia. *Geol. Ecol. Landsc.* **2020**, *5*, 65–80. [CrossRef]
85. Biswas, S.; Mukhopadhyay, B.P.; Bera, A. Delineating Groundwater Potential Zones of Agriculture Dominated Landscapes Using GIS Based AHP Techniques: A Case Study from Uttar Dinajpur District, West Bengal. *Environ. Earth Sci.* **2020**, *79*, 302. [CrossRef]

86. Magesh, N.S.; Chandrasekar, N.; Soundranayagam, J.P. Delineation of Groundwater Potential Zones in Theni District, Tamil Nadu, Using Remote Sensing, GIS and MIF Techniques. *Geosci. Front.* **2012**, *3*, 189–196. [CrossRef]
87. Mandal, U.; Sahoo, S.; Munusamy, S.B.; Dhar, A.; Panda, S.N.; Kar, A.; Mishra, P.K. Delineation of Groundwater Potential Zones of Coastal Groundwater Basin Using Multi-Criteria Decision Making Technique. *Water Resour. Manag.* **2016**, *30*, 4293–4310. [CrossRef]
88. Abdelkareem, M.; El-Baz, F. Remote Sensing of Paleodrainage Systems West of the Nile River, Egypt. *Geocarto Int.* **2017**, *32*, 541–555. [CrossRef]
89. Reddy, G.P.O.; Mouli, K.C.; Srivastav, S.K.; Srinivas, C.V.; Maji, A.K. Evaluation of Ground Water Potential Zones Using Remote Sensing Data-A Case Study of Gaimukh Watershed, Bhandara District, Maharashtra. *J. Indian Soc. Remote Sens.* **2000**, *28*, 19–32. [CrossRef]
90. Senthilkumar, M.; Gnanasundar, D.; Arumugam, R. Identifying Groundwater Recharge Zones Using Remote Sensing & GIS Techniques in Amaravathi Aquifer System, Tamil Nadu, South India. *Sustain. Environ. Res.* **2019**, *29*, 15. [CrossRef]
91. Algaydi, B.A.M.; Subyani, A.M.; Hamza, M.H.M.M. Investigation of Groundwater Potential Zones in Hard Rock Terrain, Wadi Na'man, Saudi Arabia. *Groundwater* **2019**, *57*, 940–950. [CrossRef]
92. Adeyeye, O.A.; Ikpokonte, E.A.; Arabi, S.A. GIS-Based Groundwater Potential Mapping within Dengi Area, North Central Nigeria. *Egypt. J. Remote Sens. Sp. Sci.* **2019**, *22*, 175–181. [CrossRef]
93. Andualem, T.G.; Demeke, G.G. Groundwater Potential Assessment Using GIS and Remote Sensing: A Case Study of Guna Tana Landscape, Upper Blue Nile Basin, Ethiopia. *J. Hydrol. Reg. Stud.* **2019**, *24*, 100610. [CrossRef]
94. Ahmed, J.B., II; Mansor, S. Overview of the Application of Geospatial Technology to Groundwater Potential Mapping in Nigeria. *Arab. J. Geosci.* **2018**, *11*, 504. [CrossRef]
95. Yang, J.; Zhang, H.; Ren, C.; Nan, Z.; Wei, X.; Li, C. A Cross-Reconstruction Method for Step-Changed Runoff Series to Implement Frequency Analysis under Changing Environment. *Int. J. Environ. Res. Public Health* **2019**, *16*, 4345. [CrossRef] [PubMed]
96. Bhanja, S.N.; Mukherjee, A.; Rangarajan, R.; Scanlon, B.R.; Malakar, P.; Verma, S. Long-Term Groundwater Recharge Rates across India by in Situ Measurements. *Hydrol. Earth Syst. Sci.* **2019**, *23*, 711–722. [CrossRef]
97. Alogayell, H.M.; EL-Bana, E.M.M.; Abdelfattah, M. Groundwater Quality and Suitability Assessment for Irrigation Using Hydrogeochemical Characteristics and Pollution Indices: A Case Study of North Al-Quwayiyah Governorate, Central Saudi Arabia. *Water* **2023**, *15*, 3321. [CrossRef]

hydrology

MDPI

Article

Climatic Modeling of Seawater Intrusion in Coastal Aquifers: Understanding the Climate Change Impacts

Aikaterini Lyra [1], Athanasios Loukas [2], Pantelis Sidiropoulos [2,*] and Nikitas Mylopoulos [1]

1 Laboratory of Hydrology and Aquatic Systems Analysis, Department of Civil Engineering, School of Engineering, University of Thessaly, 38334 Volos, Greece; klyra@uth.gr (A.L.); nikitas@civ.uth.gr (N.M.)
2 Laboratory of Hydraulic Works and Environmental Management, School of Rural and Surveying Engineering, Aristotle University of Thessaloniki, 54124 Thessaloniki, Greece; agloukas@topo.auth.gr
* Correspondence: pasidirop@topo.auth.gr; Tel.: +30-231-099-6105

Abstract: The study examines the impacts of climate change and sea level rise on coastal aquifers, focusing on the influence of the components of the water cycle on seawater intrusion, and the evolution of the phenomenon in the future. The simulation of coastal water resources was performed using an integrated modeling system (IMS), designed for agricultural coastal watersheds, which consists of inter-connected models of surface hydrology (UTHBAL), groundwater hydrology (MODFLOW), and seawater intrusion (SEAWAT). Climatic models for the adverse impact scenario (RCP8.5) and the medium impact scenario (RCP4.5) of climate change were used. Transient boundary head conditions were set to the coastal boundary, to dynamically represent the rise in sea level due to climate change. The response of groundwater in the coastal Almyros Basin, located in central Greece, was simulated from 1991 to 2100. The findings indicate that seawater intrusion will be advanced in the future, in both climate change scenarios. The models show varying patterns in groundwater recharge, with varying uncertainty projected into the future, and sensitivity to time in the fluctuation of the components of the water cycle.

Keywords: climate change; Med–CORDEX; integrated modeling system; groundwater; seawater intrusion; water resources

Citation: Lyra, A.; Loukas, A.; Sidiropoulos, P.; Mylopoulos, N. Climatic Modeling of Seawater Intrusion in Coastal Aquifers: Understanding the Climate Change Impacts. *Hydrology* **2024**, *11*, 49. https://doi.org/10.3390/hydrology11040049

Academic Editor: Marco Delle Rose

Received: 29 February 2024
Revised: 3 April 2024
Accepted: 4 April 2024
Published: 6 April 2024

1. Introduction

Groundwater aquifers are facing challenges related to seawater intrusion, due to climate change and anthropogenic activities. The quantity and quality of groundwater resources are being impacted by surface water depletion, rising temperatures, and sea level rise [1–3]. In the face of our changing climate, the delicate balance between freshwater and the encroaching sea has become a critical concern for coastal communities and environmental stewards worldwide [4]. As communities grapple with the uncertainty of climate change, and with the freshwater that sustains them under threat, an understanding of the seawater wedge's evolution becomes extremely important. Seawater contains salts in very high concentrations, the mass of which causes saltwater to have a higher fluid density (1.025 g/cm^3) and a higher weight than freshwater (1 g/cm^3) [5]. In seawater intrusion, saltwater infiltrates the freshwater in coastal aquifers and forms wedge-shaped plumes of contamination. The impacts of climate change and human activity are worldwide environmental concerns that need to be closely watched and further researched [6]. Sustainable groundwater management practices, aquifer recharge, and careful monitoring are essential for addressing seawater intrusion and ensuring the long-term availability of freshwater resources in coastal regions [7]. Research on seawater intrusion is more prevalent in countries facing a higher risk of this phenomenon due to factors like sea level rise, the over-pumping of groundwater, or specific geological conditions, while it is also greatly influenced by the availability of funding, research institutions, and scientific expertise within each country. The biggest contributions in this field of research, between

1970 and 2019, came from the United States with 23.9%, China with 14.8%, and Australia with 7.7%, followed by Italy, Japan, Canada, France, Germany, India, Spain, and others [8]. Seawater intrusion poses a threat to over 100 nations and regions worldwide, yet much of the coastline is not adequately monitored [9]. The causes of occurrence of seawater intrusion in coastal aquifers can vary in significance, depending on the specific circumstances of each coastal area, e.g., proximity to the sea, natural hydrogeological background, overuse of groundwater, and water table drawdown [10]. Seawater intrusion is a very important problem for coastal aquifer systems across Europe, especially along the Mediterranean coastline [11]. In recent years, the intense degradation of groundwater quality, due to salinization, has become evident in arid or semi-arid climate areas, because these areas are more vulnerable to seawater intrusion than other coastal aquifer systems in other regions of the world. A characteristic example of a highly vulnerable basin is the Mediterranean basin, and its coastal groundwaters have been documented in the literature as having been salinized to a lesser or greater extent. Chloride compounds are allowed in water for human consumption up to a limit of 250 mg/L, but in higher quantities, they are taste-perceptible and are also associated with health problems [12,13]. The maximum limits for chloride ions are also in line with the World Health Organization (WHO) [14]. The Water Framework Directive [15] in Europe, according to articles 17 (1) and (2), includes a specific Directive for Groundwater 2006/118/EC [16], wherein specific measures are established to prevent the degradation of groundwater and its impact on surface water (i.e., rivers, lakes, and wetlands). Significant importance is given to water pollution, distinguishing between substances of a natural background and those originating from other processes, and to seawater intrusion/salinization.

Climate-driven modifications in recharge and sea level rise are major drivers of saltwater intrusion, with rising sea levels affecting a large percentage of coastal watersheds [17]. Seawater intrusion simulation is a valuable tool for understanding and managing the movement of saltwater into coastal aquifers [18]. Numerical and analytical models are being developed to investigate saltwater intrusion and estimate the depth of the freshwater-to-saltwater interface. These simulations typically involve mathematical models that represent the complex hydrogeological processes governing the interaction between freshwater and seawater. The future consequences of climate change on groundwater resources are being evaluated through integrated modeling techniques, such as those joining together groundwater and climate models [19,20]. Thus, considering both sea level rise and freshwater recharge rates can enhance knowledge of the consequences of climate change on seawater intrusion [21]. When modeling seawater intrusion under climate change, several key factors are necessary to consider. These factors include the impacts of climate change on recharge and sea level rise, as well as the influence of freshwater discharges, bed slopes, and hydraulic conductivity on the length of saline intrusion [22,23]. Seawater intrusion is greatly influenced by the capacity of the aquifer to transmit water, known as its hydraulic conductivity, as well as the pattern of water flow within the main aquifers [24]. Furthermore, the rising sea levels and decreasing freshwater reserves can exacerbate the issue of saltwater intrusion [25]. Developing accurate seawater intrusion simulations requires a substantial amount of data, including aquifer properties/parameters, boundary conditions, and historical information. Seawater intrusion models make various assumptions and simplifications, such as assuming steady-state conditions or neglecting certain hydrogeological processes [26]. However, these assumptions may not hold in all situations, leading to potential inaccuracies [27]. One of the primary issues with numerical models is that, on occasion, the initial and boundary conditions must be made simpler. For instance, the boundary of the ocean can be set at a fixed level [28]. Nevertheless, seawater intrusion simulations are particularly valuable in regions with poorly monitored aquifer systems along the coastline, where limited funding for monitoring often hinders direct data collection efforts. Traditional modeling approaches often struggle to capture the variabilities of these systems, leading to inaccuracies and uncertainties in water resource management. By adopting a holistic approach, these simulations enable a comprehensive understanding of the multifaceted factors contributing to seawater intrusion, making them instrumental in formulating effective management strategies under climatic changes [29,30]. Additionally, they provide a timeseries of simulated profiles and expansion, offering a temporal dimension to the understanding of seawater intrusion dynamics and helping decision-makers plan for a more resilient and sustainable coastal future.

This study makes a contribution to the field by providing integrated modeling approaches to estimate the complex impacts of climate change and sea level rise on groundwater aquifer systems, with a specific focus on seawater intrusion. The incorporation of transient head boundary conditions in the models allows for a dynamic representation of the aquifer's behavior, providing a more realistic representation of the interactions between regional climatic forcings, sea level rise, shifts in groundwater recharge, groundwater uses, and seawater intrusion.

2. Materials and Methods

2.1. Study Area

Many coastal aquifer systems in Greece suffer from seawater intrusion and salinity degradation [31,32]. The research focuses on the Almyros basin, a coastal location in central Greece, specifically within the region of Thessaly. The basin contains a large low elevated plain in its coastal section, and its altitude varies from 0 m to roughly 1710 m, as depicted in Figure 1. The total area is 856 km^2 and is elevated by 30% to 150 m, by 57% in hilly areas to 800 m, and by 13% to mountainous areas. Climatic zones were obtained from a publicly available dataset from the global studies of Metzger et al. (2013) [33] and Sayre et al. (2014) [34]. Overlaying the contoured elevation of the study area and the climatic zones, as highlighted in Figure 1, the prevailing climatic conditions are characterized as Mediterranean; however, depending on the elevation, it is observed that specific differentiations exist. Up to 300 m, the climate is warm and semi-arid, up to 900 m, it is warm and moist, and higher than this elevation, it is cool and wet. The basin is divided into six sub-watersheds [35]. Numerous streams with intermittent flow gather into six main streams, and no other storage surface water bodies exist within the basin (Figure 1). The annual areal precipitation is 570 mm and the average temperature is 15 °C. Under the ground, there is a significant granular aquifer system that is used to cover the water supply needs, mainly agricultural and urban, of the basin. The Almyros aquifer system is mainly composed of geological materials of Neogene and Quaternary age, classified into five intercalating groups, namely clay, clay–gravel–sand, sand, clay–sand, and limestone, as highlighted and drawn in Figure 2a [36]. Limited coastal karsts a have direct hydraulic connection to sea, but not with the aquifer. Other small aquifers also exist in Mount Orthrys and to the north, but impermeable materials occur around the Almyros aquifer. Surface runoff accounts for 20% of precipitation across the Almyros Basin; 11–12% of precipitation recharges the Almyros aquifer in the northern and central regions (i.e., the watersheds of Kazani, Lahanorema, Holorema, and Xirias) and 5–6% recharges the aquifer in the southern region (the watersheds of Platanorema and Xirorema). Within the historical period of the study, agricultural activities occupy approximately 205 km^2, constituting nearly 70% of the aquifer's total area. The water uses cover the needs of the irrigated crops of wheat, alfalfa, vegetables, cotton, maize, trees, olive groves, and vineyards. However, the semi-arid hydrological status of the area, combined with the groundwater abstractions for agricultural and other uses, have led to a lowering groundwater table and seawater intrusion in the coastal lowland area, as shown and highlighted in Figure 2b. Groundwater flows from high elevated areas towards the lowlands and the sea in the central region of the aquifer, and the flow directions change on the northern coast and in the southern region, according mostly to hydrogeological permeability and to the difference in elevation between the water table and sea level. Nowadays, the Almyros basin is a basin which faces severe seawater intrusion that has evolved through the years, mostly at the northern coastline, as observed and estimated in previous studies [35,37,38] in the region and as described in the next section.

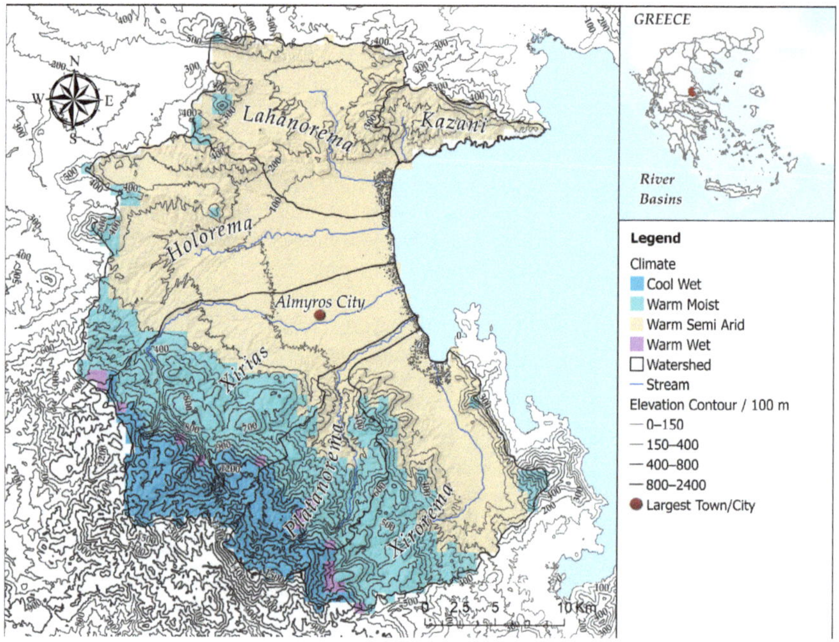

Figure 1. Map of the Almyros Basin, the climatic zones, the watersheds, the largest town/city of basin, and the contoured elevation of the wider area.

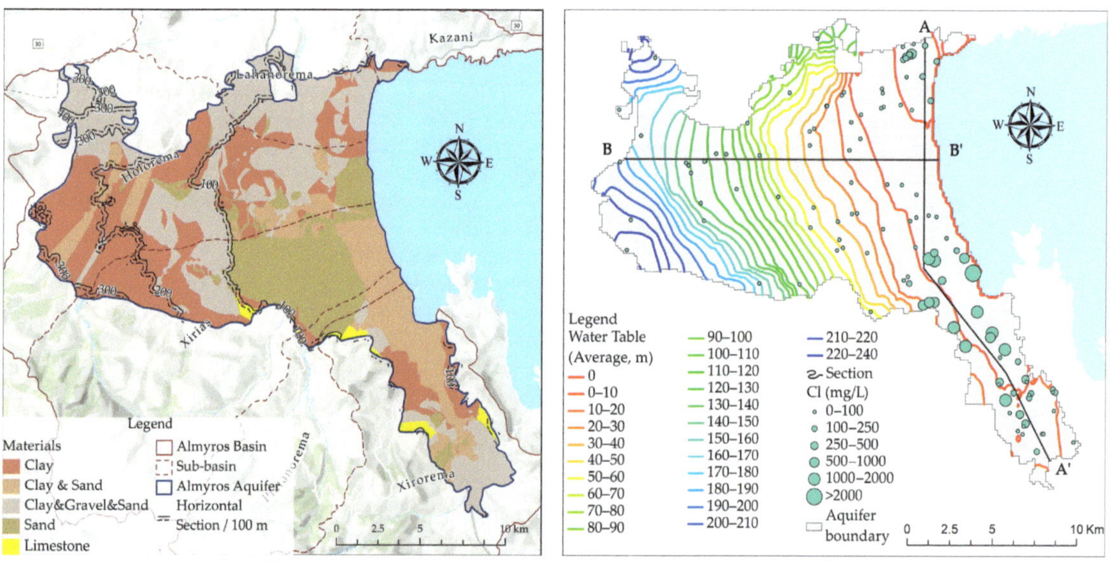

(a) (b)

Figure 2. (**a**) Hydrogeological formations within the Almyros aquifer [36]. (**b**) Mean hydraulic heads from the integrated modelling system, with the average observed levels of chloride concentrations during the historical period, spanning from 1991 to 2018, and cross-sections AA′ and BB′.

2.2. Modeling of Seawater Intrusion

2.2.1. Application of a Calibrated and Validated Integrated Modelling System (IMS)

Coastal water resources were studied using an integrated modeling system, developed in a previous study [35]. The integrated modelling system was used for simulating both surface and groundwater resources in the Almyros Basin. It consists of interlinked models which simulate surface hydrology (UTHBAL) [39], groundwater hydrology (MODFLOW) [40], and seawater intrusion (SEAWAT) [41]. A comprehensive study conducted by Lyra et al. in 2021 [35] included substantial preparatory processing, validation, and employing data from observations acquired during years of surveillance. The majority of components in the integrated modeling system underwent rigorous validation and calibration. The historical and future groundwater balance was simulated using the developed IMS. The water needs and abstractions from the groundwater, which is the only freshwater source in the study basin, for irrigation, domestic supply, and other water uses, were estimated for current and future climate conditions under two climate scenarios (RCPs 4.5 and 8.5) and 19 climate models. The groundwater balance was simulated, accounting for natural groundwater recharge (calculated using UTHBAL), groundwater abstractions, irrigation return flows (estimated as a percentage of the irrigation water applied for each crop cultivation), and the variable rise in sea level.

To accomplish this, six meteorological stations inside and around the basin were used to gather climate data for the specified region. These stations were from various sources, including Hellenic Meteorological Service and non-public meteorological stations. After collecting necessary data, these were processed and used as inputs for simulations utilizing the UTHBAL model. Precipitation by month was determined by applying the Thiessen polygon methodology, while the gradient method was utilized for the central points of the watershed. The watershed centroids were utilized in calculating the monthly temperature, through the gradient method. The UTHBAL model uses monthly sequences of areal precipitation, temperature, and potential evapotranspiration to calculate watershed runoff and groundwater recharge. Depending on the available data, the usage of the model may differ among various applications, such as being lumped, partially distributed, or entirely distributed, and comprises six parameters, which are either calibrated or estimated. For the Almyros region, the UTHBAL model was utilized as a semi-distributed model, to simulate surface hydrological processes in the six watersheds of Almyros Basin, from October 1961 to September 2018. These model parameters expressed the watershed characteristics and included the monthly melt rate factor (Cm, 6 mm/°C), the actual evapotranspiration parameter (α, 0.48), the interflow parameter (β, 0.033), the baseflow parameter (γ, 0.033), and the groundwater recharge parameter (K, 0.68), all of which play a crucial role in the functioning of the model. The values for these parameters were derived from UTHBAL's regional application in Thessaly, as the Almyros watersheds lack sufficient runoff and streamflow data for model calibration. To calculate the weighted mean curve number (CN), according to the United States Soil Conservation Service (SCS) method of considering various factors that influence the CN [42], soil composition, land use, and land cover data for every watershed were included in the calculations, which derived the values of CN_{Kazani}, at 67.93, $CN_{Lahanorema}$, at 68.11, $CN_{Holorema}$, at 68.47, CN_{Xirias}, at 60.69, $CN_{Platanorema}$, at 51.07, and $CN_{Xirorema}$, at 53.84 [35]. However, the values of CN may have large uncertainties. Groundwater recharge also takes into consideration the mean weighed return flow from irrigation from various types of crops within the study area. Using a systematic approach, the monthly runoff was determined by considering the demands of both rural and urban areas, as well as other factors. Spatial datasets on arable land and crop yields were provided by the Greek Payment Authority, while records on soil hydraulic features were integrated from the European Soil Data Centre [43]. Of the total agricultural water needs calculated with the near irrigation requirement (NIR) technique [44], approximately 9% is distributed in the Kazani watershed, 20% in the Lahanorema watershed, 21% in the Holorema watershed, 19% in the Xirias watershed, 17% in the Platanorema watershed, and 14% in the Xirorema watershed [35]. The inputs of the MODFLOW model include irrigation return flow, groundwater recharge, groundwater abstractions, net monthly sea level rise (set at the beginning of the hydrological year), and, finally, shoreline conductance.

The MOFLOW and SEAWAT groundwater models are built on a grid system, with 200 rows and 200 columns mapped and matched to the sub-watershed delineation of the surface hydrological model (UTHBAL). Each cell in this grid has a size of approximately 150 m × 150 m. The simulation of the unconfined aquifer was accomplished through the application of the layer property flow package. A 'No-Flow Boundary' setting was implemented for the western region of the study area, because there are impenetrable geological formations nearby that prevent the flow of fluids. In contrast, a transient head boundary was utilized in the eastern region, to take into account the changing sea level. Different hydrogeological zones were simulated with varying hydraulic conductivity values, ranging from 0.1 to 18.7 m/day, with an mean equal to 2.3 m/day, while the maximum velocities run in the northern coastal region, with clayish and mostly sandy materials [35]. Borehole public data, used to pinpoint locations of water pumping installations, were also a significant input of the groundwater model [45]. Useful field measurements of water levels and chloride levels in groundwater were provided by the Hellenic Survey of Geology and Mineral Exploration, Magnesia Prefecture, Regional Government of Thessaly, and earlier research investigations. The modelling system was calibrated and validated against water levels, and the MODFLOW model demonstrated exceptional performance during both the calibration (1991–2009) and validation (2013–2015) phases. The calibration process yielded high scores of statistical measures, such as Nash–Sutcliffe efficiency (NSE), R^2, and index of agreement (IA) values of 0.98, 0.98, and 0.99, respectively. The MODFLOW model demonstrated significant efficiency even throughout the verification phase, with NSE, R^2, and IA ratings of 0.99. The seawater intrusion model was also calibrated and validated against chloride concentration observations for the Almyros aquifer. The SEAWAT model demonstrated exceptional performance during both the calibration (1991–2004) and validation (2005–2007) phases. The calibration process yielded high Nash–Sutcliffe efficiency (NSE), R^2, and index of agreement (IA) values of 0.92, 0.94, and 0.98, respectively. During the validation phase, the SEAWAT model maintained its high efficacy, with NSE, R^2, and IA scores of 0.89, 0.95, and 0.98, respectively. These scores clearly highlight the high efficacy and consistency of the integrated modeling system.

2.2.2. Climate Change Scenarios

In order to accurately simulate the Almyros aquifer's water resources and seawater intrusion in light of potential climate shifts, it was imperative to utilize datasets projecting future alterations in temperature, precipitation, and sea level. Climatic models were acquired from Med-CORDEX (Coordinated Regional Climate Downscaling Experiment for the Mediterranean region), a repository for regional climate model data focusing on the Mediterranean region. Specifically, the representative concentration pathways (RCP) scenarios 8.5 (RCP8.5) and 4.5 (RCP4.5) were selected for analysis. RCP8.5 represents a higher greenhouse gas emission trajectory, reflecting a scenario of increased emissions and greater climate change impacts. On the other hand, RCP4.5 represents a more moderate emission trajectory. The Med-CORDEX models used in the study are given in Table 1.

After obtaining climate models from Med-CORDEX for the RCP8.5 and RCP4.5 scenarios, a crucial step was taken in guaranteeing the precision and dependability of the data. The bias correction technique of Quantile Empirical Mapping, a highly regarded method, was utilized to refine and harmonize the model outputs, which was undertaken in earlier studies [36,37], the results of which were used in the present simulations. Critical efficacy scores are summarized in Appendix A. The use of bias-corrected future climatic data is integral in addressing any systematic errors or inconsistencies between the climate model projections and actual observations. By implementing quantile mapping, the accuracy of climatic models was improved, ensuring they accurately reflect historical climate patterns and serve as a reliable basis for projecting future scenarios in the Almyros basin.

Table 1. Simulations from the Med–CORDEX utilized for analysis.

Realm	GCM Model ǀ CMIP5 Ensemble	Regional Climate Model	Climate Projection
MED/11	CNRMCM5r8i1p1	1.CNRMALADIN52v1	RCP8.5/RCP4.5
	ICTPRegCM4r1i1p1	2.ICTPRegCM43v1*	RCP8.5
MED44	CMCCCM.r1i1p1	3.CMCCCLM4819.v1	RCP8.5/RCP4.5
	CNRMCM5_r8i1p1	4.CNRMALADIN52.v1	RCP8.5/RCP4.5
	HadGEM2ES.r1i1p1	5.ELURegCM43v1	RCP8.5/RCP4.5
	HadGEM2ES.r1i1p1	6.ICTPRegCM43.v4*	RCP4.5
	HadGEM2ES.r1i1p1	7.ICTPRegCM43.v7*	RCP8.5
	IPSLCM5AMR.r1i1p1	8.LMDLMDZ4NEMOMED8.v1	RCP8.5/RCP4.5
	MPIESMLR.r1i1p1	9.GUFCCLM4818.v1	RCP8.5/RCP4.5
	MPIESMMR.r1i1p1	10.ICTPRegCM43.v7*	RCP8.5
MED44i	CNRMCM5.r1i1p1	11.ENEAPROTHEUS.v2*	RCP4.5
	MPIESMLR.r1i1p1	12.UNIBELGRADEEBU.v1*	RCP8.5
	MPIESMLR.r1i1p1	13.UNIBELGRADEEBUPOM2c.v1*	RCP8.5

* Available only for one RCP.

2.2.3. Sea Level Rise and Transient Head Boundary Conditions

Sea level rise in the Mediterranean Sea is a pressing issue which has garnered significant attention in recent years. As global temperatures continue to rise, the resulting melting of polar ice caps and glaciers has led to an increase in sea levels worldwide. The Mediterranean region is particularly vulnerable to sea level rise, as it is surrounded by densely populated coastal areas that are at risk of flooding and erosion. However, it has a lower rate of sea level rise compared to other regions. The sea level rise is also influenced by other factors, such as atmospheric pressure anomalies, variations in evaporation and precipitation, the North Atlantic oscillation, Eastern Mediterranean transient, and the balancing effect of salinification near the coasts [46,47].

According to gridded satellite historical datasets, the average sea level at the coastline of Almyros was 0.05 m between 1993 and 2018 (25 years) [48]. The annual rate of increase was calculated to be 3.8 mm (the Figure 3a), which was also verified by the estimations of the European Environmental Agency (EEA) [49] for the period of 1992 to 2013. However, the sea level rise was considered to be zero during the period of 1991 to 2018, because the available timeseries of the sea level did not cover the time span of historical simulations. The formation of climate scenarios had to also include the projected change in the average sea level, since the Almyros watershed and the aquifer system are located adjacent to the sea (Figure 2). The sea level is estimated to increase in the future, with a higher rise in the RCP8.5 climate scenario and a lower rise in the RCP4.5 climate scenario. The sea level rise for the climatic scenario RCP4.5 is projected to be between 0.3 and 0.4 m between 2081 and 2100 according to the EEA [50], and, for RCP8.5, is projected to be 0.5–0.6 m between 2081 and 2100 [51], respectively. These estimated projected values of sea level rise were equally distributed over time for the period 2019–2100 (Figure 3b). The sea level rise was considered to happen at the start of every hydrological year (i.e., each October).

For RCP4.5, the sea level rise was calculated as 3.6 mm/y, for the total sea level rise to reach 0.3 m in 2100, and for RCP8.5 as 6.3 mm/y, for the total sea level rise to reach approximately 0.5 m in 2100.

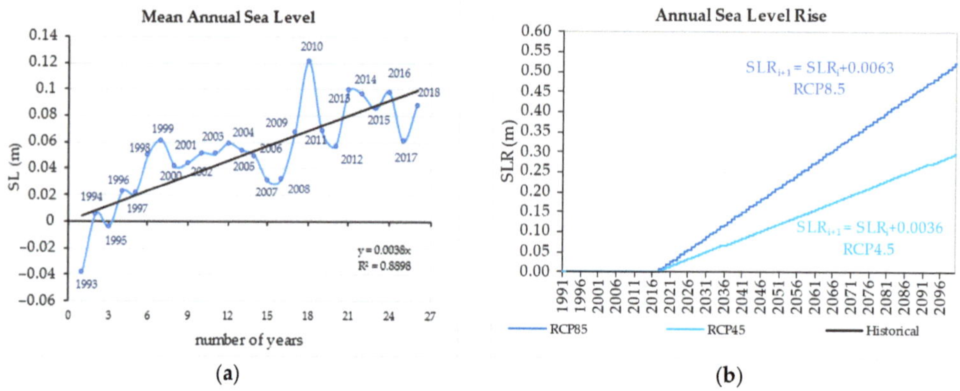

Figure 3. (**a**) Mean annual sea level for the period 1991–2018 according to [48]. (**b**) Simulated sea level rise for the period 1991–2100.

2.3. Evaluation of Results

A ternary diagram is a comprehensive visual tool used in environmental science to depict the composition of a three-component system [52]. This diagram features an equilateral triangle, with each point representing one of the three components, and the sides representing all possible combinations of these components. Every point in a ternary diagram precisely represents the combination or content of the three elements. The distance between the point and the triangle's sides indicates the relative weights of each element. Importantly, the sum of these distances is constant, illustrating the fact that the three components together make up the entirety of the system. The ternary diagram has been used to assess seawater intrusion in coastal aquifers, a growing concern in many regions due to the development of coastal areas and increasing demands for freshwater. By analyzing the inflows and outflows of coastal aquifers using the ternary diagram, the analysis of the water balance of aquifer systems can provide valuable information for the dynamics of water movement and the key components that influence water availability and seawater intrusion.

3. Results

The studies by Lyra and associates (2021) [35] and Lyra and Loukas (2022) [37] gave rise to the development of the integrated modeling system, utilized to model and project the potential effects of climatic changes on the Almyros basin and its aquifer system, in Thessaly, Greece, covering both the worst-case (RCP8.5) and moderate (RCP4.5) climate scenarios. The findings of the simulation process implemented for the evaluation of seawater intrusion and climate change scenarios and climatic models are provided and analyzed in the following paragraphs.

3.1. Projected Water Balance—Seawater Intrusion

The results of the simulations of the integrated modelling system regarding the water balance for the two climatic scenarios and climatic models are depicted in Figure 4. The ternary diagram was employed for analyzing the components of the water budget of the aquifer system, particularly in the context of seawater intrusion, groundwater recharge, irrigation return flow, groundwater abstractions, and aquifer discharge. The grouping of the variables was firstly considered, namely which components held the inflows and the outflows of the aquifer system. The aquifer outflows consisted of groundwater abstractions, while the aquifer inflows were considered separately. Water inflows from the sea are displayed on the left side of the triangle, while water inflows from groundwater recharge and irrigation return flows are the displayed on the right side of the triangle. Then, the original values of the results were normalized as percentages. The normalized percentages of the components of the water budget were displayed as averaged annual values of the three components for the periods of 1991 to 2000, 2001 to 2018, 2019 to 2050, 2051 to 2080, and 2081 to 2100. These factors are represented by the three equilateral triangle corners, and all potential combinations of the three are shown on each side. The ideal distribution would be 0% seawater intrusion, 50% groundwater recharge and irrigation return flows, and 50% groundwater abstractions and aquifer discharge, for freshwater to not be contaminated with seawater, and for permanent groundwater reserves to not be consumed for water needs. Through the placement of the proportional relations of the components of the groundwater budget as points within the triangle, the diagram allows for a thorough examination of how seawater intrusion, groundwater recharge, and groundwater abstractions interact and impact the distribution of freshwater in the aquifer system.

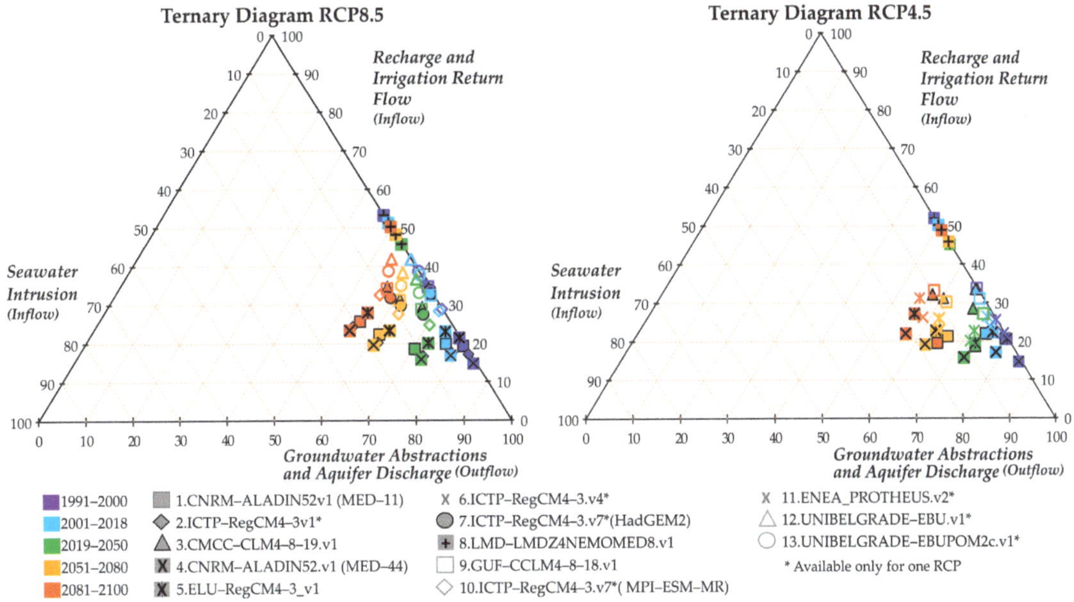

Figure 4. Ternary plots of the water balance components under climate scenarios RCP8.5 and RCP4.5 during various time periods. The time periods covered are 1991–2000, 2001–2018, 2019–2050, 2051–2080, and 2081–2100. The integrated modeling system (IMS) was utilized for the simulations.

From the ternary plots, it is observed that, in both climatic scenarios, RCP8.5 and RCP4.5, the aquifer outflows constitute 55% to 85% of the water balance for all periods. The outflows consist mostly of groundwater abstractions while the aquifer discharge is negligible. Hence, this is why the points of the diagram are accumulated at the right side of large component values. Moreover, it is observed that the groundwater recharge and irrigation return flow form a more distinct relation with seawater intrusion, because, when the aquifer recharge is high, the seawater intrusion is quite low and almost zero in some climate models. As the recharge of water in the aquifer decreases, the presence of seawater in the water balance increases, leading to the salinization of the groundwater. Notably, as the time progresses to 2100, groundwater abstractions are reduced due to altered seasonality; the groundwater recharge and irrigation return flow decrease in 2019–2050 but increase in 2050–2100; and the seawater intrusion increases from 2019 to 2100. Hence, the groundwater recharge and irrigation return flow play the most critical roles in the evolution of seawater intrusion under climate change when groundwater is abstracted, even with a decreasing trend. According to RCP8.5, the analyzed models show varying patterns in groundwater recharge and irrigation return flow over distinct periods. For example, the 1.CNRMALADIN52v1 (MED11) model displays a fluctuation of 19.2% to 25.7% for these factors, suggesting a sensitivity to time. Similarly, the 8.LMDLMDZ4NEMOMED8.v1 model depicts a gradual decrease from 53.3% to 50.3%. Seawater intrusion also exhibits upward trends throughout the projected timeframes, as illustrated by multiple climate models. In the initial period of 1991–2000, all models showed relatively low percentages, ranging from 0.1% to 0.7%. However, as the decades progress, there was a clear escalation in seawater intrusion percentages. Between the years 2001 and 2018, there was a steady increase in values for seawater intrusion, ranging from 0.1% to 4.3%, across multiple models. This trend is expected to continue for the foreseeable future, with projected percentages for 2019–2050 ranging from 1.6% to 11.0% and, for 2051–2080, escalating to 3.5% to 19.1%. The latter part of the century (2081–2100) shows a continuing upward trend, with expected percentages of seawater intrusion between 4.1% and 22.2%. Interestingly, the 4.CNRMALADIN52.v1 (MED44) model consistently predicts the highest percentage of seawater intrusion across all time periods. In contrast, most models show a consistent decrease in groundwater abstractions and aquifer discharges across different time periods. For example, when observing changes over time, it is evident that the 1.CNRMALADIN52v1 (MED11) model experienced a significant decrease, from 80.2% to 55.5%. This highlights a notable decrease in both groundwater usage and aquifer discharge. On the other hand, the 4.CNRMALADIN52.v1 (MED44) model initially had the highest percentages, at 84.6%, but, over time, it experienced a noteworthy decline to 54.5% by 2081–2100. In contrast, the 8.LMDLMDZ4NEMOMED8.v1 model displayed an increase from 46.7% to 54.2%, indicating a positive trend in groundwater outflows.

During the initial period of RCP4.5 (1991–2000), the variability among climatic models was evident, as the values of groundwater recharge and irrigation return flow ranged from 20.6% to 51.9%. As time progresses, fluctuations and trends become apparent. Many models show a decreasing trend in recharge in the later periods, although there are some fluctuations. For example, the 4.CNRMALADIN52v1 (MED44) model experienced a decline from 15.8% to 19.2% in the initial period but then shows a slight increase to 22.0% in the final period. A similar trend can be observed in the 8.LMDLMDZ4NEMOMED8v1 model, which displays a steady decrease from 51.9% to 48.7%, indicating a potential impact of climate change on the recharge of groundwater and return flow from irrigation. From 1991–2000 to 2081–2100, there were significant fluctuations in the rates of seawater intrusion, according to different models. One consistent trend is seen in Model 8.LMDLMDZ4NEMOMED8v1, which maintains consistently low percentages throughout the time period, with a peak of 0.1% in the later years. However, Model 4.CNRMALADIN52v1 (MED44) shows a worrisome increase, starting at 0.7% in 1991–2000 and reaching 21.2% in 2081–2100. The same unpredictability is seen in the trends of groundwater abstractions and aquifer discharge from 1991–2000 to 2081–2100, as shown by various climatic models. One example is the decline seen in Model 1.CNRMALADIN52.v1, which shows a decline from 78.8% to 64.8%, indicative of a reduction in groundwater abstractions and aquifer discharge. The 3.CMCCCCLM4819v1 model consistently exhibits a decline from 66.5% to 57.6%, indicating a persistent downward trend throughout the evaluated periods. On the contrary, the model 8.LMDLMDZ4NEMOMED8v1 displays a contrasting trend, with an increase from 48.0% to 51.2%.

3.2. Projected Chloride Concentrations with the SEAWAT Model under Climate Change

The SEAWAT model was applied to simulate the flow of variable density and the movement of chlorides in the Almyros coastal aquifer at a monthly time step for the two climate scenarios RC8.5 and RCP4.5, for the 19 climatic models. The time periods under examination included 1991 to 2000, 2001 to 2018, 2019 to 2050, 2051 to 2080, and 2081 to 2100.

3.2.1. Groundwater Model Cross-Sections

Figures 5 and 6 provide a thorough examination of the impacts of sea level rise and climate change on the Almyros aquifer system, presenting intricate details and analyses. Figure 5 captures the changes in sections AA′ and BB′ (presented in Figure 2b) for the eleven bias-corrected models under the RCP8.5 scenario of adverse impacts, while Figure 6 focuses on the medium impacts scenario, RCP4.5, and displays the corresponding sections for the eight bias-corrected models. All the SEAWAT model simulations for chloride concentrations for the historical period of 1991–2018 showed a consistent pattern across the various climatic scenarios and models. As time progresses towards 2100, the results reveal distinct changes and fluctuations, particularly in the levels of chloride concentrations.

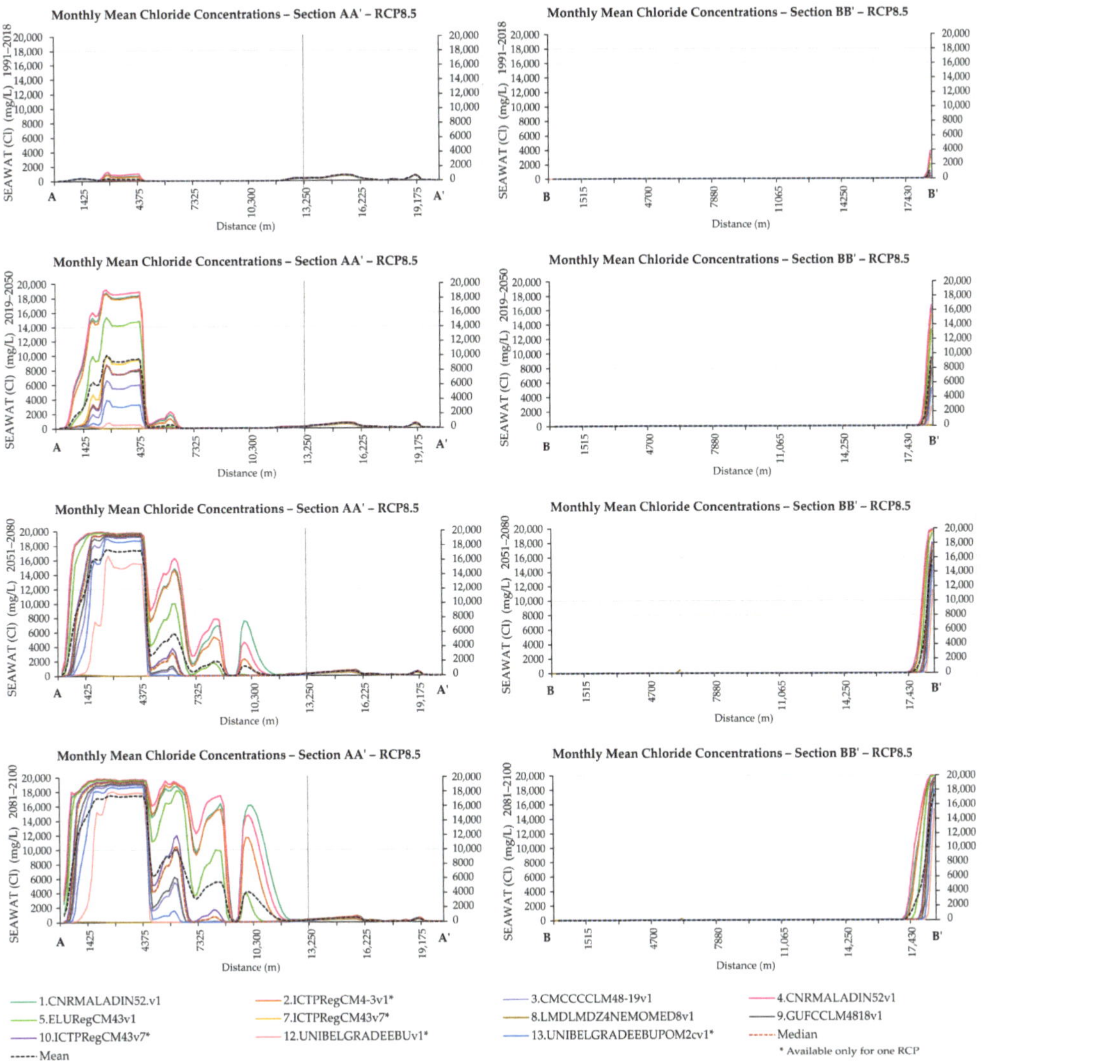

Figure 5. Section AA′ on the left side and Section BB′ on the right illustrate the chloride concentrations in milligrams per liter (mg, L) across the following four time periods: 1991–2018, 2019–2050, 2051–2080, and 2081–2100. These concentrations are derived from the bias-corrected Med-CORDEX models under the RCP8.5 scenario.

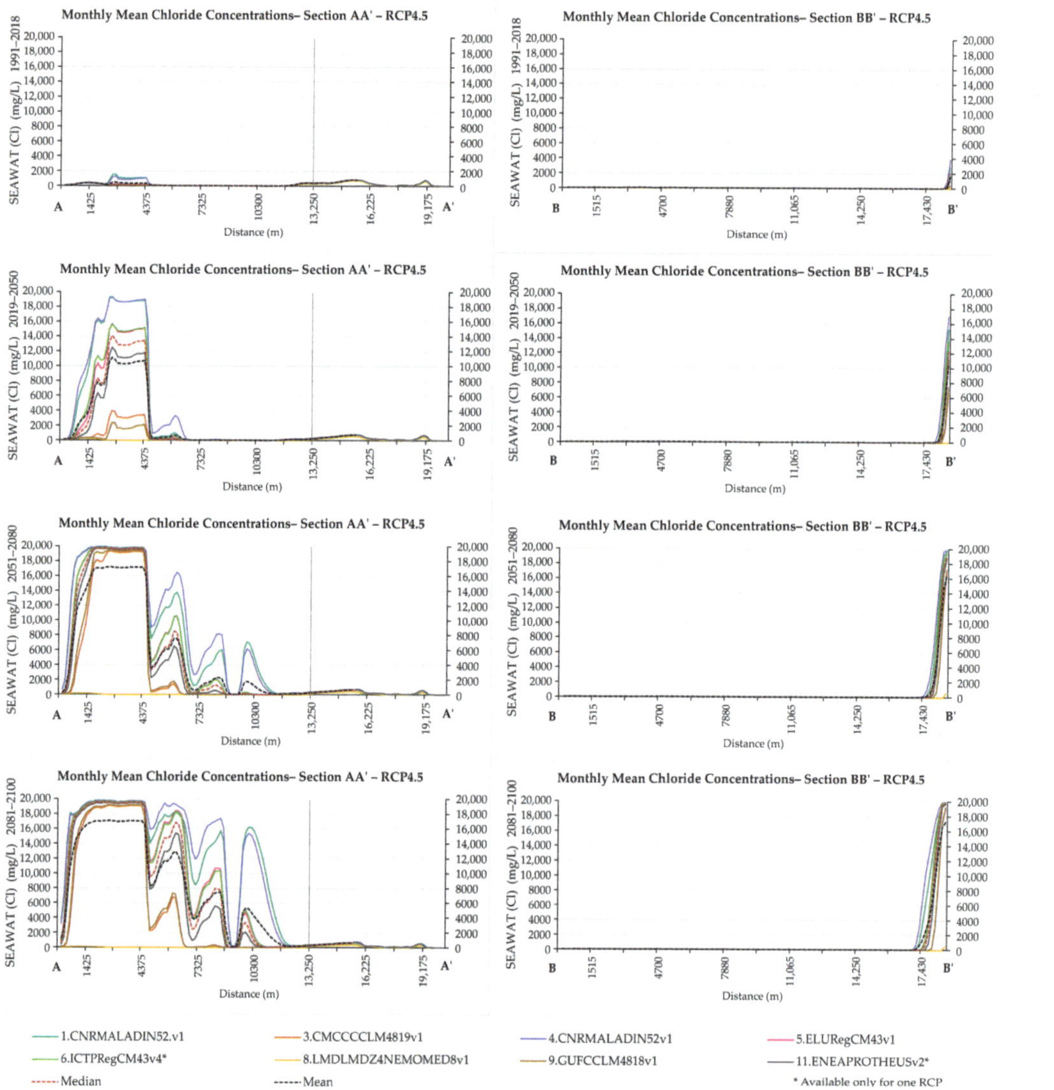

Figure 6. Section AA′ on the left side and Section BB′ on the right illustrate the chloride concentrations in milligrams per liter (mg, L) across the following four time periods: 1991–2018, 2019–2050, 2051–2080, and 2081–2100. These concentrations are derived from the bias-corrected Med-CORDEX models under the RCP4.5 scenario.

Concerning the incorporation of the climatic scenario RCP8.5 in Figure 5, in Section AA′, the highest chloride concentrations are found on the northern coast, when using the bias-corrected climate models 4.CNRMALADIN52.v1 (MED44), 1.CNRMALADIN52v1 (MED11), and 2.ICTPRegCM43v1* in 2019–2050, and also when using 5.ELURegCM43v1 up to 2100. The greater differentiations among the models present in the areas of sandy and clayish materials in the central region of the coastal zone, originated from the same climate models. Chloride concentrations in Section BB′ verify the aforementioned distributions of the effects of climate models and show the shape of the seawater wedge in the aquifer system towards the inland, which advances with time.

In the case of the climate scenario RCP 4.5 in Figure 6, in Section AA′, the highest chloride concentrations are produced on the northern coast, when using the bias-corrected climatic models 4.CNRMALADIN52v1 (MED44) and 1.CNRMALADIN52.v1 (MED11), the values of which are similar to the values of RCP8.5. However, the model 5.ELURegCM43v1 presents lowered chloride concentrations, compared to RCP8.5. The greater differentiations among the models again present in the areas of sandy and clayish materials in the middle of the coastal region, originated from the same climatic models. In Section BB′, the seawater wedge appears similar to the distributions of RCP8.5, with no significant variations in the maximum values of accumulated chlorides from seawater intrusion.

For these reasons, the extent to which seawater intrusion expands inland was calculated by taking the tolerance limit for crops, 100 mg/L, as a threshold value. Table 2 depicts the statistics of the creation of the seawater wedge in the aquifer system of Almyros for RCP8.5 climatic scenario and model, for the time periods from 1991 to 2100.

Table 2. Maximum seawater intrusion and the formation of a saltwater wedge inland, in km, under RCP8.5.

Med-CORDEX Model	1991–2018	2019–2050	2051–2080	2081–2100
1.CNRMALADIN52v1 (MED11)	0.303	0.625	1.106	1.304
2.ICTPRegCM43v1*	0.301	0.612	1.068	1.427
3.CMCCCLM4819.v1	0.113	0.366	0.598	0.635
4.CNRMALADIN52.v1 (MED44)	0.310	0.629	1.215	1.575
5.ELURegCM43v1	0.272	0.474	0.878	1.155
7.ICTPRegCM43.v7*(HadGEM2)	0.133	0.443	0.629	0.782
8.LMDLMDZ4NEMOMED8.v1	—	—	—	0.099
9.GUFCCLM4818.v1	0.124	0.383	0.605	0.673
10.ICTPRegCM43.v7*(MPIESMMR)	0.151	0.428	0.642	0.793
12.UNIBELGRADEEBU.v1*	—	0.268	0.413	0.459
13.UNIBELGRADEEBUPOM2c.v1*	—	0.309	0.469	0.599

* Available only for one RCP.

The table presents data on seawater intrusion and the formation of saltwater wedges inland under the RCP8.5 scenario, categorized by different Med-CORDEX models and time periods. The values in the table represent the distance of intrusion in kilometers for each model during the following four time intervals: 1991–2018, 2019–2050, 2051–2080, and 2081–2100. The values indicate the extent of seawater intrusion, with higher values representing a greater intrusion distance. The maximum seawater intrusion is observed when using the model 4.CNRMALADIN52v1 (MED44) during the last time interval, reaching 1.575 km, while the minimum intrusion is found when using the 8.LMDLMDZ4NEMOMED8.v1 model during the 2081–2100 period, with a distance of 0.099 km. It should be noted that the climatic models 8.LMDLMDZ4NEMOMED8.v1, 12.UNIBELGRADEEBU.v1*, and 13.UNIBELGRADEEBUPOM2c.v1* present a complete lack of seawater intrusion during the historical periods, while under 8.LMDLMDZ4NEMOMED8.v1 model continues to present a complete lack of seawater intrusion until 2080. The steeper increases happen under the climatic models 1.CNRMALADIN52v1 (MED11), 2.ICTPRegCM43v1*, 3.CMCCCLM4819.v1, and 4.CNRMALADIN52.v1 (MED44), while the rest of the models present milder trends. Table 3 shows the statistics of the creation of the seawater wedge in the aquifer system of Almyros for the RCP4.5 climatic scenario and model, for the time periods from 1991 to 2100. The 1.CNRMALADIN52v1 (MED11) model exhibits a minimum intrusion of 0.297 km in the period 1991–2018, while the maximum intrusion increases to 1.450 km in the later period of 2081–2100. The model 8.LMDLMDZ4NEMOMED8.v1 indicates the minimum seawater intrusion of 0.131 km and a maximum intrusion of 0.138 km in the latter two periods. The steepest increases happen for the climate model 4.CNRMALADIN52v1 (MED44), while the rest of the models also show steep trends in the formulation of the seawater wedge, except for 8.LMDLMDZ4NEMOMED8.v1.

Table 3. Maximum seawater intrusion and formation of saltwater wedge inland, in km, under RCP4.5.

Med-CORDEX Model	1991–2018	2019–2050	2051–2080	2081–2100
1.CNRMALADIN52v1 (MED11)	0.297	0.598	1.066	1.450
3.CMCCCLM4819.v1	0.126	0.316	0.615	0.740
4.CNRMALADIN52.v1 (MED44)	0.310	0.689	1.222	1.572
5.ELURegCM43v1	0.277	0.489	0.901	1.211
6.ICTPRegCM43.v4*	0.264	0.571	0.892	1.199
8.LMDLMDZ4NEMOMED8.v1	—	—	0.138	0.131
9.GUFCCLM4818.v1	0.138	0.316	0.615	0.737
11.ENEAPROTHEUS.v2*	0.178	0.456	0.769	0.948

* Available only for one RCP.

3.2.2. Spatial Analysis of Seawater Intrusion under Climate Change

The spatial distributions of chloride concentrations simulated with the integrated modeling system (IMS) and the SEAWAT model are displayed in the following figures. The IMS climatic outputs that were chosen to be presented are the largest and smallest seawater intrusion levels, according to the results presented in Table 2 (RCP8.5) and Table 3 (RCP4.5) for each of the climate scenarios. The simulated results were averaged for their monthly values for the time frames of both the historical period, 1991 to 2018, and the future periods, from 2019 to 2050, 2051 to 2080, and 2081 to 2100. Figures 7–11 display the progression of chloride concentrations in (mg/L) across the following four distinct time periods: 1991–2018, 2019–2050, 2051–2080, and 2081–2100. The progression through the four panels from (a) to (d) visually communicates the simulated chloride concentration over time, indicating areas that might experience significant environmental changes due to seawater intrusion. Each map is segmented into different colored areas, each representing a specific range of chloride concentration, from 0–100 mg/L up to 15,000–20,000 mg/L, highlighting areas of varying salinity levels, and providing a clear visual representation of how chloride levels change over time and space.

For RCP8.5, the simulated data were derived from the bias-corrected 4.CNRMALADIN52.v1 (MED44) model, which yields the maximum seawater intrusion and is illustrated in Figure 7. The average chloride concentrations during the period of 1991–2018 are aggregated mostly in the northern coastal region and the southern coastal strait of the aquifer system (Figure 7a). The values of chloride concentrations range from 100 mg/L to the colored class of 6000 to 10,000 mg/L, with local points exceeding 10,000 mg/L. The progression to the next time period (2019–2050) shows significant increases along the whole coastline (Figure 7b). The aggregation of chlorides falls in the colored class of 15,000 to 20,000 mg/L, and the groundwater at the coastal area is salinized almost up to the concentration of chlorides in seawater. Seawater intrusion appears most intense and fierce in the north, and in the central southern coastal areas. Moving on to the subsequent time frame (2051–2080), the chloride concentrations reach the levels of seawater and take over 1.5 km of the whole coastline and coastal area inland (Figure 7c). Seawater intrusion shows variations in expansion among the northern, the central, and the southern coastal areas. During the final time period (2081–2100), seawater intrusion shows extreme intensification (Figure 7d). The aggregation of chlorides reaches the levels of seawater and expands over 2 km inland, while the course also shifts into the southern part of the aquifer system. For RCP4.5, the simulated data are derived from the bias-corrected 4.CNRMALADIN52.v1 (MED44) model, which yields the maximum seawater intrusion and is illustrated in Figure 8.

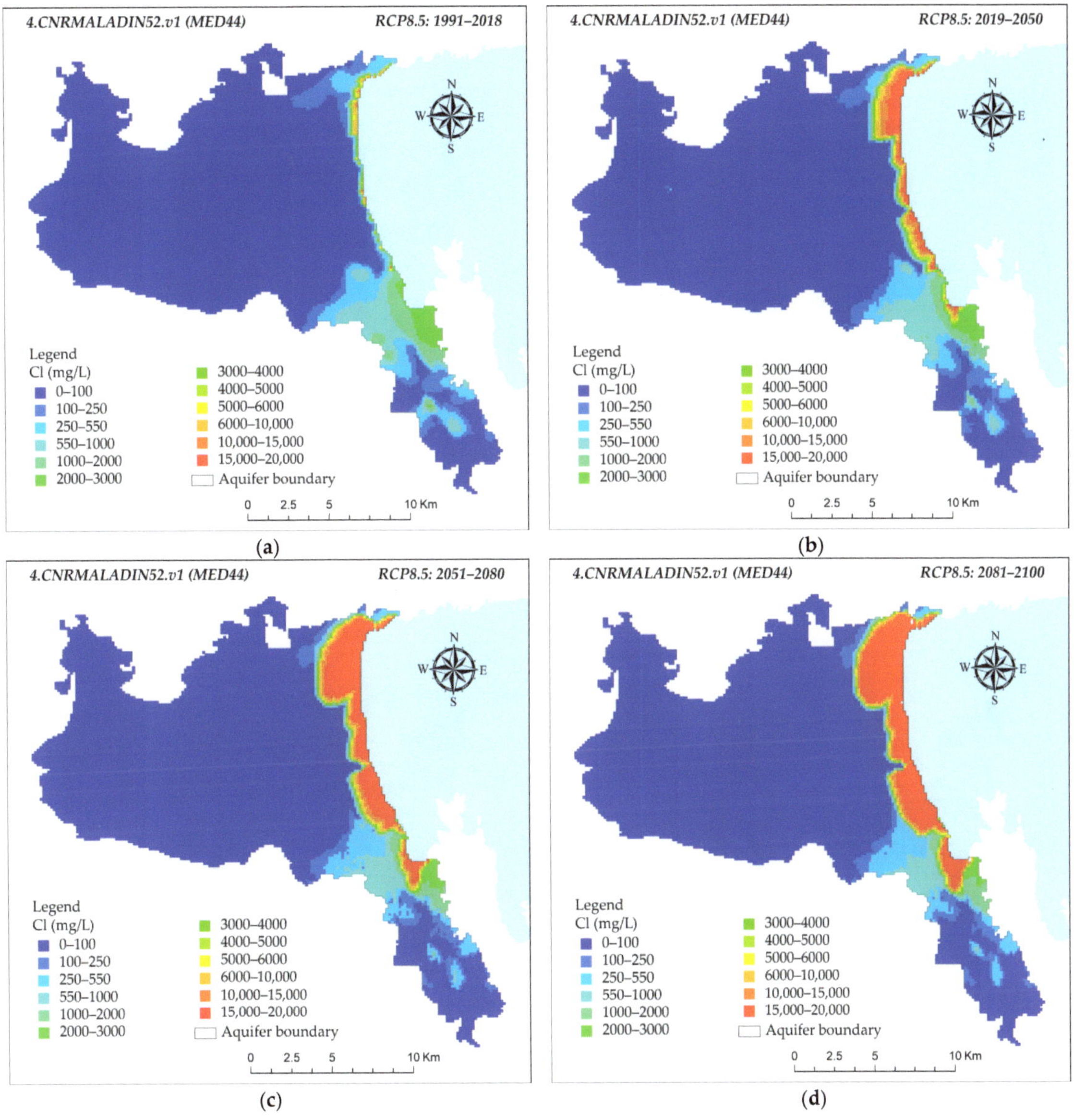

Figure 7. Chloride concentrations in milligrams per liter (mg/L) across the following four time periods: (**a**) 1991–2018, (**b**) 2019–2050, (**c**) 2051–2080, and (**d**) 2081–2100. These concentrations are derived from the bias-corrected 4.CNRMALADIN52.v1 (MED44) model under climate scenario RCP8.5.

The spatial distributions of simulated chlorides, the pace of seawater intrusion intensification, and its progression are evident in the four panels from (a) to (d). As graphically verified through cross-checking, the chloride concentrations closely resemble those obtained under RCP8.5, with negligible differences in inland expansion. For both the RCP8.5 and PCP4.5 scenarios, the simulated chlorides are derived from the bias-corrected 8.LMDLMDZ4NEMOMED8.v1 model, which yields the minimum seawater intrusion and is illustrated in Figure 9. Seawater intrusion is detected for the future periods from 2051 to 2100, according to Tables 2 and 3.

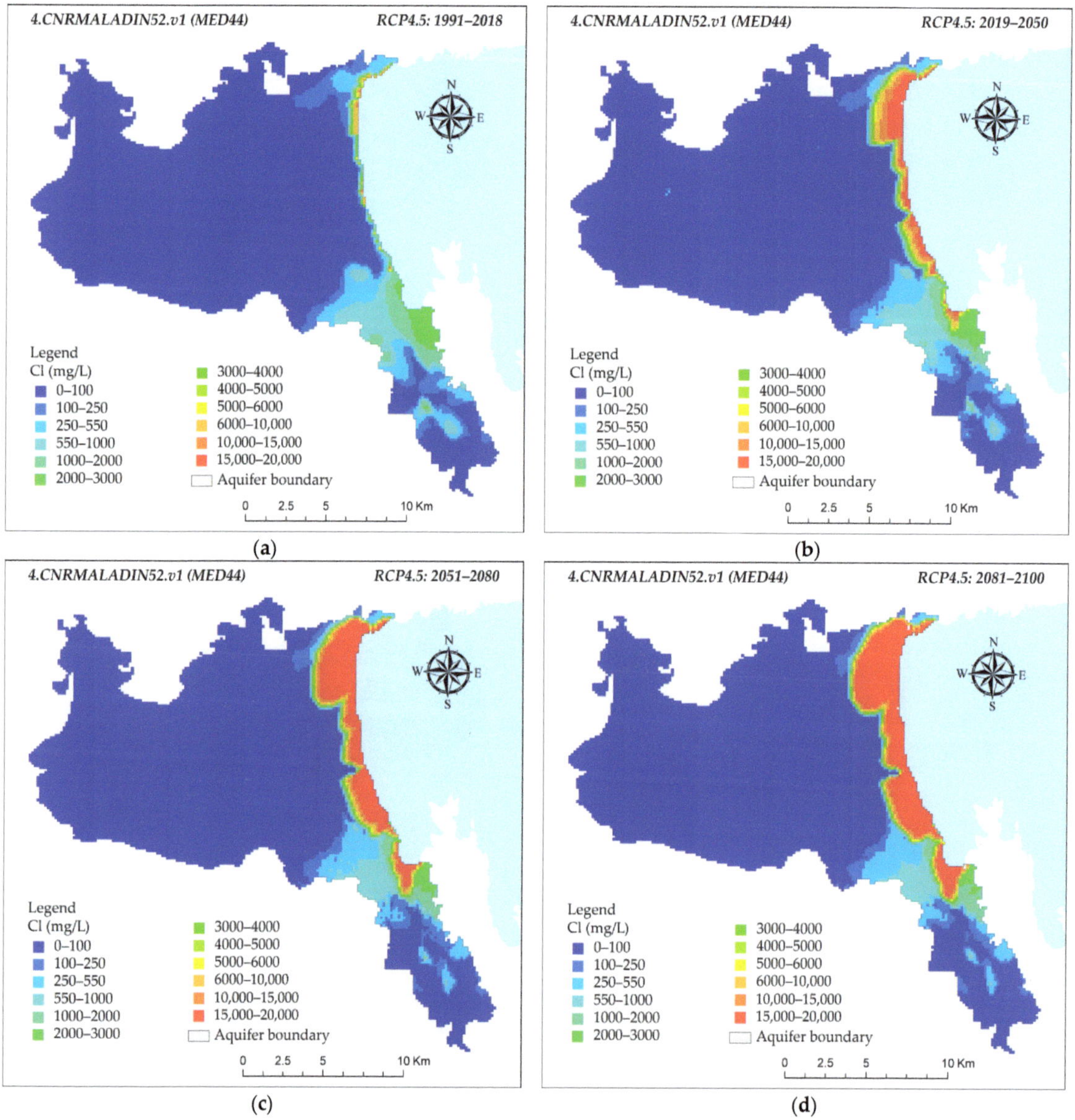

Figure 8. Chloride concentrations in milligrams per liter (mg/L) across the following four time periods: (**a**) 1991–2018, (**b**) 2019–2050, (**c**) 2051–2080, and (**d**) 2081–2100. These concentrations are derived from the bias-corrected 4.CNRMALADIN52.v1 (MED44) model under climate scenario RCP4.5.

The panels of Figure 9a,b illustrate the simulated chloride levels for RCP8.5, in the periods of 2051 to 2080 and 2081 to 2100, respectively. Chloride concentrations aggregate in the northernmost coastal region and in the southern coastal strait of the aquifer. However, the chlorides appear diluted in both time periods under 8.LMDLMDZ4NEMOMED8.v1 model, in the southern coastal strait, reaching mostly the colored class of 550 to 1000 mg/L. Seawater intrusion levels for RCP4.5 present alleviating shifts in the southern coastal strait of the aquifer but increase in the northernmost coastal region. The phenomenon is not intense under the climatic model 8.LMDLMDZ4NEMOMED8.v1, for either of the two scenarios of climate change.

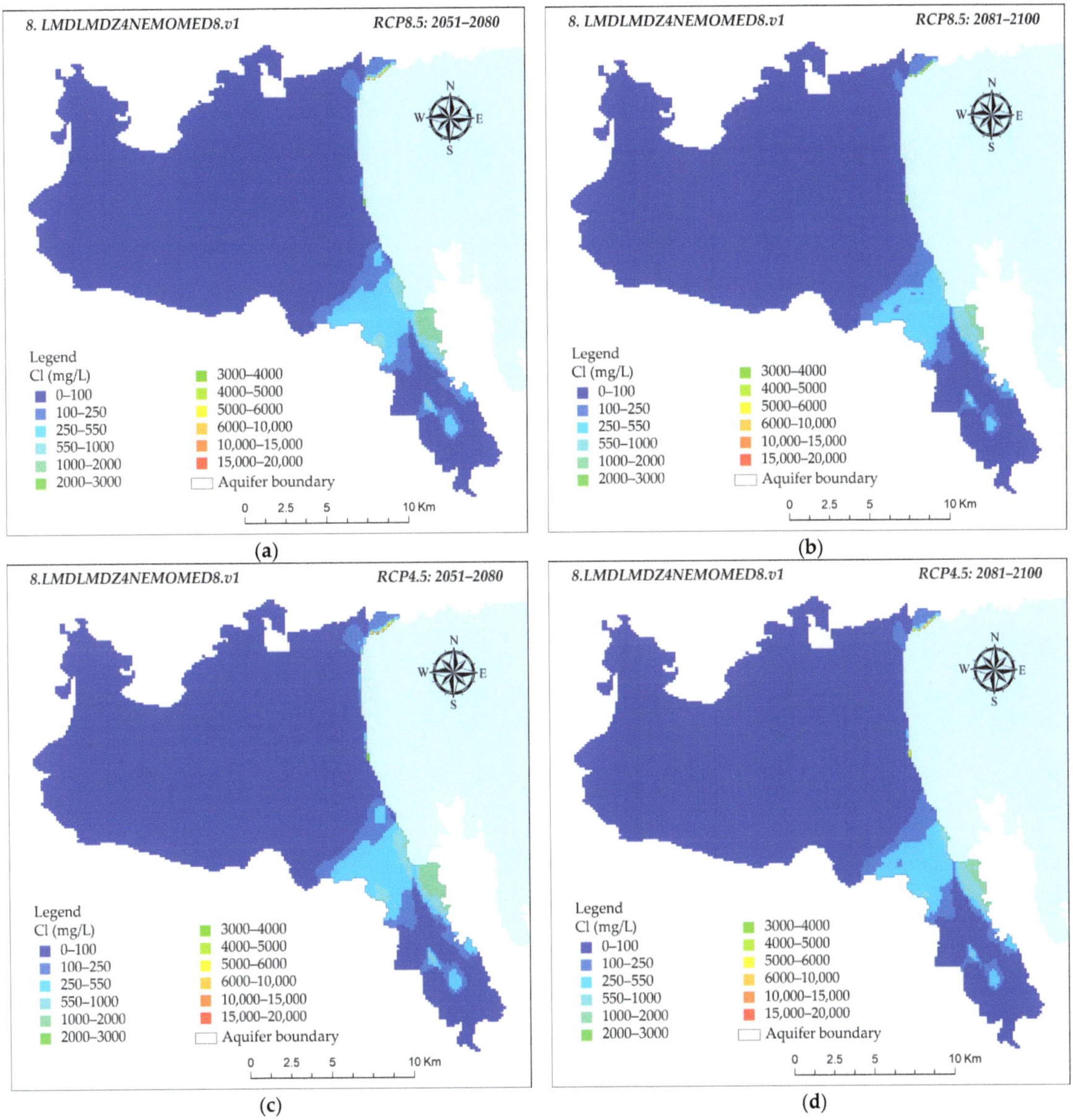

Figure 9. Chloride concentrations in milligrams per liter (mg/L) across the following time periods: (**a**,**c**) 2051–2080 and (**b**,**d**) 2081–2100. These concentrations are derived from the bias-corrected 8.LMDLMDZ4NEMOMED8.v1 model under climate scenarios RCP4.5 and RCP8.5.

For RCP8.5, the simulated data are derived from the bias-corrected 12.UNIBEL-GRADEEBU.v1* model, which yields significant seawater intrusion levels and is illustrated in Figure 10. The average chloride concentrations during the period of 1991–2018 are aggregated mostly in the southern coastal strait of the aquifer system (Figure 10a).

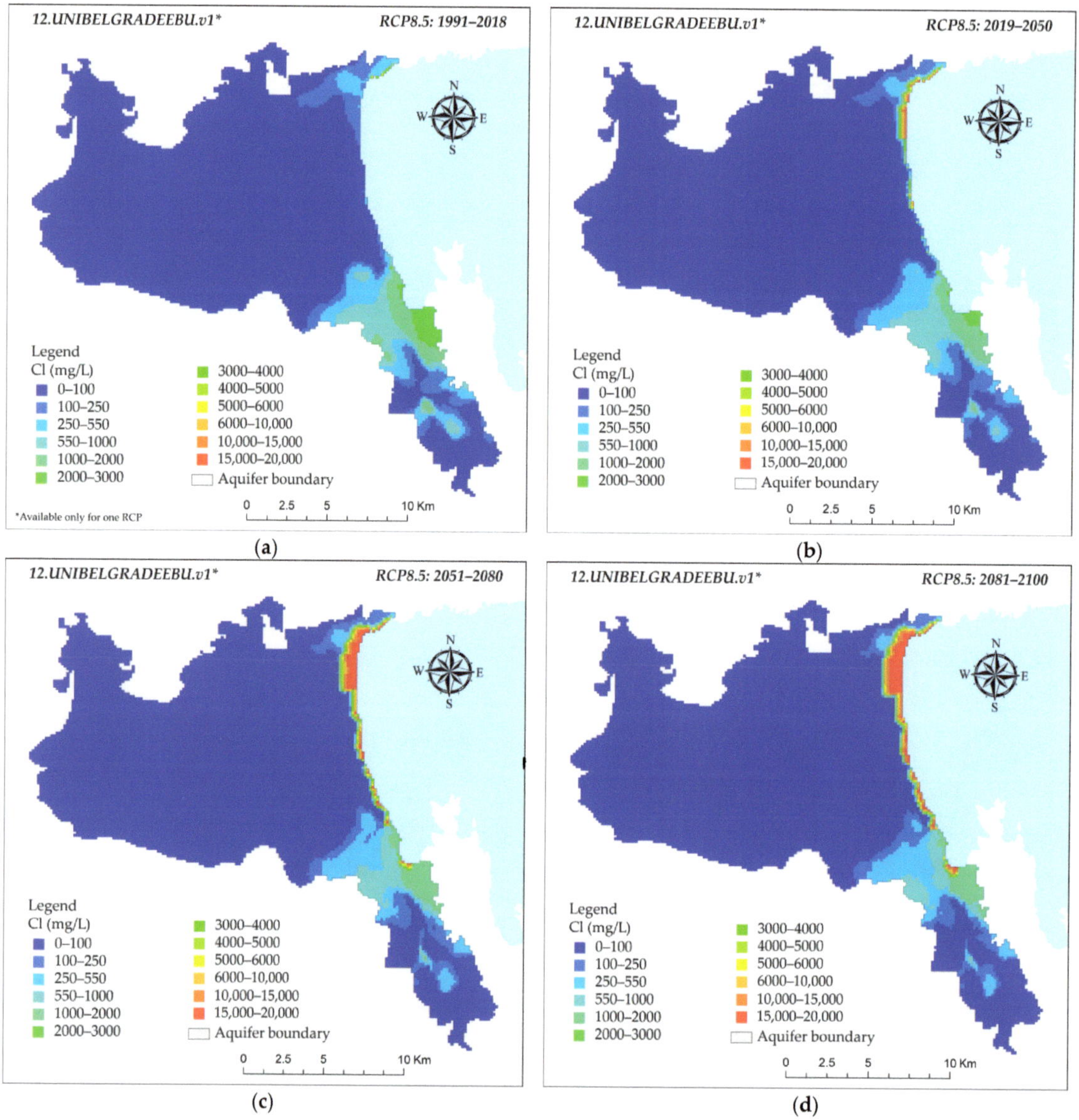

Figure 10. Chloride concentrations in milligrams per liter (mg/L) across the following four time periods: (**a**) 1991–2018, (**b**) 2019–2050, (**c**) 2051–2080, and (**d**) 2081–2100. These concentrations are derived from the bias-corrected 12.UNIBELGRADEEBU.v1* under climate scenario RCP8.5. (*Available only for one RCP).

The values of chloride concentrations range from 550 mg/L to the colored class of 3000 to 4000 mg/L, with local points exceeding 15,000 mg/L. The progression to the next time period (2019–2050) shows significant increases along the northern coastline (Figure 10b). The aggregation of chlorides falls in the colored class of 15,000 to 20,000 mg/L, and the groundwater at the coastal area is salinized almost up to the concentration of chlorides in seawater near the coast. Moving on to the subsequent time frame (2051–2080), the chloride concentrations reach the levels of seawater and take over approximately 1 km of the northern coastal area inland (Figure 10c). The salinized southern coastal strait of the aquifer shows minor shrinkage, although the aggregated chlorides remain at same levels as during 2019 to 2050, while local exceedances of 15,000 mg/L are detected near the coast. During the final time period (2081–2100), seawater intrusion shows notable intensification (Figure 10d). The aggregation of chlorides reaches the chloride levels of seawater and expands over 1.5 km inland. Seawater intrusion expands along the coastline, reaching chloride concentration levels more than 15,000 mg/L, but the inward expansion is quite short, compared to previous models.

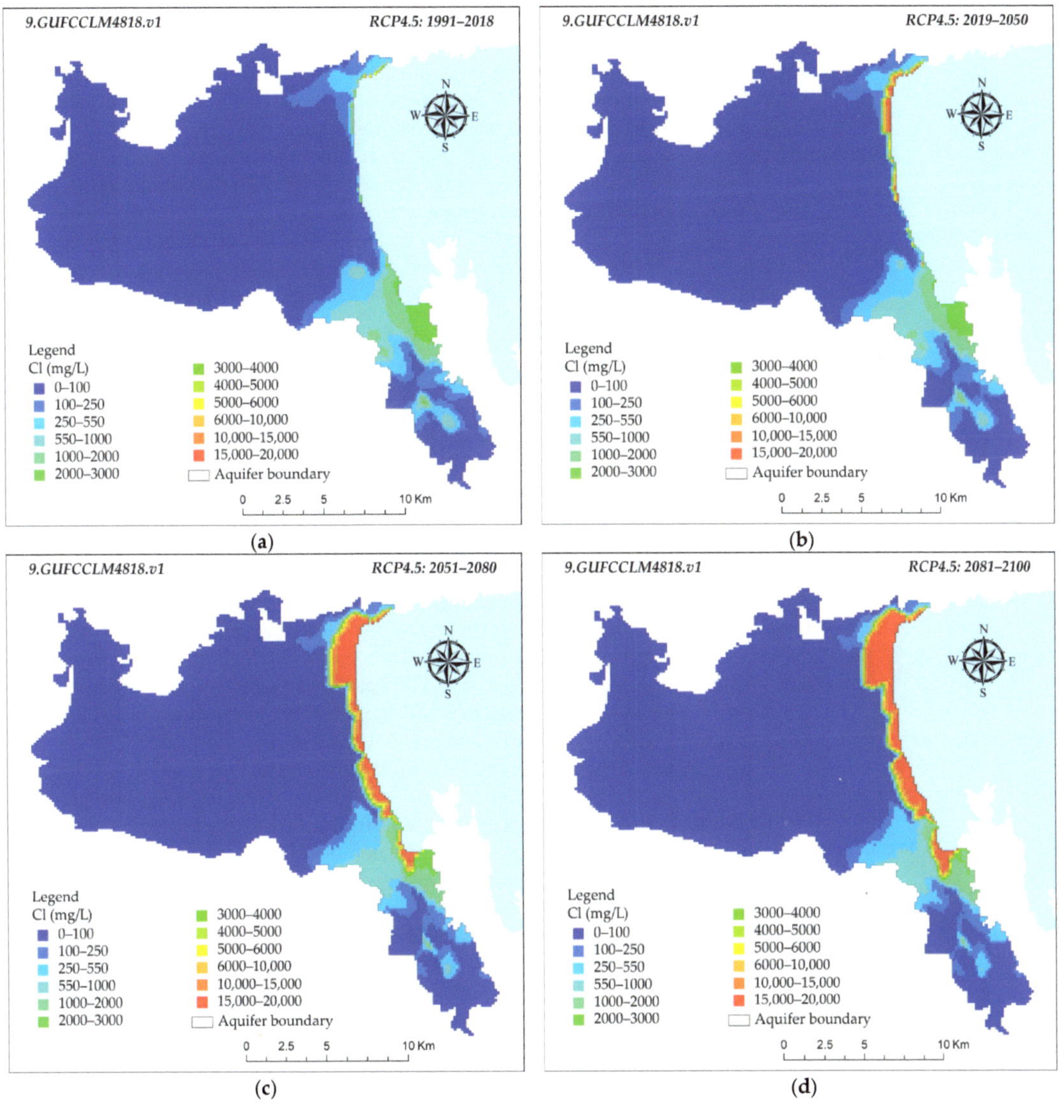

Figure 11. Chloride concentrations in milligrams per liter (mg/L) across the following four time periods: (**a**) 1991–2018, (**b**) 2019–2050, (**c**) 2051–2080, and (**d**) 2081–2100. These concentrations are derived from the bias-corrected 9.GUFCCLM4818.v1 model under climate scenario RCP4.5.

For RCP4.5, the simulated chlorides are derived from the bias-corrected 9.GUFC-CLM4818.v1* model, which yields significant seawater intrusion levels and is illustrated in Figure 11. The average chloride concentrations during the period of 1991–2018 are aggregated mostly in the southern coastal strait of the aquifer system (Figure 11a). The values of chloride concentrations range from 550 mg/L to the colored class of 4000 to 5000 mg/L, while no local increases occur. The progression to the next time period (2019–2050) shows significant increases along the northern coastline and down along the central coast (Figure 11b). Seawater intrusion reaches the chloride concentration of seawater in the north, over 15,000 mg/L, and the inward expansion is limited, compared to the results of the previous model, with more adverse impacts. The chloride concentrations seem relatively stable in the southern coastal strait of the aquifer. Moving on to the subsequent time frame (2051–2080), the seawater intrusion levels show variations in expansion among the northern, central, and southern coastal areas. Seawater intrusion appears most intense and fierce in the northern, central, and southern coastal areas. The chloride concentrations reach the levels of seawater and take over approximately 2 km of the inland northern coastal area, and 1 km of the inland southern coastal area (Figure 11c). The phenomenon appears to quickly exacerbate during this time frame. During the final time period (2081–2100), seawater intrusion shows evident intensification (Figure 11d). The chloride concentrations take over more than 2 km of the northern coastal area, approximately 1 km of the central coastal region, and 2 km of the southern coastal area. Remarkably, an inland shift of seawater is detected in the southern strait of the aquifer. The phenomenon appears exacerbated and adverse during the last time frame of the century.

4. Discussion

This research examined water reserves, particularly focusing on groundwater, and the components of water balance and seawater intrusion for two climatic scenarios, RCP8.5, and RCP4.5, alongside the projected climate data from nineteen (19) climatic models within the Med-CORDEX database. The scopes of this research were as follows: (i) to provide integrated modeling approaches to assess the complex impacts of climate change and sea level rise on groundwater aquifer systems, with a specific focus on seawater intrusion; (ii) to incorporate transient head boundary conditions in the models for a dynamic representation of the aquifer's behavior; (iii) to provide a more realistic and sophisticated representation of the interactions between regional climatic forcings, sea level rise, changes in recharge, groundwater uses, and seawater intrusion; and (iv) to examine the components of the water cycle that affect and cause the exacerbation of seawater intrusion. The assessment process employs the integrated modeling system [37] specifically developed for agricultural coastal watersheds. This system comprises linked models for surface water hydrology [32] and groundwater models of aquifer hydrology [40] and seawater intrusion [41].

The integrated modelling system simulations reveal that the water balance of the aquifer system is affected by seawater intrusion, groundwater recharge, and groundwater abstractions. In both climatic scenarios, aquifer outflows constitute 55% to 85% of the water balance, mostly consisting of groundwater abstractions. Groundwater recharge and irrigation return flow form a distinct relation with seawater intrusion, with high recharge levels causing low seawater intrusion, which drops to almost zero in some climate models.

As the time progresses to the year 2100, the advancement of seawater intrusion in the coastal aquifer is strongly controlled by the combined effects of the rate at which groundwater is replenished by natural sources (recharge), the water pumped out of the aquifer (abstractions), and the water applied to crops (irrigation) that seeps back into the groundwater system (return flow), especially in light of changing climate conditions.

Seawater intrusion poses a substantial challenge, impacting groundwater recharge and irrigation return flow. Different climate models under the RCP8.5 scenario show various trends. However, most models predict that seawater intrusion will rise over time. From 1991–2000, seawater intrusion percentages are relatively low, but as decades progress, they increase. The projected percentages for 2019–2050 and 2051–2080 are expected to rise, while the last timeframe of the century (2081–2100) showcases a continuing upward trend. Most models show a consistent decrease in groundwater abstractions and aquifer discharge, with the highest percentage predicted by the 4.CNRMALADIN52.v1 model. Groundwater recharge and irrigation return flow experience variability during the initial period of RCP4.5 (1991–2000). As time progresses, fluctuations and trends become apparent, with many models showing a decreasing trend in recharge. From 1991–2000 to 2081–2100, significant fluctuations in seawater intrusion rates are observed, with one consistent trend maintaining low percentages. Variability is also observed in the groundwater abstractions and aquifer discharge trends, depending on the increase or decrease rate of precipitation and temperature indicated by the various climatic models. Climate change may impact groundwater recharge and return flow, which are crucial components of the water cycle.

The SEAWAT model simulations for chloride concentrations In the historical period of 1991–2018 show consistent patterns across various climatic scenarios and models. As time progresses towards 2100, the results reveal distinct changes and fluctuations, particularly in the levels of chloride. The highest chloride concentrations are produced at the northern coast using bias-corrected climate models 4.CNRMALADIN52.v1 (MED44) and 1.CNRMALADIN52.v1 (MED11), while the seawater wedge in the aquifer system advances inland over time. The extent to which seawater intrusion expands inland was calculated by taking the tolerance limit for crops, 100 mg/L, as a threshold value. The statistics for the creation of the seawater wedge in the aquifer system of Almyros, categorized by different Med-CORDEX models and time periods, show steeper increases in the formation of the seawater wedge, except for the model 8.LMDLMDZ4NEMOMED8.v1.

The spatial distributions of chloride concentrations, as simulated by the integrated modeling system (IMS) and SEAWAT model, were averaged monthly for both the historical period, of 1991 to 2018, and the future periods, from 2019 to 2100. Comparing the spatial distributions of chloride concentrations under the two climate scenarios, RCP8.5 and RCP4.5, respectively, it is observed that the phenomenon of seawater intrusion appears mostly exacerbated and high-paced, with great expansion across the coastal lowlands. Interestingly, seawater intrusion is more high-paced in RCP8.5 than RCP4.5. In the RCP8.5 scenario, from 1991 to 2018, chloride concentrations are concentrated primarily in the northern coastal region and the southern coastal strait, ranging from 100 mg/L to 6000–10,000 mg/L. From 2019 to 2050, there are substantial increases in chloride concentrations along the entire coastline, peaking at 15,000–20,000 mg/L, with intense seawater intrusion observed in the northern, central, and southern coastal areas. Moving into 2051–2080, chloride concentrations reach seawater levels, extending over 1.5 km of the coastline, and displaying variations in expansion across the northern, central, and southern coastal areas. The period of 2081–2100 witnesses the extreme intensification of seawater intrusion, with chloride concentrations reaching seawater levels and expanding over 2 km inland. In the RCP4.5 scenario, chloride concentration trends closely mirror those observed under RCP8.5, displaying minimal differences in inland expansion. The simulations, derived from the bias-corrected 8.LMDLMDZ4NEMOMED8.v1 model, reveal a scenario of minimal seawater intrusion. Notably, chloride concentrations in the southern coastal strait remain diluted, ranging from 550 to 1000 mg/L for both scenarios, indicating consistent patterns of less severe impacts. From 1991 to 2018, chloride concentrations in the RCP8.5 scenario, modeled by 12.UNIBELGRADEEBU.v1*, aggregate predominantly in the southern coastal strait, ranging from 550 mg/L to the colored class of 3000 to 4000 mg/L. Between 2019 and 2050, there are significant increases along the northern coastline, peaking at 15,000 to 20,000 mg/L, leading to seawater salinization. By 2051–2080, chloride concentrations reach seawater levels, extending over approximately 1 km of the northern coastal area inland. In the final period (2081–2100), there is a notable intensification of seawater intrusion, with chloride concentrations expanding over 1.5 km inland.

The results indicate consistent chloride concentration patterns across RCP8.5 and RCP4.5 scenarios, affirming the models' robustness and the sensitivity of the simulated results to emission scenarios. However, different models (namely 8.LMDLMDZ4NEMOMED8.v1* and

12.UNIBELGRADEEBU.v1*) demonstrate varied seawater intrusion levels, underscoring the critical role of model selection. Temporally, seawater intrusion exhibits a progressively worsening trend, particularly in the latter half of the century under RCP8.5, emphasizing the need for climate change mitigation. Spatially, variations in impact intensity, particularly in the northern, central, and southern coastal areas, underscore the localized vulnerability to seawater intrusion. One common limitation of climate models in general is their inherent uncertainty, due to the complexity of the Earth's climate system and the numerous variables that influence it. Nevertheless, the consistent chloride concentration patterns observed provide valuable perspectives, despite the inherent uncertainties associated with climate modeling. The range of differences between simulations (for different climate models) is a key tool for understanding climate change dynamics under specific emission scenarios, highlights the challenges associated with climate modeling, and underscores the need for continuous improvement and validation processes to enhance the credibility of climate projections. Climate models are constantly being updated and refined as new data become available, so it is important to view their predictions as projections rather than definitive outcomes.

5. Conclusions

The present research highlights the significant impact of climate change on groundwater coastal aquifer systems, particularly in relation to seawater intrusion. The integrated modeling system was employed to simulate the ways in which climate change may affect water resources and seawater intrusion in the Almyros basin and its groundwater system. The incorporation of transient head boundary conditions allows for a dynamic representation of aquifer behavior, offering a more accurate depiction of the ever-changing coastal environment under climate change. The findings emphasize the crucial role of groundwater recharge and irrigation return flow in influencing the evolution of seawater intrusion under different climatic scenarios, with varying trends and uncertainty projected for the future. Sea level rise and climate change have a substantial impact on vulnerable coastal aquifers, and computational models are essential for evaluating how these factors alter the processes of seawater intrusion. Seawater intrusion was analyzed under two climate scenarios, RCP8.5 and RCP4.5, using 19 bias corrected climatic models. In RCP8.5, the model 4.CNRMALADIN52v1 (MED44) has the highest intrusion at 1.575 km, while the model 8.LMDLMDZ4NEMOMED8.v1 has the lowest at 0.099 km during the 2081–2100 period. Some models show no seawater intrusion, either historically or until 2080. In RCP4.5, the model 1.CNRMALADIN52v1 (MED11) has the highest intrusion at 1.450 km in the later period, while the model 8.LMDLMDZ4NEMOMED8.v1 has the lowest, at 0.131 km. Most models exhibit increasing trends in seawater intrusion, occurring on the northern coast and expanding to other coastal areas during the period from 1991 to 2100. The variations among the model results highlight the following: (i) the significance of model selection, (ii) the inherent uncertainties of climate modelling, (iii) the necessity for climate change mitigation actions, and (iv) the vulnerability of coastal aquifers to seawater intrusion. It is important to consider multiple models and scenarios, due to chloride concentration trends, the intensification of seawater intrusion in the second half of the century, and uncertainties in predicting the exact extent of seawater intrusion under climate change conditions.

This information is crucial for policymakers seeking to mitigate the impacts of sea level rise on coastal communities. Thus, action and measures should be taken to address the exacerbation of seawater intrusion now and in the future, such as the following: (i) groundwater management and monitoring, (ii) the use of managed aquifer recharge (MAR), (iii) land-use/crop cultivation planning, (iv) the application of agricultural best management practices (BMPs) for irrigation (e.g., deficit irrigation), (v) the application of climate change adaptation measures, (vi) the development of surface water storage (e.g., reservoirs) and/or the use of marginal water sources (e.g., treated wastewater and brackish water), (vi) the provision of incentives to adopt water-efficient technologies and practices, (vii) community and stakeholder involvement, (viii) emergency response plans to address sudden increases in salinity or other indicators of seawater intrusion, ensuring swift and effective actions, and, last but not least, (ix) research and modelling to account for the propagated uncertainties of climate change to the modelling simulations. Several of the aforementioned measures and actions have been simulated and showcased in the Ph.D. thesis authored by Lyra [38]. By integrating these measures into a comprehensive and adaptive management approach, coastal regions can better protect their freshwater resources from the threats of seawater intrusion.

Author Contributions: Conceptualization, methodology, writing—original draft, supervision, and writing—review and editing: A.L. (Athanasios Loukas); conceptualization, methodology, software, writing—original draft, writing—review, formal analysis, investigation, and data curation: A.L. (Aikaterini Lyra); writing—original draft preparation, formal analysis, writing—review and editing, and data curation: P.S.; writing—original draft and writing—review and editing: N.M. All authors have read and agreed to the published version of the manuscript.

Funding: This research was co-financed by Greece and the European Union (European Social Fund—ESF) through the Operational Programme ("Human Resources Development, Education and Lifelong Learning" in the context of the project "Strengthening Human Resources Research Potential via Doctorate Research" (MIS-5000432), implemented by the State Scholarships Foundation (IKY).

Data Availability Statement: The article encompasses the original contributions made in the study; for additional inquiries, please contact the corresponding authors.

Acknowledgments: The simulations employed in this study were obtained from the Med-CORDEX repository (www.medcordex.eu).

Conflicts of Interest: The authors declare no conflicts of interest. The funders had no involvement in the study's design, data collection, analysis, interpretation, article writing, or publication decision.

Appendix A

Table A1. Statistics of efficiency of bias correction of climate models for the historical period of 1970 to 2000 [32].

Climate	Model	Prec–NSE	Prec–VE	Temp–NSE
RCP 8.5	1.CNRMALADIN52v1	0.720	0.610	0.940
	2.ICTPRegCM43v1*	0.730	0.710	0.940
	3.CMCCCLM4819.v1	0.910	0.790	0.930
	4.CNRMALADIN52.v1	0.640	0.570	0.930
	5.ELURegCM43v1	0.880	0.780	0.940
	7.ICTPRegCM43.v7*(HadGEM2)	0.960	0.870	0.940
	8.LMDLMDZ4NEMOMED8.v1	0.730	0.580	0.930
	9.GUFCCLM4818.v1	0.950	0.830	0.040
	10.ICTPRegCM4–3.v7*	0.950	0.870	0.940
	12.UNIBELGRADEEBU.v1*	0.940	0.850	0.940
	13.UNIBELGRADEEBUPOM2c.v1*	0.970	0.900	0.940
RCP 4.5	1.CNRMALADIN52v1	0.750	0.640	0.940
	3.CMCCCLM4819.v1	0.910	0.790	0.940
	4.CNRMALADIN52.v1	0.640	0.570	0.930
	5.ELURegCM43v1	0.880	0.780	0.940
	6.ICTPRegCM43.v4*	0.900	0.810	0.940
	8.LMDLMDZ4NEMOMED8.v1	0.730	0.580	0.930
	9.GUFCCLM4-8-18.v1	0.950	0.830	0.740
	11.ENEAPROTHEUS.v2*	0.920	0.820	0.780

* Available only for one RCP.

References

1. Nemer, Z.; Khaldaoui, F.; Benaissa, Z.; Belaroui, A.; Tebbouche, M.Y.; Ydri, A. Hydrogeophysical investigation of aquifer parameters and seawater intrusion: A case study from Eastern Mitidja plain, Algeria. *Geomech. Geophys. Geo-Energy Geo-Resour.* **2023**, *9*, 60. [CrossRef]
2. EL Hamidi, M.J.; Larabi, A.; Faouzi, M. Numerical Modeling of Saltwater Intrusion in the Rmel-Oulad Ogbane Coastal Aquifer (Larache, Morocco) in the Climate Change and Sea-Level Rise Context (2040). *Water* **2021**, *13*, 2167. [CrossRef]

3. Zhao, J.; Lin, J.; Wu, J.; Wu, J. Impact of climate change on multi-objective management of seawater intrusion in coastal karst aquifers in Zhoushuizi district of Dalian City, China. *Hydrogeol. J.* **2021**, *29*, 2329–2346. [CrossRef]
4. IPCC. *Climate Change 2014—Impacts, Adaptation and Vulnerability: Part B: Regional Aspects: Working Group II Contribution to the IPCC Fifth Assessment Report: Volume 2: Regional Aspects*; Cambridge University Press: Cambridge, UK, 2014; pp. 1787–1820, ISBN 9781107683860. Available online: https://www.ipcc.ch/report/ar5/wg2/ (accessed on 28 January 2024).
5. Bear, J.J.; Cheng, H.D.A. Seawater Intrusion. In *Theory and Applications of Transport in Porous Media*; Bear, J., Cheng, A.H.D., Eds.; Springer: Dordrecht, The Netherlands, 2010; pp. 593–636. [CrossRef]
6. Wang, X.; Geng, X.; Sadat-Noori, M.; Zhang, Y. Groundwater-Seawater Exchange and Environmental Impacts. *Front. Water* **2022**, *4*, 928615. [CrossRef]
7. Hussain, M.S.; Abd-Elhamid, H.F.; Javadi, A.A.; Sherif, M.M. Management of Seawater Intrusion in Coastal Aquifers: A Review. *Water* **2019**, *11*, 2467. [CrossRef]
8. Cao, T.; Han, D.; Song, X. Past, present, and future of global seawater intrusion research: A bibliometric analysis. *J. Hydrol.* **2021**, *603*, 126844. [CrossRef]
9. Werner, A.D.; Bakker, M.; Post, V.E.A.; Vandenbohede, A.; Lu, C.; Ataie-Ashtiani, B.; Simmons, C.T.; Barry, D.A. Seawater intrusion processes, investigation and management: Recent advances and future challenges. *Adv. Water Resour.* **2013**, *51*, 3–26. [CrossRef]
10. Lepuri, S.; Loukas, A.; Lyra, A. Seawater Intrusion Vulnerability Assessment Using the GALDIT and the Modified GALDIT–AHP Methods: Application in the Coastal Almyros Aquifer, Thessaly, Greece. *Environ. Sci. Proc.* **2023**, *25*, 15. [CrossRef]
11. Mastrocicco, M.; Colombani, N. The Issue of Groundwater Salinization in Coastal Areas of the Mediterranean Region: A Review. *Water* **2021**, *13*, 90. [CrossRef]
12. EU. *Report from the Commission: In Accordance with Article 3.7 of the Groundwater Directive 2006/118/EC on the Establishment of Groundwater Threshold Values*; European Commission: Brussels, Belgium, 2010.
13. EU. Directive (EU) 2020/2184 of the European Parliament and of the Council of 16 December 2020 on the quality of water intended for human consumption (recast) (Text with EEA relevance). In *Revised EU Drinking Water Directive (EU) 2020/2184*; Official Journal of the European Union: Luxembourg, 2020. Available online: https://eur-lex.europa.eu/eli/dir/2020/2184/oj (accessed on 28 January 2024).
14. WHO. *Guidelines for Drinking-Water Quality: First Addendum to the Fourth Edition*; World Health Organization: Geneva, Switzerland, 2017.
15. EU. Directive 2000/60/EC of the European Parliament and of the Council establishing a framework for Community action in the field of water policy—European Environment Agency. *Off. J. Eur. Union* **2000**, *L327*, 1–73.
16. EC. Groundwater Directive (GWD) 2006/118/EC. Directive 2006/118/EC of the European Parliament and of the Council of 12 December 2006 on the Protection of Groundwater against Pollution and Deterioration. *Off. J. Eur. Union* **2006**, *L372*, 19–31. Available online: http://eur-lex.europa.eu/LexUriServ/LexUriServ.do?uri=OJ:L:2006:372:0019:0031:EN:PDF (accessed on 28 January 2024).
17. Yang, H.; Sun, H.; Liu, T.; Yang, X.; Yang, F.; Jiao, J. Characterization of seawater intrusion based on machine learning and implications for offshore management under shared socioeconomic paths. *J. Hydrol.* **2023**, *623*, 129862. [CrossRef]
18. Halder, S. Development and Management of Coastal Aquifer System Through Seawater Intrusion Modelling. In *Groundwater Development and Management: Issues and Challenges in South Asia*; Sikdar, P.K., Ed.; Springer International Publishing: Cham, Switzerland, 2019; pp. 209–225. [CrossRef]
19. Abd-Elaty, I.; Zeleňáková, M.; Krajníková, K.; Abd-Elhamid, H.F. Analytical Solution of Saltwater Intrusion in Costal Aquifers Considering Climate Changes and Different Boundary Conditions. *Water* **2021**, *13*, 995. [CrossRef]
20. Akbarpour, S.; Niksokhan, M.H. Investigating effects of climate change, urbanization, and sea level changes on groundwater resources in a coastal aquifer: An integrated assessment. *Environ. Monit. Assess.* **2018**, *190*, 579. [CrossRef] [PubMed]
21. Chun, J.A.; Lim, C.; Kim, D.; Kim, J.S. Assessing Impacts of Climate Change and Sea-Level Rise on Seawater Intrusion in a Coastal Aquifer. *Water* **2018**, *10*, 357. [CrossRef]
22. Menten, G.; Melo, W.; Pinho, J.; Iglesias, I.; Antunes do Carmo, J. Simulation of Saltwater Intrusion in the Minho River Estuary under Sea Level Rise Scenarios. *Water* **2023**, *15*, 2313. [CrossRef]
23. Adams, K.H.; Hamlington, B.; David, C.; Reager, J.; Sawyer, A.; Buzzanga, B.; Fredericks, J. Estimating Saltwater Intrusion Using Remote-Sensing Datasets and Analytical Approaches. Copernicus Meetings. 2023. Available online: https://meetingorganizer.copernicus.org/EGU23/EGU23-3641.html?pdf (accessed on 1 February 2024).
24. Mini, P.; Singh, D.; Sarangi, A. Predictive Simulation of Seawater Intrusion Control Measures in a Coastal Aquifer. *J. Sustain. Dev. Energy Water Environ. Syst.* **2022**, *10*, 1090399. [CrossRef]
25. Abd-Elhamid, H.F.; Abdel-Aal, G.M.; Fahmy, M.; Sherif, M.; Zeleňáková, M.; Abd-Elaty, I. Experimental and Numerical Study to Investigate the Impact of Changing the Boundary Water Levels on Saltwater Intrusion in Coastal Aquifers. *Water* **2022**, *14*, 631. [CrossRef]
26. Cherubini, C.; Sathish, S.; Pastore, N. Dynamics of Coastal Aquifers: Conceptualization and Steady-State Calibration of Multilayer Aquifer System-Southern Coast of Emilia Romagna. *Water* **2023**, *15*, 2384. [CrossRef]
27. Paldor, A.; Frederiks, R.S.; Michael, H.A. Dynamic Steady State in Coastal Aquifers Is Driven by Multi-Scale Cyclical Processes, Controlled by Aquifer Storativity. *Geophys. Res. Lett.* **2022**, *49*, e2022GL098599. [CrossRef]
28. Abdoulhalik, A.; Ahmed, A.A.; Abdelgawad, A.M.; Hamill, G.A. Towards a Correlation between Long-Term Seawater Intrusion Response and Water Level Fluctuations. *Water* **2021**, *13*, 719. [CrossRef]

29. Elshall, A.S.; Arik, A.D.; El-Kadi, A.I.; Pierce, S.; Ye, M.; Burnett, K.M.; Wada, C.A.; Bremer, L.L.; Chun, G. Groundwater sustainability: A review of the interactions between science and policy. *Environ. Res. Lett.* **2020**, *15*, 093004. [CrossRef]
30. Felisa, G.; Panini, G.; Pedrazzoli, P.; Di Federico, V. Combined Management of Groundwater Resources and Water Supply Systems at Basin Scale Under Climate Change. *Water Resour. Manag.* **2022**, *36*, 915–930. [CrossRef]
31. NCESD. *Greece, State of the Environment Report, Summary*; National Center of Environment and Sustainable Development: Athens, Greece, 2018; pp. 70–71, ISBN 978-960-99033-3-2.
32. MEECC. *River Basin Management Plans*; Hellenic Democracy, Government Gazette: Thessaly, Greece, 2014.
33. Metzger, M.J.; Bunce, R.G.H.; Jongman, R.H.G.; Sayre, R.; Trabucco, A.; Zomer, R. A high-resolution bioclimate map of the world: A unifying framework for global biodiversity research and monitoring. *Glob. Ecol. Biogeogr.* **2013**, *22*, 630–638. [CrossRef]
34. Sayre, R.; Dangermond, J.; Frye, C.; Vaughan, R.; Aniello, P.; Breyer, S.P.; Cribbs, D.; Hopkins, D.; Nauman, R.; Derrenbacher, W.; et al. *A New Map of Global Ecological Land Units—An Ecophysiographic Stratification Approach*; Association of American Geographers, U. S. Geological Survey, GEO BON: Washington, DC, USA, 2014. Available online: https://pubs.usgs.gov/publication/70187380 (accessed on 28 January 2024).
35. Lyra, A.; Loukas, A.; Sidiropoulos, P.; Tziatzios, G.; Mylopoulos, N. An Integrated Modeling System for the Evaluation of Water Resources in Coastal Agricultural Watersheds: Application in Almyros Basin, Thessaly, Greece. *Water* **2021**, *13*, 268. [CrossRef]
36. Lyra, A.; Loukas, A.; Sidiropoulos, P.; Vasiliades, L. Climate Change Impacts on Nitrate Leaching and Groundwater Nitrate Dynamics Using a Holistic Approach and Med-CORDEX Climatic Models. *Water* **2024**, *16*, 465. [CrossRef]
37. Lyra, A.; Loukas, A. Simulation and Evaluation of Water Resources Management Scenarios Under Climate Change for Adaptive Management of Coastal Agricultural Watersheds. *Water Resour. Manag.* **2022**, *37*, 2625–2642. [CrossRef]
38. Lyra, A. Simulation and Management of Degraded Water Resources of Coastal Watersheds. Ph.D. Thesis, University of Thessaly, School of Engineering, Department of Civil Engineering, Thessaly, Greece, 2023. [CrossRef]
39. Loukas, A.; Mylopoulos, N.; Vasiliades, L. A Modeling System for the Evaluation of Water Resources Management Strategies in Thessaly, Greece. *Water Resour. Manag.* **2007**, *21*, 1673–1702. [CrossRef]
40. Harbaugh, A.W.; Banta, E.R.; Hill, M.C.; McDonald, M. *Modflow-2000, The U. S. Geological Survey Modular Ground-Water Model-User Guide To Modularization Concepts and the Ground-Water Flow Process*; USGS: Reston, VA, USA, 2000; p. 134.
41. Guo, W.; Langevin, C.D. *User's Guide to SEAWAT; A Computer Program for Simulation of Three-Dimensional Variable-Density Ground-Water Flow*; 06-A7; USGS: Reston, VA, USA, 2002. [CrossRef]
42. US Soil Conservation Service. *National Engineering Handbook, Section 4, Hydrology*; US Soil Conservation Service: Washington, DC, USA, 1972.
43. ESDAC. European Soil Data Centre, Joint Research Centre, European Commission. Available online: https://esdac.jrc.ec.europa.eu (accessed on 21 January 2021).
44. Dastane, N. *Effective Rainfall, FAO Irrigation and Drainage Paper No. 25*; Food and Agriculture Organization of the United Nations: Rome, Italy, 1974.
45. MEE-GSNEW. Ministry of Environment and Energy, Secretariat of Natural Environment and Water. Available online: http://lmt.ypeka.gr/public_view.html (accessed on 28 January 2024).
46. Orlić, M.; Pasarić, M.; Pasarić, Z. Mediterranean Sea-Level Variability in the Second Half of the Twentieth Century: A Bayesian Approach to Closing the Budget. *Pure Appl. Geophys.* **2018**, *175*, 3973–3988. [CrossRef]
47. Tsimplis, M.N.; Marcos, M.; Somot, S. 21st century Mediterranean sea level rise: Steric and atmospheric pressure contributions from a regional model. *Glob. Planet Chang.* **2008**, *63*, 105–111. [CrossRef]
48. E.U. Copernicus-Marine-Service-Information. Sea Level Daily Gridded Data from Satellite Observations for the Mediterranean Sea from 1993 to 2020. Copernicus Climate Change Service (C3S). Available online: https://data.marine.copernicus.eu/product/SEALEVEL_EUR_PHY_L4_MY_008_068/description (accessed on 28 January 2024).
49. EEA; CLS. Trend in Absolute Sea Level in European Seas Based on Satellite Measurements (1992–2013). In Ocean Monitoring and Forecasting (MyOcean). 2017. Available online: https://www.eea.europa.eu/data-and-maps/figures/sea-level-changes-in-europe-october-1992-may-1 (accessed on 28 January 2024).
50. EEA; ICDC. RCP4.5 Total Sea Level Rise Projections. 2017. Available online: https://www.eea.europa.eu/data-and-maps/figures/projected-change-in-sea-level (accessed on 28 January 2024).
51. EEA. Projected Change in Relative Sea Level, 2081–2100. 2019. Available online: https://www.eea.europa.eu/data-and-maps/figures/projected-change-in-relative-sea-level (accessed on 28 January 2024).
52. Weisstein, E. Ternary Diagram. 2014. Available online: https://mathworld.wolfram.com (accessed on 28 January 2024).

hydrology

MDPI

Article

Groundwater Flow Model Calibration Using Variable Density Modeling for Coastal Aquifer Management

Martha Perdikaki *, Efthymios Chrysanthopoulos, Konstantinos Markantonis and Andreas Kallioras

School of Mining and Metallurgical Engineering, National Technical University of Athens, 15772 Athens, Greece; echrysanthopoulos@metal.ntua.gr (E.C.); markantonis@metal.ntua.gr (K.M.); kallioras@metal.ntua.gr (A.K.)
* Correspondence: mperdikaki@metal.ntua.gr

Abstract: The paper investigates the mechanism of seawater intrusion and the performance of free and open-source codes for the simulation of variable density flow problems in coastal aquifers. For this purpose, the research focused on the Marathon Watershed, located in the northeastern tip of Attica, Greece. For the simulation of the groundwater system, MODFLOW, MT3DMS and SEAWAT codes were implemented, while sensitivity analysis and calibration processes were carried out with UCODE. Hydraulic head calibration was performed on the MODFLOW model, and TDS concentration was validated in the SEAWAT model. The calibrated parameters of the MODFLOW model were obtained for the variable density flow simulation with SEAWAT. The MODFLOW and SEAWAT hydraulic head outputs were analyzed and compared to one another. The outcome of this analysis is that SEAWAT produced slightly better results in terms of the hydraulic heads, concluding that parameter transferability can take place between the two models. For the purpose of the seawater intrusion assessment, the use of the SEAWAT code revealed that the aquifer is subjected to passive and passive–active seawater intrusion during wet and dry seasons, respectively. Finally, an irregular shape of a saltwater wedge is developed at a specific area associated with the hydraulic parameters of the aquifer.

Keywords: MODFLOW; SEAWAT; UCODE; seawater intrusion; sensitivity analysis; calibration

Citation: Perdikaki, M.; Chrysanthopoulos, E.; Markantonis, K.; Kallioras, A. Groundwater Flow Model Calibration Using Variable Density Modeling for Coastal Aquifer Management. *Hydrology* **2024**, *11*, 59. https://doi.org/10.3390/hydrology11040059

Academic Editors: Pantelis Sidiropoulos, Augustina Clara Alexander and Peiyue Li

Received: 11 January 2024
Revised: 17 April 2024
Accepted: 18 April 2024
Published: 22 April 2024

1. Introduction

The salinization of coastal aquifers due to seawater intrusion has been investigated for many years. The diversity and complexity of groundwater formations form different mechanisms and patterns of seawater intrusion, complicating the management and mitigation of the phenomenon [1]. The mechanism of seawater intrusion (SWI) and the saltwater wedge has been investigated through several methods. The most common methods that are examined in the literature are geophysical investigation, e.g., refs. [2–4], laboratory experiments, e.g., refs. [5,6], mathematical models, e.g., refs. [7,8] or a combination of the above.

Different models and numerical solutions are utilized for the simulation of seawater intrusion. Dokou and Karatzas [9] used the finite element FEFLOW code [10] to model SWI in a karstic aquifer, while Stoeckl et al. [11] used the same code for investigating the effects of post-pumping in the SWI mechanism. Na et al. [12] employed the TOUGH2 multi-phase transport simulator [13] to model an experimental setup and test the influence of seawater density in a coastal aquifer under confined conditions. The SEAWAT [14] three-dimensional finite difference code by the United States Geological Survey (USGS) is used by many researchers for the investigation of seawater intrusion dynamics, e.g., refs. [15–17]. The simplest approach of the sharp interface with the SWI2 package of MODFLOW [18] is also utilized, while other researchers have applied the MODFLOW and MT3DMS [19] codes to simulate seawater encroachment on the mainland [20].

The sensitivity analysis and calibration process are very important when either simulating a natural system or in cases of experimental setups where observation data are obtained, e.g., refs. [21,22]. In the field of seawater intrusion, the calibration process is much more complicated since the modeler has to deal with several issues, such as the use of concentration data, tidal data or head data that are affected by variable density [23]. Due to the complexity and the high computational requirements of variable density flow codes along with the uncertainty of head measurements in aquifers affected by seawater intrusion [23], many researchers simplify this procedure by utilizing constant density models, e.g., refs. [24–26].

For the purposes of this research, the mechanism of seawater intrusion was investigated through numerical modeling in the case of the Marathon coastal aquifer in Greece. The MODFLOW [18], MT3DMS [19] and SEAWAT codes [14] of the United States Geological Survey (USGS) were employed for this process. The implementation of the two groundwater flow codes (MODFLOW and SEAWAT) served two different purposes: (i) to explore seawater intrusion mechanism in the aquifer and (ii) to assess parameter transferability between MODFLOW and SEAWAT, i.e., the calibration of the MODFLOW parameters using observed head values and afterward, the use of the exact same dataset without calibration in the SEAWAT model. The main outcome of the research is the improvement of the groundwater flow model by comparing the performance of the SEAWAT code against MODFLOW in terms of the hydraulic heads. With respect to the comparison of the results of MODFLOW and SEAWAT in terms of the hydraulic heads, it is to be expected that SEAWAT will perform better after calibration as it entails a variable density factor. However, the parameter transferability, as studied herein, constitutes a research question. Furthermore, the SEAWAT variable density flow model revealed the irregular shape of the saltwater wedge at the eastern part of the plain, affected by the geological heterogeneity in the study area.

2. Materials and Methods

A graphical representation of the methodological framework utilized in the present research is presented in Figure 1. The following sections describe, in detail, each step of the methodology.

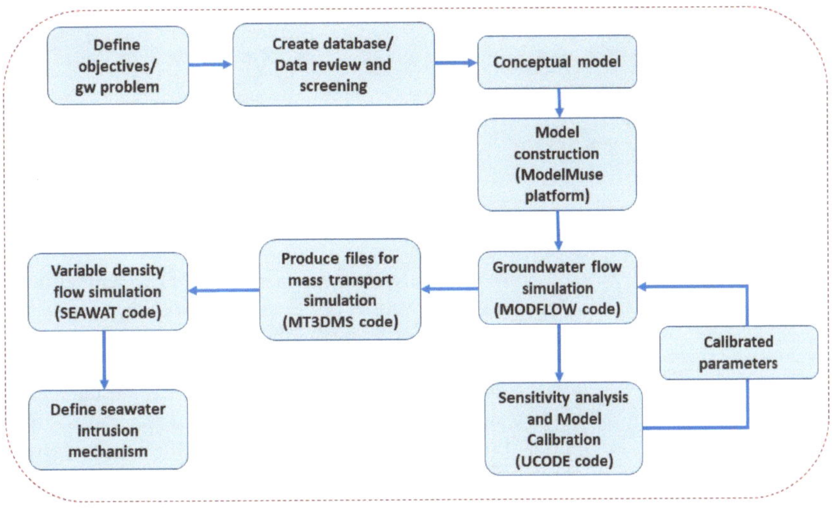

Figure 1. Methodological framework.

2.1. Study Area

The Marathon Plain is located at the NE of Attica (Greece) and is formed within a hilly marble formation at the north, NE and SW, with the Mediterranean Sea at the south (Figure 2). The plain is approximately 40 km^2, and it forms a typical coastal Mediterranean catchment. It is an area of great interest due to historical events (the battle of Marathon in 490 BC) and its coastal ecosystem (Schinias National Park), forming a resort available for several activities at the coastal zone of Attica. The climatic condition is semi-arid, with dry and warm summers and mild winters. According to weather data of the National Observatory of Athens meteorological station in the area (http://penteli.meteo.gr/stations/neamakri/, accessed on 22 January 2024), the average annual temperature is 18.3 °C, while a wide range of temperature changes has been observed during the year from −1.8 °C in January to 43.3 °C in August. The average annual precipitation is around 500 mm, occurring mainly between October and March, with the highest precipitation recorded in December. During the dry season, the monthly rainfall does not exceed 30 mm, while the driest month is August, with a range of 0 up to a few millimeters of precipitation.

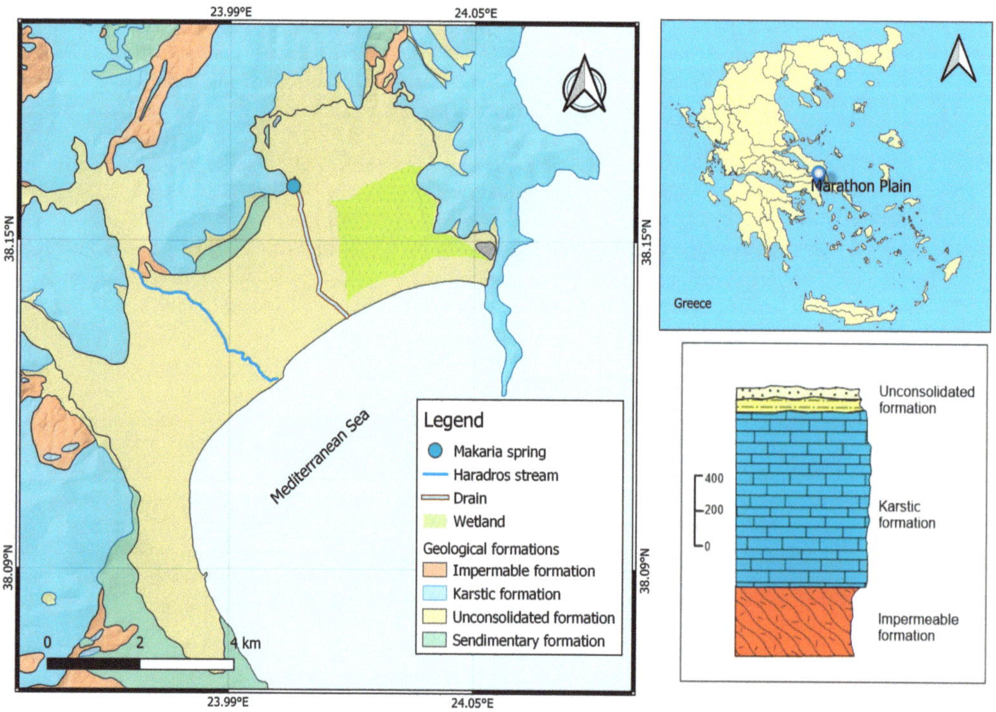

Figure 2. Study area—Marathon hydrosystem, with a hydrogeological cross-section.

The groundwater system consists of two aquifer units: a karstic aquifer developed in marble and a granular aquifer that overlies the marble formation developed in terrestrial, alluvial deposits (Figure 2). The impermeable bedrock of the karstic aquifer is composed of a schist layer.

Along with the aquifer formations, the Marathon hydrosystem is formed by three major hydrologic compounds: the Haradros ephemeral stream, the karstic spring of Makaria that discharges in a drain canal and a naturally occurring wetland at the eastern part of the plain (Figure 2).

The land use in Marathon and the surrounding area is mainly agricultural, domestic and touristic. As a result of these activities, water demand highly varies all year round, with peak demand during summertime. The groundwater in the area is a valuable source of freshwater, especially for the numerous farms and greenhouses that are irrigated entirely from groundwater resources. The uncontrolled abstraction—especially during the dry season—leads to the phenomena of aquifer quantity, quality deterioration and seawater encroachment on the mainland [27,28]. Figure 3 presents the TDS concentration in the study area (October 2018).

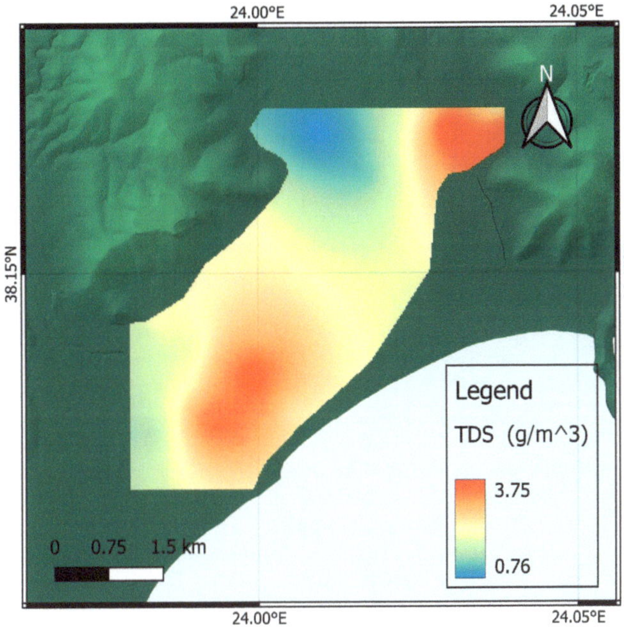

Figure 3. TDS concentration in Marathon Plain.

2.2. Investigation of Seawater Intrusion Pattern and Mechanism through Modeling

2.2.1. Modeling and Calibration Codes

Three different mathematical codes were utilized to investigate the seawater intrusion in the alluvial aquifer of Marathon. Firstly, the finite difference 3D code for groundwater movement in porous media, -MODFLOW-2005 [18], was applied. The partially differential equation that is executed by the code is defined as:

$$\frac{\partial}{\partial x}\left(K_{xx}\frac{\partial h}{\partial x}\right)+\frac{\partial}{\partial y}\left(K_{yy}\frac{\partial h}{\partial y}\right)+\frac{\partial}{\partial z}\left(K_{zz}\frac{\partial h}{\partial z}\right)-W=S_s\frac{\partial h}{\partial t} \quad (1)$$

where K_{xx}, K_{yy} and K_{zz} are values of hydraulic conductivity along the x, y and z coordinate axes (L/T); h is the potentiometric head (L); W is a volumetric flux per unit volume, representing sources and/or sinks of water, with $W < 0.0$ for flow out of the groundwater system and $W > 0.0$ for flow into the system (T − 1); *SS* is the specific storage of the porous material (L − 1); and t is time (T) [18].

Finally, the SEAWAT v.4 code [14] was used in order to analyze groundwater movement under the effect of the variable density flow that is caused by seawater intrusion. The code integrates MODFLOW-2000 [29] and MT3DMS [19] codes simulating groundwater movement and contaminant transport, with modifications to address the problem of variable density flow in the aquifers. Further calibration for the hydraulic properties of the aquifer was not applied. The implementation of the SEAWAT code in this research serves two main purposes: (i) to explore the seawater intrusion mechanism in the several layers of the aquifer and (ii) to examine the performance of the model against the already calibrated MODFLOW model in terms of the hydraulic heads.

The ModelMuse platform of the USGS [30] was used to construct and simulate the model and visualize the model results. The SEAWAT code has not been integrated into ModelMuse yet, thus limiting the ability to simulate variable density flow problems using the finite difference method. Nevertheless, the generated files of both MODFLOW-2005 and MT3DMS can be used with minor modifications and additions so that it is possible to run the SEAWAT code externally (off-platform) [31].

2.2.2. Sensitivity Analysis and Calibration Code

For the purpose of this research, an existing model [26] was used, in which the sensitivity analysis and calibration were performed. The conceptualization of the model was not changed, but the model was updated with new hydraulic head measurements. For calibration purposes, we obtained hydraulic head data on a monthly timestep after a field campaign that lasted one hydrological year (October 2018–September 2019). The hydraulic head values were calculated by water level measurements (obtained with a water level meter) and absolute ground elevations (obtained with Differential GPS). For further information, interested readers are referred to [26]. The sensitivity analysis took place in an automated way using the UCODE_2014 [32] coupled with MODFLOW [33]. The code calculates parameters' sensitivity with fit-independent statistics in order to identify the importance of the observations for each parameter and the importance of a parameter for the simulation, as well as the importance of every observation for the simulation [34]. The parameter sensitivity levels are calculated considering composite scaled sensitivities (CSS) that indicate the available information that observations offer to estimate a specific parameter. In conjunction with CSS, parameter correlation coefficients (PCC) are utilized to calculate the correlation between two parameters. In cases of extreme correlation ($|PCC| > 0.95$), the parameters cannot be calculated independently during the parameter estimation process [34]. In the parameter estimation process, UCODE_2014 creates an objective function that is defined as a sum of the squared and weighted residuals of each observation minus the corresponding simulated value. The group of parameters that produces the minimum of the objective function forms the best prediction for the simulation [34].

2.2.3. Conceptual Model and Boundary Conditions

The conceptualization of the Marathon hydrosystem occurred through an extensive field investigation and review of past scientific reports [35–37]. The research was focused on the eastern part of the plain—at the east of Haradros stream—due to the high chloride concentration in the area, as it was detected in previous field research and hydrochemical analysis [28].

The extent and boundaries of the model are formed within the boundaries of the alluvial deposits of the plain (Figure 2). The thickness of the deposits was determined by the geophysical explorations of Melissaris and Stavropoulos [35]. Based on the data, the alluvial aquifer was divided into three sub-layers due to different grain size distribution with different hydraulic characteristics. The upper and lower layers comprise two different clay and sand–gravel zones, while the middle layer is composed of sand–gravel material. The extent of each zone is shown in Figure 4. The occurrence of different zones determines the hydrodynamic conditions and hydrochemical status of the aquifer. The thickness of each layer varies, although the formation does not exceed 60 m throughout the plain. Since no information is available for the extension of the aquifer offshore, the boundary of the system was defined along the coastline.

The coastal aquifer was simulated under transient conditions for 12 stress periods on a daily timestep. Each stress period represents a month of the hydrologic year, spanning from October 2018 to September 2019. The active area of the model is 19.67 km^2, divided into a grid of 50 m $\times$ 50 m. This results in 7868 active cells per layer.

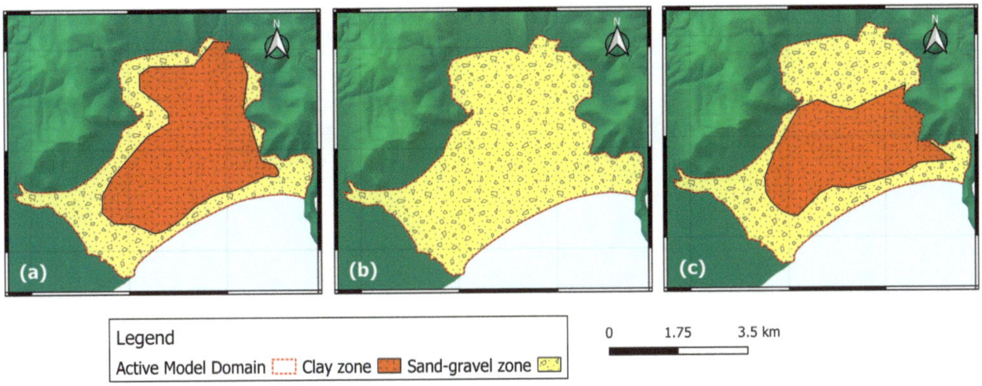

Figure 4. Layer zonation in the (**a**) upper aquifer layer, (**b**) middle aquifer layer and (**c**) lower aquifer layer.

The major hydrologic compounds of the area are represented by different boundary conditions (Figure 2), as listed below and presented in Figure 5:

1. Dirichlet boundary conditions

- Coastline (Time-Variant Specified-Head package—CHD) at 0 m;
- Makaria Spring (Time-Variant Specified-Head package—CHD) at 2 m;
- Coastal Wetland (Time-Variant Specified-Head package—CHD) ranging from −1.2 m to 0.6 m to obtain observations in the vicinity of the wetland.

2. Neuman boundary conditions

- Abstraction wells and boreholes (Well package–WEL). Due to a lack of pumping rate data from the wells and drills, the pumping rate was determined based on data from the study of the Ministry of Agriculture [35] with an annual pumping rate of about 2×10^6 m^3;
- Percolation from precipitation (Recharge package—RCH). The distribution of rainwater percolation throughout the year was calculated through a model by employing different hydrological methods [38]. The model includes the Soil Conservation Service Curve Number (SCS-CN) approach [39], the Penman–Monteith method [40]) for Potential Evapotranspiration (PET), while the Soil Moisture Balance (SMB) [41] method was utilized to estimate the Actual Evapotranspiration and rainfall percolation in the aquifer.

3. Cauchy boundary conditions

- Inflows from the surrounding marble (General Head Boundary–GHB). For the boundary head of the GHB, several hydraulic head measurements were gathered during the field campaign (from October 2028 to September 2019). The hydraulic head of the karstic formation in the area varied from 2.5 m (during the dry season) up to 4 m (during the wet season);
- Drain canal (Drain package—DRN). For the elevation of the drain canal in the area, a Digital Elevation Model (DEM) was utilized.
- Haradros stream (River package—RIV). Since the stream has an ephemeral flow, a river stage of up to 1 m above the river bottom was utilized, according to the sparse observation data that were obtained during the field campaigns.

The Head Observation package (HOB) was also utilized in order to assign the observed values to the model.

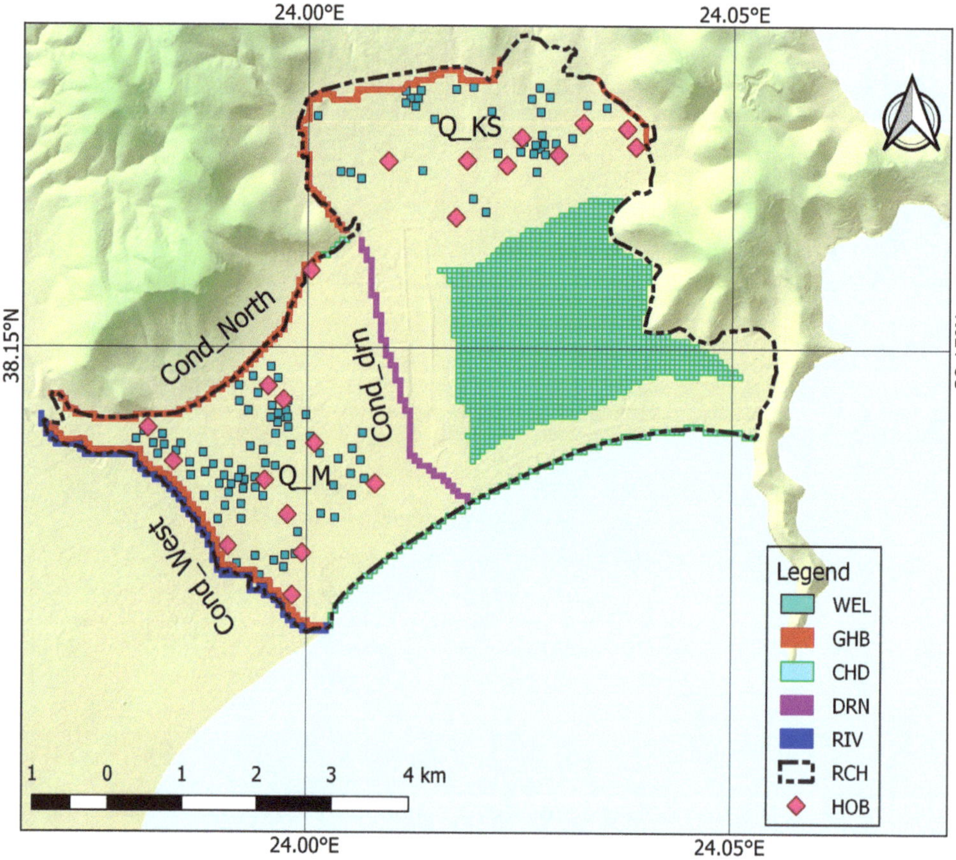

Figure 5. Model boundary conditions, sinks and sources.

To develop the variable density flow model, the MODFLOW and MT3DMS models were employed. The contaminant transport model was integrated into the flow model with common spatial and temporal discretization, enabling the Sink & Source Mixing Package (SSM). For the simulation, a type of pollutant was defined at the coastline as a constant concentration boundary condition representing the Mediterranean Sea. Initial conditions for the contaminant transport model were set from the chemical analysis data in terms of total dissolved solids (TDSs).

2.2.4. Parametrization

An important task during the conceptualization of a flow model is the determination of the type and value of the parameters and their spatial distribution. For the model of Marathon, parametrization was applied in order to define the parameters of the aquifers' different zones and boundary conditions based on the collected data. Furthermore, the use of parameters is required for both the sensitivity analysis and calibration processes. In Table 1, the selected parameters for the aquifer of Marathon, as well as their initial values retrieved from past research [26,35,42], are presented.

Table 1. Model parameters.

Parameter	Parameter Name	Initial Value
Hydraulic conductivity of clay zone	HK1	0.2 (m/day)
Hydraulic conductivity of sand–gravel zone	HK2	3.312 (m/day)
Specific yield for the upper layer	Sy	0.02
Conductance of the northern GHB boundary (marble)	Cond_North	32,210 (m^2/day)
Conductance of the western GHB boundary (adjacent aquifer)	Cond_West	7395.8 (m^2/day)
Conductance of the DRN boundary	Cond_drn	8.64 (m^2/day)
Specific storage of clay zone	SS_Par1	8.24E−3 (m^{-1})
Specific storage of sand–gravel zone	SS_Par2	8.24E−3 (m^{-1})
Pumping rate (WEL package) at the W of the plain	Q_M	−60 (m^3/day)
Pumping rate (WEL package) at the NE of the plain	Q_KS	−170 (m^3/day)

3. Results

3.1. Sensitivity Analysis of MODFLOW Parameters

The results of the sensitivity analysis for the MODFLOW model were analyzed based on the CSS and PCC indices calculated by UCODE2014.

For the alluvial aquifer model, the results of the sensitivity analysis suggest that the model is highly influenced by the parameters related to the characteristics of the middle layer and sand–gravel zones, as well as the pumping parameters. The high values of the CSS index for the parameters HK2, SS_Par2, Q_KS and Q_M, as shown in Figure 6a, are partly due to the effect of pumping that takes place mainly in the middle layer formed entirely of sand–gravel. The specific yield (Sy) of the upper layer is less important for the simulation, while the parameters concerning the clay zones determine the result to an even lesser extent. The parameters concerning the boundary conditions and the model outputs (Cond_North, Cond_West and Cond_drn) have no significant sensitivity. A high correlation is observed between the Kato Souli pumping parameter (Q_KS) and the hydraulic properties of the sand–gravel (HK2, SS_Par2) and a smaller one with the Marathon pumping (Q_M) related to the effect of pumping on the middle layer, as described above and shown in Figure 6b.

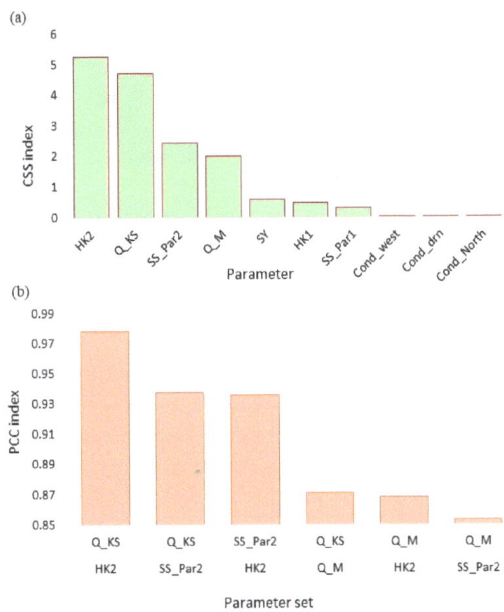

Figure 6. Sensitivity analysis results: (**a**) Composite Scaled Sensitivities—CSS; (**b**) Parameter Correlation Coefficient—PCC.

3.2. MODFLOW Model Calibration and Results

Parameter estimation for the MODFLOW model was performed with respect to the sensitivity analysis results. The parameters selected for the UCODE2014 parameter estimation process concern the aquifer characteristics HK2, SS_Par2, Sy, HK1 and SS_Par1. The parameters Q_M and Q_KS were not included in the model calibration, and the initial values were used for the entire simulation. These parameters were primarily utilized to investigate the impact of withdrawals to the aquifer. The calibrated parameters are presented in the following table (Table 2).

Table 2. Parameter estimation results.

Parameter	Calibrated Value
HK2	12.92 (m/day)
SS_Par2	7.79E−05 (m^{-1})
SY	0.03927
HK1	0.8061 (m/day)
SS_Par1	0.000542 (m^{-1})

The model improved significantly with respect to the measured hydraulic head after the calibration process. Furthermore, the statistics of the differences observed vs. the simulated hydraulic head, shown in Figure 7, present an acceptable agreement between the simulated and observed hydraulic head values.

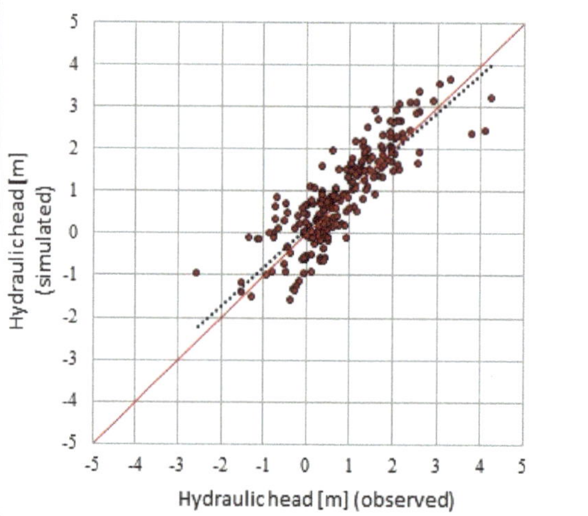

Observation points	216
Mean error difference	-4.27E-02
Standard deviation of error	5.91E-01
Min error difference	1.18E-03
Max error difference	1.69E+00
Absolute mean error difference	4.71E-01

Figure 7. Observed vs. simulated hydraulic head values for the MODFLOW model.

3.3. Parameter Fitting in SEAWAT Model

The parameters obtained from the MODFLOW-2005 calibration were used in the variable density flow model, SEAWAT. The Observation (OBS) Process, as described by Langevin et al. [43], was activated manually to test the fitting of the parameters in the SEAWAT model. The simulation results, in relation to the observed hydraulic head values, are presented in Figure 8. An acceptable agreement between the simulated and observed hydraulic head values was achieved in the SEAWAT model by utilizing the hydraulic parameters that were previously calibrated in the MODFLOW model. In fact, there is a slightly better match between the observed and simulated values with the SEAWAT model, as indicated by the metrics in Figure 8. The use of the SEAWAT model against the

MODFLOW model in terms of the hydraulic head produced a greater match between the simulated and measured values and improved the results of the already calibrated flow model (MODFLOW-2005) without further calibration. The main difference between the two models is reflected in the near-the-shoreline observed vs. simulated values where the root-mean-square error (RMSE) is equal to 0.58 for the MODFLOW model, while for the SEAWAT model, the RMSE decreases to 0.50.

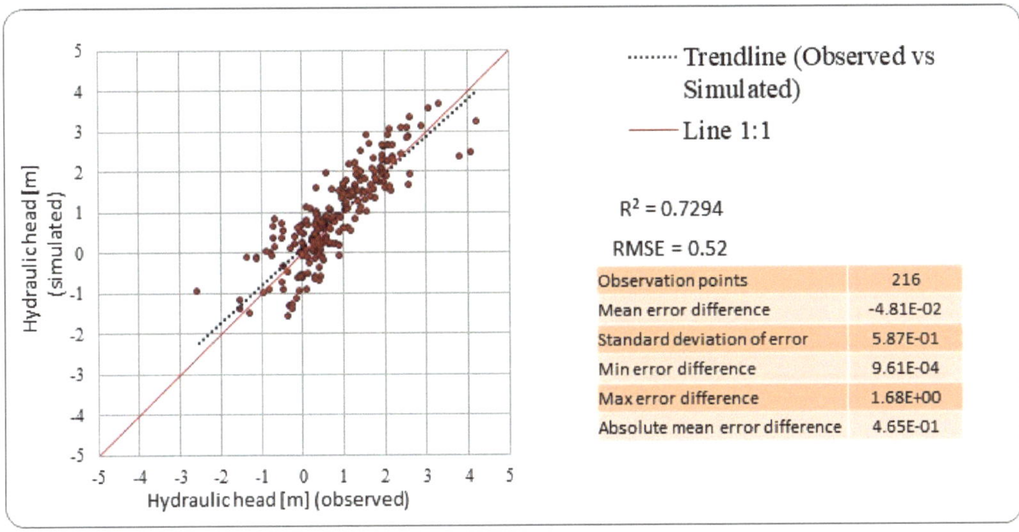

Observation points	216
Mean error difference	-4.81E-02
Standard deviation of error	5.87E-01
Min error difference	9.61E-04
Max error difference	1.68E+00
Absolute mean error difference	4.65E-01

Figure 8. Observed vs. simulated hydraulic head values for the SEAWAT model.

In addition, the simulation of the variable density flow model with the SEAWAT model produced different results in the distribution of the piezometric curves with respect to the MODFLOW model. The maps of iso-piezometric curves for the two models are presented in Figures 9 and 10, for the simulated hydrological year (2018–2019). The main difference between the two models lies in the near-shoreline piezometric contours that are shifted further inland. This shift is due to the effect of the transition zone that cannot be captured by a non-variable density model. The difference between the two simulations is more pronounced during the period of extensive pumping, i.e., between July 2019 and September 2019.

Regarding the transport model parameters, no automatic calibration was performed due to the limited data. However, the manual calibration of longitudinal dispersivity and porosity was applied by assigning different parameter values utilized in the existing research [44–49]. The parameter calibration process revealed that no significant improvement in simulated–observed TDS values occurred. Nevertheless, there was an acceptable agreement between the observed vs. simulated TDS values (R^2 = 0.9).

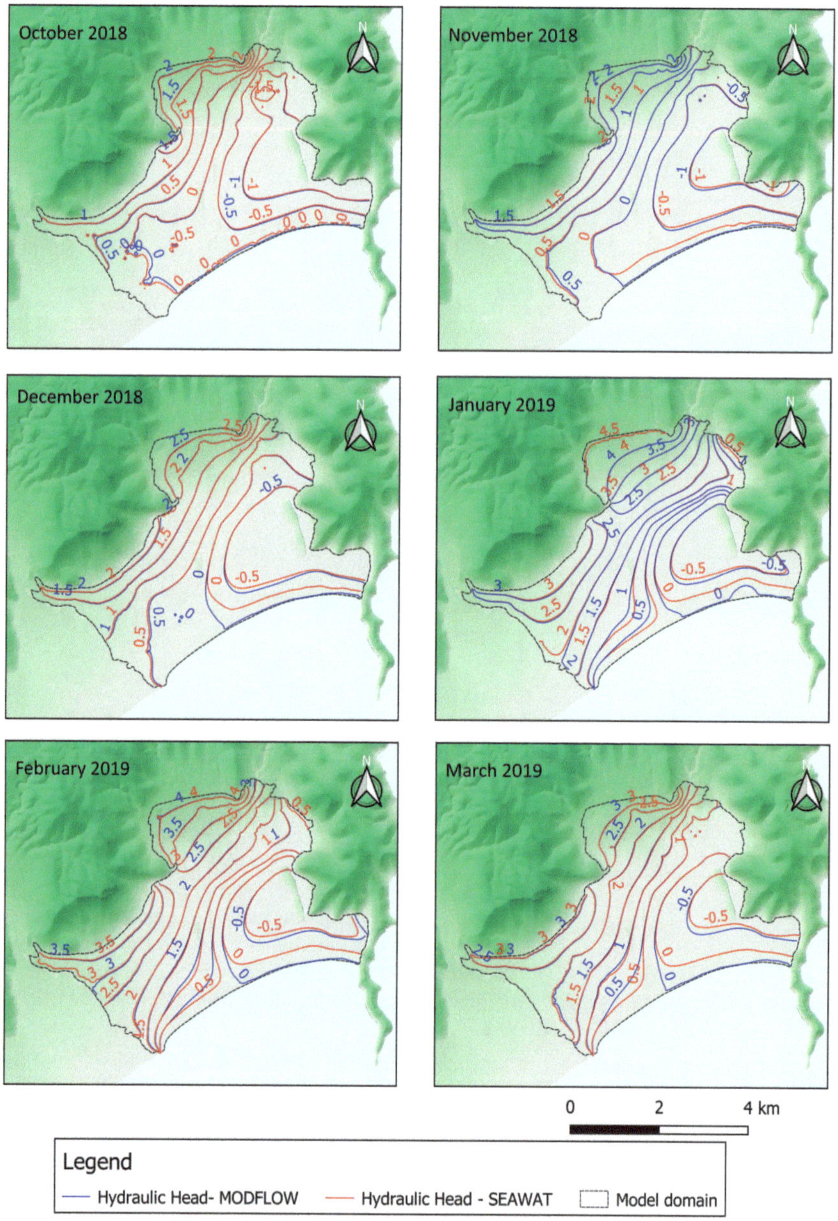

Figure 9. MODFLOW vs. SEAWAT hydraulic head in meters—wet season.

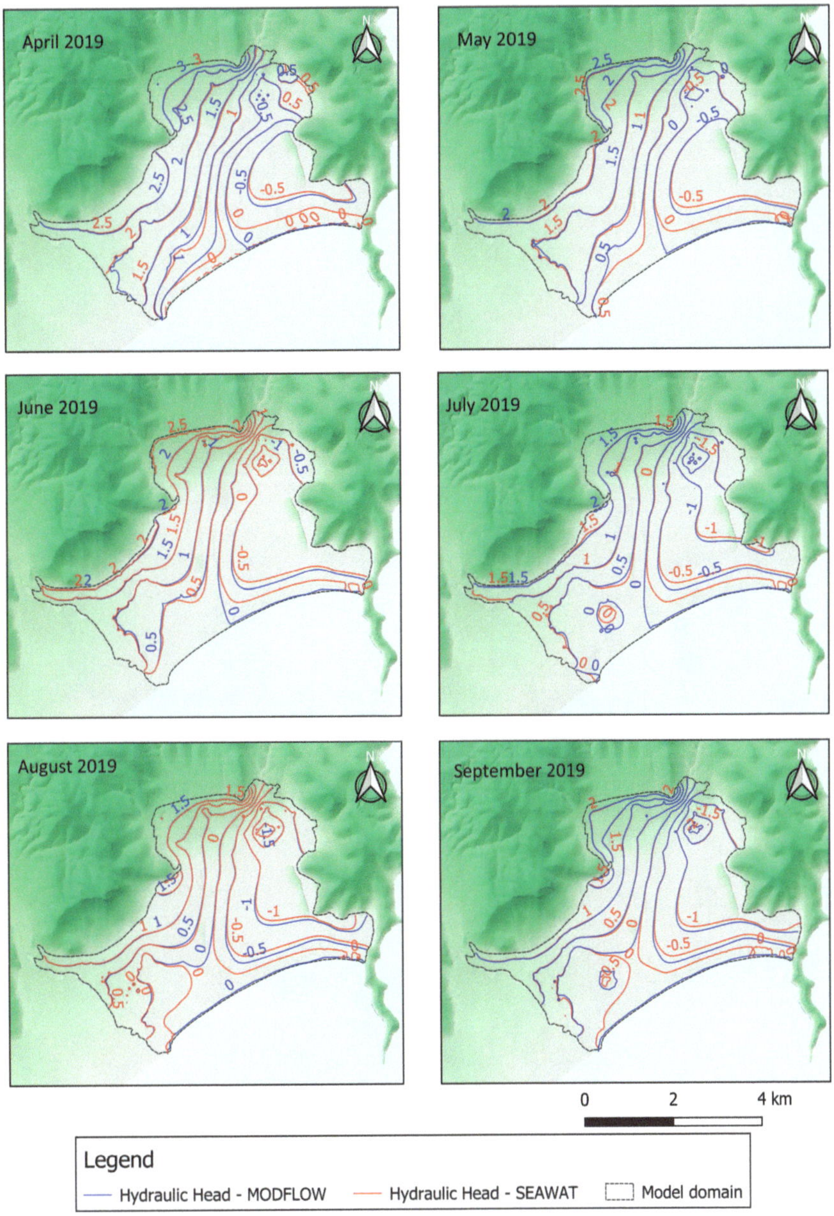

Figure 10. MODFLOW vs. SEAWAT hydraulic head in meters—dry season.

3.4. The Mechanism of Seawater Intrusion

In order to investigate further the mechanism of seawater intrusion in Marathon, the model was simulated for a longer period to achieve the natural mixing between the aquifer concentration and the seawater. The main goal was to investigate and determine the shape and the extent of the saltwater wedge in the coastal granular formation. According to Mabrouk et al. [50], two approaches are commonly used to determine the distribution of salinity in the aquifer: (a) assigning the initial salinity distribution spatially based on existing salinity data or (b) starting the simulation with a fully saline or fully freshwater aquifer and simulating for an extended period until the groundwater system reaches a dynamic equilibrium of a freshwater–seawater interface. Gomaa et al. [51], in their research in a coastal granular aquifer of Egypt, concluded that in the second case, freshwater occupied the development area, and the result did not match the salinity data. In contrast, the results of the first approach were in agreement with the observed values. Based on this conclusion and considering the existing salinity data, the first approach was adopted for the case of Marathon.

A reliable steady-state SEAWAT model was initially developed. To achieve steady-state conditions, the SEAWAT model was run for a "warm-up" period of 3000 timesteps. It must also be noted that the model was also run for a "warm-up" period of 10,000 timesteps, but the results revealed that steady-state conditions for the study area were achieved after 2000 timesteps. To this end, in order to reduce the computational burden, it was chosen to proceed with the modeling exercise using the initial model with the 3000 timesteps. The variable density flow was then simulated for 1827 timesteps under transient flow conditions for 5 hydrological years (2015–2020).

In Figure 11, cross-sections with the distribution of total dissolved solids (TDS) are plotted in the most affected areas of the aquifer as a result of the SEAWAT simulation.

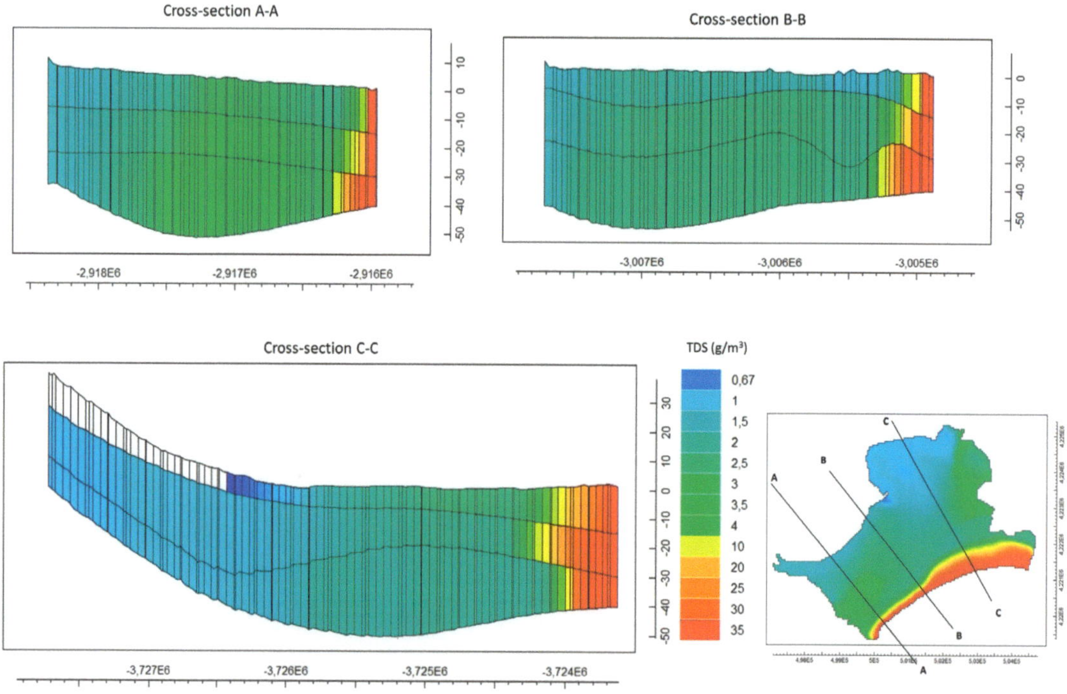

Figure 11. Aquifer TDS concentration in cross-section as a result of SEAWAT model.

In cross-section A-A′, a typical seawater intrusion wedge is formed. At this location, the extent of the freshwater–saltwater interface increases with depth toward inland. In the central part of the plain, the TDS concentration increases with depth up to 4 g/m^3. The main cause of high TDS concentration is attributed to the cone of depression that appears during and after the dry period, a fact that is also validated by the chemical analysis and simulation results. During that season, extensive pumping led to upward movement of the freshwater–saltwater interface. The cone of depression is also validated by the hydraulic head in the area during the period of extensive pumping, i.e., from July to October (Figure 10). The upward movement of the interface is reflected in cross-section A-A as the TDS concentration is higher in the bottom layer (4 g/m^3), and it is gradually decreased in the top layer (2.5 g/m^3). Salinity is reduced toward the north where the karstic marble formation of lower salinity recharges the alluvial aquifer.

In cross-section B-B, the interface is moved further inland across the coastline, while in the northern part, the TDS concertation is lower than in cross-section A-A. The decrease in salinity is due to the seizing of groundwater withdrawals since the specific area is no longer cultivated.

Cross-section C-C presents the TDS concentration in the area of the wetland, where seawater intrusion extends further inland than in the western part of the plain. The extended transition zone is caused by the low topography of the wetland, as also stated by Margoni [37], in conjunction with the negative values of the hydraulic head that lower the hydraulic pressure in the aquifer, leading to extended seawater intrusion. In addition, the aquifer geological heterogeneity near the coastline (Figure 4) could have an effect on the extent and irregular shape of the transition zone.

4. Discussion

During the present research, an important issue arose related to the calibration process of the model since three different codes were implemented for the simulation of the coastal aquifer. Generally, for hydraulic parameter calibration, the MODFLOW model is utilized without taking into account the effects of the variable density flow that the SEAWAT code incorporates. Abd-Elaty and Zelenakova [47], in their research on the Nile Delta aquifer, calibrated the SEAWAT model with the trial-and-error method with respect to the TDS concentration. Ansarifar et al. [25] simulated the effect of sea level fluctuation with a calibrated MODFLOW model. Lyra et al. [24] utilized the MODFLOW model and PEST analysis to simulate and calibrate the hydraulic heads of a coastal aquifer, while solute transport parameters were calibrated through trial and error in the SEAWAT model. Nasiri et al. [52] in their research on seawater intrusion mitigation in coastal aquifers, calibrated hydraulic heads in the MODFLOW model and TDS concentration in the MT3DMS model before simulating the variable density flow with SEAWAT. In our research, the hydraulic heads were calibrated in the MODFLOW model and the TDS concentration in the SEAWAT model. In addition to methods used in the aforementioned studies, the OBS process was activated in the SEAWAT model [43]. This procedure revealed that the hydraulic parameters calibrated in MODFLOW produced better results for the observed vs. simulated head values of the SEAWAT model without further calibration by capturing the hydraulic head differences that the constant density flow model (MODFLOW) could not capture. Furthermore, this research reveals that the distribution of the hydraulic heads with SEAWAT differs from the piezometry produced by the MODFLOW model, especially in the areas of high TDS concentration (near the shoreline and at the wetland), where variable density effects are more pronounced. SEAWAT, the density-variant model, can be compared with MODFLOW, the density-invariant model, in which concentration or temperature changes do not affect the density. The results of the two cases are very similar or identical (Figures 7 and 8) for the upstream part, where it can be observed that density is likely an insignificant part of the system behavior in that part of the two models. However, there is a significant difference between the two cases within the coastal zone, where it is proved that density is an important factor. As Post and Simmons [53] quote, the convective groundwater systems are mixed convection systems where forced convection (flow driven by pressure or hydraulic head gradients) and free convection (flow driven by density variations) co-exist.

As for the solute transport model parameters, in our case study, longitudinal dispersivity was of low importance for the TDS concentration, a result that is in accordance with Shoemaker's research. Further validation of the model results can be carried out by obtaining hydraulic head and flow data near the coastline since effective locations for the calibration of SEAWAT with hydraulic head observations are the toe of the transition zone and areas near the coastline [54]. Unfortunately, during field research in the area, it was not possible to locate wells near the coastline, possibly due to the seizing of groundwater abstractions as a result of seawater intrusion. Furthermore, for the specific case study, the vertical discretization was decided with respect to the different hydrogeological zones of the alluvial aquifer, as indicated in Figure 4. Further vertical discretization would also improve the model. In order to apply vertical discretization and validate the produced results, an advanced research infrastructure (e.g., multilevel piezometers) is essential. As a result, this topic is reserved for future work.

Another outcome of the present research is the investigation of the seawater intrusion mechanism in the alluvial aquifer of Marathon. Considering the distribution and the shape of the hydraulic head in the aquifer (Figures 9 and 10), the results of the simulation reveal that the aquifer is subjected to passive–active seawater intrusion, as analyzed by Werner [55]. In this situation, groundwater flows both inland and toward the sea. The phenomenon is more pronounced in the vicinity of the low-topography wetland, while at the central part of the plain, there is a temporal variability of the phenomenon. During the wet season, the hydraulic gradient is seaward, and a typical case of passive seawater intrusion occurs, while during the period of extensive pumping (July–October), the negative values of the hydraulic head increase the landward movement of seawater.

The extent and shape of the transition zone, as affected by the heterogeneity that characterizes the coastal formations, has been investigated through mathematical models by many researchers (e.g., refs. [56–58]), with the main conclusion that in heterogeneous porous formations, the freshwater–saltwater transition zone intensifies as a result of irregular flow conditions [1]. Across the wetland area, the transition zone does not form the typical saltwater wedge. In the middle layer of the aquifer, a much wider transition zone develops than in the lower layer, as it is affected by the different hydraulic properties of the aquifer's zones. At the bottom layer, the zone of lower hydraulic conductivity (Figure 4c) prevents a faster development of seawater encroachment. The results are in accordance with Michael et al.'s [59] conclusion that geological heterogeneity has such a strong influence on the flow conditions that the salinity distribution and the shape of the freshwater–saltwater interface have an irregular and complex form, with no resemblance to the wedge-shaped interface consideration.

5. Conclusions

The present study involves the evaluation of variable density flow and assessment of the seawater intrusion mechanism in a coastal aquifer through modeling. The study area is a typical coastal Mediterranean hydrosystem located in Marathon, Greece.

The main conclusions derived from the use of MODFLOW and SEAWAT models to simulate groundwater flow are as follows: (i) the sensitivity analysis performed on the MODFLOW model with the UCODE_2014 code indicates that the most sensitive parameters are the hydraulic conductivity (HK) of the main aquifer layer (middle layer of the model), as well as the parameter of pumping (Q), concluding that the parameters related to the particular characteristics of the aquifer and the total water use (withdrawals) in the watershed have the greatest influence on the simulation of the aquifer, (ii) the use of the SEAWAT code to simulate flow under variable density conditions produced a greater match between the simulated and observed values of the hydraulic heads and, in this sense, further improved the results of the already calibrated flow model (MODFLOW-2005) and (iii) the distribution of hydraulic heads in the area produced with the MODFLOW model differ significantly with those produced with the SEAWAT model in the areas of high TDS concentration. Additionally, it appears that longitudinal dispersivity is not sensitive to the simulation. Regarding the "porosity" parameter, it appears that this parameter is more important for the simulation results in the variable density flow model than the "longitudinal dispersivity" parameter. However, the model results seem to be affected at very small porosity values (i.e., 0.05).

Parameter transferability seems to work, as even without calibration, the results of the SEAWAT model indicate a slightly better performance; thus, parameter transferability can take place between the two models. The limitations of this study lie in the lack of observed hydraulic head data near the coastline that would further validate the proposed framework since the most effective observation locations for the calibration of seawater intrusion models are those that are located near the coastline.

Finally, the mechanism of seawater intrusion in the area was assessed through the TDS concentration results produced from the SEAWAT model. The results indicate that the cone of depression at the central part of the plain is responsible for the high TDS concentration throughout the year. Furthermore, in the eastern part of the plain, the heterogeneity that characterizes the coastal formation in terms of the hydraulic parameters creates an irregular shape of the saltwater wedge in the specific area.

Author Contributions: Conceptualization, M.P., K.M. and A.K.; methodology, M.P., K.M. and A.K.; software, M.P., E.C., K.M. and A.K.; validation, M.P., E.C., K.M. and A.K.; formal analysis, M.P.; investigation, M.P., K.M. and A.K.; data curation, M.P.; writing—original draft preparation, M.P. and E.C.; writing—review and editing, M.P., E.C., K.M. and A.K.; visualization, M.P.; supervision, A.K. All authors have read and agreed to the published version of the manuscript.

Funding: This research was partially funded by the Special Account for Research Funding (E.L.K.E.) of the National Technical University of Athens (N.T.U.A.).

Data Availability Statement: The datasets generated and/or analyzed during the current study are available from the corresponding author upon reasonable request.

Acknowledgments: The authors would like to thank all the anonymous reviewers whose comments have greatly improved this manuscript.

Conflicts of Interest: The authors declare no conflicts of interest.

References

1. Jiao, J.; Post, V. *Coastal Hydrogeology*; Cambridge University Press: Cambridge, UK, 2019; Volume 44. ISBN 9781139344142.
2. Apostolopoulos, G.; Pouliaris, C.; Karizonis, S.; Perdikaki, M.; Kallioras, A. Integrated Analysis of the Coastal Aquifer System of Thorikos Valley, Attica, Greece. In Proceedings of the NSG2021 1st Conference on Hydrogeophysics, Bordeaux, France and Online, 29 August–2 September 2021; European Association of Geoscientists & Engineers: Utrecht, The Netherlands, 2021; pp. 1–5.
3. Cosentino, P.; Capizzi, P.; Fiandaca, G.; Martorana, R.; Messina, P.; Pellerito, S. Study And Monitoring Of Salt Water Intrusion In The Coastal Area Between Mazara Del Vallo And Marsala (South-Western Sicily). In *Methods and Tools for Drought Analysis and Management*; Springer: Dordrecht, The Netherlands, 2007; pp. 303–321.
4. Águila, J.F.; McDonnell, M.C.; Flynn, R.; Hamill, G.A.; Ruffell, A.; Benner, E.M.; Etsias, G.; Donohue, S. Characterizing Groundwater Salinity Patterns in a Coastal Sand Aquifer at Magilligan, Northern Ireland, Using Geophysical and Geotechnical Methods. *Environ. Earth Sci.* **2022**, *81*, 231. [CrossRef]
5. Wilson, E.A.; Wells, A.J.; Hewitt, I.J.; Cenedese, C. The Dynamics of a Subglacial Salt Wedge. *J. Fluid Mech.* **2020**, *895*, A20. [CrossRef]
6. Narayanan, D.; Eldho, T.I. Validation of Boundary Layer Solution for Assessing Mixing Zone Dynamics of Saltwater Wedge in a Coastal Aquifer. *J. Hydrol.* **2023**, *617*, 128899. [CrossRef]
7. Torres-Bejarano, F.; González-Martínez, J.; Rodríguez-Pérez, J.; Rodríguez-Cuevas, C.; Mathis, T.J.; Tran, D.K. Characterization of Salt Wedge Intrusion Process in a Geographically Complex Microtidal Deltaic Estuarine System. *J. S. Am. Earth Sci.* **2023**, *131*, 104646. [CrossRef]
8. Ke, S.; Chen, J.; Zheng, X. Influence of the Subsurface Physical Barrier on Nitrate Contamination and Seawater Intrusion in an Unconfined Aquifer. *Environ. Pollut.* **2021**, *284*, 117528. [CrossRef] [PubMed]
9. Dokou, Z.; Karatzas, G.P. Saltwater Intrusion Estimation in a Karstified Coastal System Using Density-Dependent Modelling and Comparison with the Sharp-Interface Approach. *Hydrol. Sci. J.* **2012**, *57*, 985–999. [CrossRef]
10. Diersch, H.-J.G. *FEFLOW*; Springer: Berlin/Heidelberg, Germany, 2014; Volume 9783642387. ISBN 978-3-642-38738-8.
11. Stoeckl, L.; Walther, M.; Morgan, L.K. Physical and Numerical Modelling of Post-Pumping Seawater Intrusion. *Geofluids* **2019**, *2019*, 7191370. [CrossRef]
12. Na, J.; Chi, B.; Zhang, Y.; Li, J.; Jiang, X. Study on the Influence of Seawater Density Variation on Sea Water Intrusion in Confined Coastal Aquifers. *Environ. Earth Sci.* **2019**, *78*, 669. [CrossRef]
13. Xu, T.; Sonnenthal, E.; Spycher, N.; Pruess, K. *TOUGHREACT User's Guide: A Simulation Program for Non-Isothermal Multiphase Reactive Geochemical Transport in Variably Saturated Geologic Media, V1.2.1*; Lawrence Berkeley National Laboratory: Berkeley, CA, USA, 2008.
14. Langevin, C.D.; Thorne, D.T., Jr.; Dausman, A.M.; Sukop, M.C.; Guo, W. *SEAWAT Version 4: A Computer Program for Simulation of Multi-Species Solute and Heat Transport*; U.S. Geological Survey Techniques and Methods Book 6; U.S. Geological Survey: Reston, VA, USA, 2007; 39p.

15. Dunlop, G.; Palanichamy, J.; Kokkat, A.; EJ, J.; Palani, S. Simulation of Saltwater Intrusion into Coastal Aquifer of Nagapattinam in the Lower Cauvery Basin Using SEAWAT. *Groundw. Sustain. Dev.* **2019**, *8*, 294–301. [CrossRef]
16. Abd-Elhamid, H.F.; Abd-Elaty, I.; Sherif, M.M. Effects of Aquifer Bed Slope and Sea Level on Saltwater Intrusion in Coastal Aquifers. *Hydrology* **2019**, *7*, 5. [CrossRef]
17. Gopinath, S.; Srinivasamoorthy, K.; Saravanan, K.; Prakash, R.; Karunanidhi, D. Characterizing Groundwater Quality and Seawater Intrusion in Coastal Aquifers of Nagapattinam and Karaikal, South India Using Hydrogeochemistry and Modeling Techniques. *Hum. Ecol. Risk Assess. Int. J.* **2019**, *25*, 314–334. [CrossRef]
18. Harbaugh, A.W. *MODFLOW-2005, The U.S. Geological Survey Modular Ground-Water Model—The Ground-Water Flow Process*; U.S. Geological Survey Techniques and Methods; U.S. Geological Survey: Reston, VA, USA, 2005; p. 253. ISBN 6-A16.
19. Zheng, C.; Wang, P. *MT3DMS: A Modular Three-Dimensional Multispecies Transport Model for Simulation of Advection, Dispersion, and Chemical Reactions of Contaminants in Groundwater Systems Documentation and User 's Guide*; US Army Engineer Research and Development Center: Vicksburg, MS, USA, 1999.
20. Akhtar, J.; Sana, A.; Tauseef, S.M.; Tanaka, H. Numerical Modeling of Seawater Intrusion in Wadi Al-Jizi Coastal Aquifer in the Sultanate of Oman. *Hydrology* **2022**, *9*, 211. [CrossRef]
21. Chang, S.W.; Clement, T.P. Experimental and Numerical Investigation of Saltwater Intrusion Dynamics in Flux-controlled Groundwater Systems. *Water Resour. Res.* **2012**, *48*, W09527. [CrossRef]
22. Khadim, F.K.; Dokou, Z.; Lazin, R.; Moges, S.; Bagtzoglou, A.C.; Anagnostou, E. Groundwater Modeling in Data Scarce Aquifers: The Case of Gilgel-Abay, Upper Blue Nile, Ethiopia. *J. Hydrol.* **2020**, *590*, 125214. [CrossRef]
23. Carrera, J.; Hidalgo, J.J.; Slooten, L.J.; Vázquez-Suñé, E. Computational and Conceptual Issues in the Calibration of Seawater Intrusion Models. *Hydrogeol. J.* **2010**, *18*, 131–145. [CrossRef]
24. Lyra, A.; Loukas, A.; Sidiropoulos, P.; Tziatzios, G.; Mylopoulos, N. An Integrated Modeling System for the Evaluation of Water Resources in Coastal Agricultural Watersheds: Application in Almyros Basin, Thessaly, Greece. *Water* **2021**, *13*, 268. [CrossRef]
25. Ansarifar, M.-M.; Salarijazi, M.; Ghorbani, K.; Kaboli, A.-R. Simulation of Groundwater Level in a Coastal Aquifer. *Mar. Georesources Geotechnol.* **2020**, *38*, 257–265. [CrossRef]
26. Perdikaki, M.; Makropoulos, C.; Kallioras, A. Participatory Groundwater Modeling for Managed Aquifer Recharge as a Tool for Water Resources Management of a Coastal Aquifer in Greece. *Hydrogeol. J.* **2022**, *30*, 37–58. [CrossRef]
27. Psychoyou, M.; Mimides, T.; Rizos, S.; Sgoubopoulou, A. Groundwater Hydrochemistry at Balkan Coastal Plains—The Case of Marathon of Attica, Greece. *Desalination* **2007**, *213*, 230–237. [CrossRef]
28. Perdikaki, M.; Manjarrez, R.C.; Pouliaris, C.; Rossetto, R.; Kallioras, A. Free and Open-Source GIS-Integrated Hydrogeological Analysis Tool: An Application for Coastal Aquifer Systems. *Environ. Earth Sci.* **2020**, *79*, 16. [CrossRef]
29. Harbaugh, B.A.W.; Banta, E.R.; Hill, M.C.; Mcdonald, M.G. *MODFLOW-2000, The U.S. Geological Survey Modular Graound-Water Model—User Guide to Modularization Concepts and the Ground-Water Flow Process*; U.S. Geological Survey: Reston, VA, USA, 2000; p. 130.
30. Winston, R.B. *ModelMuse: A Graphical User Interface for MODFLOW-2005 and PHAST*; U.S. Geological Survey Techniques and Methods 6–A29; U.S. Geological Survey: Reston, VA, USA, 2009; p. 52.
31. Perdikaki, M. *Modelling and Optimized Management in Coastal Aquifer Systems*; National Technical University of Athens: Athens, Greece, 2022.
32. Poeter, E.P.; Hill, M.C.; Lu, D.; Tiedeman, C.R.; Mehl, S. *UCODE_2014, with New Capabilities to Define Parameters Unique to Predict Ions, Calculate Weights Using Simulated Values, Estimate Parameters with SVD, Evaluate Uncertainty with MCMC, and More*; Integrated Groundwater Modeling Center (IGWMC), of the Colorado School of Mines: Golden, CO, USA, 2014.
33. Banta, E.R. ModelMate—A Graphical User Interface for Model Analysis. In *Techniques and Methods 6–E4*; U.S. Geological Survey: Reston, VA, USA, 2000.
34. Poeter, E.; Hill, M.; Banta, E.; Mehl, S.; Christensen, S. *UCODE 2005 and Six Other Computer Codes for Universal Sensitivity Analysis, Calibration, and Uncertainty Evaluation*; U.S. Geological Survey: Reston, VA, USA, 2005.
35. Melissaris, P.; Stavropoulos, X. *Hydrogeological Study of the Marathon Plain, Atticaq*; Ministry of Agricultural Development: Department of Geology–Hydrogeology: Athens, Greece, 1999.
36. Margoni, S. *Research of the Environmental Evolution Processes of the Wetlands and Plain of Marathon during the Holocene with the Use of Geographic Information Systems*; Aristotle University of Thessaloniki: Thessaloniki, Greece, 2006.
37. Lozios, S. *Tectonic Analysis of the Metamorphic Units of Northeastern Attica*; National and Kapodistrian University of Athens: Athens, Greece, 1994.
38. Perdikaki, M.; Kallioras, A. Estimating Groundwater Recharge from Precipitation in a Coastal Mediterranean Aquifer: The Case of Marathon. In Proceedings of the 12th World Congress of EWRA on Water Resources and Environment, Thessaloniki, Greece, 27 June–1 July 2023.
39. Mckeever, V.; Owen, W.; Rallison, R.; Engineers, H. National Engineering Handbook Section 4—Hydrology; Soil Conservation Service: 1972. Available online: https://docslib.org/doc/4839451/national-engineering-handbook-section-4-hydrology (accessed on 22 January 2024).
40. Allen, R. Penman–Monteith Equation. In *Encyclopedia of Soils in the Environment*; Elsevier: Amsterdam, The Netherlands, 2005; Volume 4, pp. 180–188.

41. Thornthwaite, C.W.; Mather, J.R. Instructions and Tables for Computing Potential Evapotranspiration and the Water Balance. Drexel Institute of Technology. Laboratory of Climatology. *Publ. Climatol.* **1957**, *10*, 181–289.
42. Siemos, N. *Evaluation of Water Resources in Attica and Islands of Argosaronic Gulf*; Institute of Geology and Mineral Exploration: Athens, Greece, 2010.
43. Langevin, C.D.; Shoemaker, W.B.; Guo, W. *MODFLOW-2000, the U.S. Geological Survey Modular Ground-Water Model-Documentation of the SEAWAT-2000 Version with the Variable-Density Flow Process (VDF) and the Integrated MT3DMS Transport Process (IMT)*; U.S. Geological Survey: Reston, VA, USA, 2003.
44. Gelhar, L.W.; Welty, C.; Rehfeldt, K.R. A Critical Review of Data on Field-Scale Dispersion in Aquifers. *Water Resour. Res.* **1992**, *28*, 1955–1974. [CrossRef]
45. Schulze-Makuch, D. Longitudinal Dispersivity Data and Implications for Scaling Behavior. *GroundWater* **2005**, *43*, 443–456. [CrossRef] [PubMed]
46. Oude Essink, G.H.P.; Van Baaren, E.S.; De Louw, P.G.B. Effects of Climate Change on Coastal Groundwater Systems: A Modeling Study in the Netherlands. *Water Resour. Res.* **2010**, *46*, W00F04. [CrossRef]
47. Abd-Elaty, I.; Zelenakova, M. Saltwater Intrusion Management in Shallow and Deep Coastal Aquifers for High Aridity Regions. *J. Hydrol. Reg. Stud.* **2022**, *40*, 101026. [CrossRef]
48. Tran, D.A.; Tsujimura, M.; Pham, H.V.; Nguyen, T.V.; Ho, L.H.; Le Vo, P.; Ha, K.Q.; Dang, T.D.; Van Binh, D.; Doan, Q.-V. Intensified Salinity Intrusion in Coastal Aquifers Due to Groundwater Overextraction: A Case Study in the Mekong Delta, Vietnam. *Environ. Sci. Pollut. Res.* **2022**, *29*, 8996–9010. [CrossRef] [PubMed]
49. Siarkos, I. *Developing a Methodological Framework Using Mathematical Simulation Models in Order to Investigate the Operation of Coastal Aquifer Systems: Application in the Aquifer of N. Moudania*; Aristotle University of Thessaloniki, School of Civil Engineering: Thessaloniki, Greece, 2015.
50. Mabrouk, M.; Jonoski, A.; Oude Essink, G.; Uhlenbrook, S. Assessing the Fresh–Saline Groundwater Distribution in the Nile Delta Aquifer Using a 3D Variable-Density Groundwater Flow Model. *Water* **2019**, *11*, 1946. [CrossRef]
51. Gomaa, S.M.; Hassan, T.M.; Helal, E. Assessment of Seawater Intrusion under Different Pumping Scenarios in Moghra Aquifer, Egypt. *Sci. Total Environ.* **2021**, *781*, 146710. [CrossRef]
52. Nasiri, M.; Moghaddam, H.K.; Hamidi, M. Development of Multi-Criteria Decision Making Methods for Reduction of Seawater Intrusion in Coastal Aquifers Using SEAWAT Code. *J. Contam. Hydrol.* **2021**, *242*, 103848. [CrossRef]
53. Post, V.; Simmons, C. *Variable Density Groundwater Flow*; The Groundwater Project: Guelph, ON, Canada, 2022. ISBN 9781774700464.
54. Shoemaker, W.B. Important Observations and Parameters. *GroundWater* **2004**, *42*, 829–840.
55. Werner, A.D. On the Classification of Seawater Intrusion. *J. Hydrol.* **2017**, *551*, 619–631. [CrossRef]
56. Abarca, E.; Vázquez-Suñé, E.; Carrera, J.; Capino, B.; Gámez, D.; Batlle, F. Optimal Design of Measures to Correct Seawater Intrusion. *Water Resour. Res.* **2006**, *42*, W09415. [CrossRef]
57. Kerrou, J.; Renard, P. A Numerical Analysis of Dimensionality and Heterogeneity Effects on Advective Dispersive Seawater Intrusion Processes. *Hydrogeol. J.* **2010**, *18*, 55–72. [CrossRef]
58. Pool, M.; Post, V.E.A.; Simmons, C.T. Effects of Tidal Fluctuations and Spatial Heterogeneity on Mixing and Spreading in Spatially Heterogeneous Coastal Aquifers. *Water Resour. Res.* **2015**, *51*, 1570–1585. [CrossRef]
59. Michael, H.A.; Scott, K.C.; Koneshloo, M.; Yu, X.; Khan, M.R.; Li, K. Geologic Influence on Groundwater Salinity Drives Large Seawater Circulation through the Continental Shelf. *Geophys. Res. Lett.* **2016**, *43*, 10782–10791. [CrossRef]

hydrology

MDPI

Article

A Simulation–Optimization Model for Optimal Aquifer Remediation, Using Genetic Algorithms and MODFLOW

Yiannis N. Kontos

School of Civil Engineering, Aristotle University of Thessaloniki, 54124 Thessaloniki, Greece; ykontos@civil.auth.gr

Abstract: This paper investigates the optimal remediation process in an aquifer using Modflow 6 software and genetic algorithms. A theoretical confined aquifer has been polluted over a long period of time by unnoticed leakage in a pipeline conveying leachate from an adjacent landfill to a wastewater treatment plant. When the extended leakage and groundwater pollution are discovered, the optimal planning of the remediation strategy is investigated using the pump-and-treat method or/and hydrodynamic control of the pollution. The practical goal is to find the optimal locations and flow rates of two additional pumping wells, which will pump the polluted water or/and control pollution, protecting an existing drinking water pumping well, securing its fully operational mode even during the remediation process with the minimum possible cost, simply represented by the pumped water volume of the additional wells. The remediation process is considered complete when the maximum concentration in the aquifer drops below a certain limit. The Modflow software (handled by the Flopy Python package) simulates the flow field and advective–dispersive mass transport, and a genetic algorithm is used as the optimization tool. The coupled simulation–optimization model, Modflow-GA, complemented by a sophisticated post-processing results analysis, provides optimal and alternate sub-optimal remediation strategies for the decision makers to select from.

Keywords: polluted aquifer; groundwater pollution; groundwater resources management; optimization; optimal remediation strategies; genetic algorithms; pump-and-treat; pollution control; MODFLOW; Flopy

Citation: Kontos, Y.N. A Simulation–Optimization Model for Optimal Aquifer Remediation, Using Genetic Algorithms and MODFLOW. *Hydrology* **2024**, *11*, 60. https://doi.org/10.3390/hydrology11050060

Academic Editors: Augustina Clara Alexander and Pantelis Sidiropoulos

Received: 8 March 2024
Revised: 21 April 2024
Accepted: 21 April 2024
Published: 24 April 2024

1. Introduction

1.1. Groundwater Pollution

Groundwater pollution poses a significant global water quality issue. It arises from diverse sources, such as agriculture, industry, urban and municipal waste, landfills, or poorly designed/constructed septic systems. Groundwater is then introduced with nitrate, heavy metals, organic chemicals, pathogens, and emerging pollutants. Approximately 20% of globally recorded 2000–2013 disease outbreaks (GIDEON; [1]) were water-related diseases, mainly caused by contaminated drinking water supplies and often associated with private/household wells and small groundwater supplies [2,3]. Disposal of municipal solid waste poses a global challenge; landfills are the primary method for waste storage due to their simplicity and cost-effectiveness [4]. In many countries, waste management practices are often unscientific and unplanned, leading to significant environmental issues [5]. This paper deals with the optimal remediation of a theoretical aquifer polluted by leachate from a similar case. Landfills function as biological reactors where refuse undergoes transformation, resulting in gas and leachate production, the latter being a liquid stream carrying soluble products from waste [6]. Leachate composition varies but typically contains dissolved organics, inorganic salts, ammonia, heavy metals, and xenobiotic organic compounds from various sources [7]. Monitoring landfill leachate and tools for the production of optimal remediation strategies are crucial to ensure that landfill operations do not harm human health or the environment.

1.2. Remediation of Aquifers Polluted by Household and Municipal Waste

The treatment and remediation of various types of waste, particularly household and municipal solid waste, require diverse technologies due to their heterogeneous nature. Composting and anaerobic digestion, permeable reactive barriers, electrokinetic technology, micro and nanobubbles, and solubilizing agents have been developed for different waste materials. Biologically enhanced degradation, employing fungi, bacteria, and actinomycetes, is another approach for treating contaminated soils and landfill leachates. These methods are expensive and require highly experienced and specialized personnel and equipment/materials. Often, optimal remediation and effective management of contaminated aquifers demand sophisticated approaches for pollution control or remediation [8]. These methods often involve setting up and operating a network of wells to either manage the dispersion of contaminants through altering groundwater levels and flow paths (hydraulic control—HC) or to lessen the concentration of pollutants by partial or complete extraction, aiming to achieve predetermined concentration levels or overall mass reduction, followed by appropriate treatment (pump-and-treat—PAT) [9]. Pollution control strategies can be further categorized based on the manner in which pollution plumes are intercepted or contained [10], including concentration control (adhering to maximum allowable concentration at specific checkpoints) [11], hydraulic control (maintaining certain head difference, gradient, or velocity at designated points) [12], and advective control [13].

Optimal aquifer remediation problems belong to the class of constrained, nonlinear, stochastic, multi-objective optimization problems. State-of-the-art optimization methods include analytical approaches [14], metaheuristics, such as harmony search algorithms [15], and genetic algorithms (GAs) [16–18]. Moreover, chromatic graphs [19] and even game theory are used [20], while the rapid development of machine learning is also mirrored [21]. The current methodology is based on our own previous similar work concerning optimal pollution control and pump-and-fertilize strategies in a nitro-polluted aquifer, using GAs and Modflow [22].

1.3. Current Research Problem

This paper deals with the optimal management of a theoretical polluted aquifer, specifically, the optimal remediation of a confined aquifer. The aquifer is assumed to have been polluted over a long period of time by unnoticed leakage in two sections of an underground pipeline conveying leachate from a landfill to a wastewater treatment plant. Upon delayed discovery of the extended pollution leak, it is decided to immediately shutdown the landfill's drainage network operation. Simulation of the pollution's spread is needed in order to predict the aquifer's current state, and an optimal remediation strategy must be planned. Optimal remediation refers to the investigation of the optimal locations and flow rates of two additional pumping wells (AWs) to protect an existing freshwater supply pumping well. AWs should secure the water supply well's fully operational mode throughout the remediation process at minimum cost. Optimization entails minimization of the operational cost of the AWs' system; simply represented here by the total volume of water pumped by AWs. The protection of the existing supply well means that the pollutant concentration does not exceed a certain threshold. Another constraint derives from relevant rules in effect at certain regions of the Greek district, e.g., the Regional Unit of Chalkidiki, Prefecture of Macedonia; these dictate a minimum distance of any well to another or the landfill. The remediation process is assumed to be concluded when the maximum pollutant concentration of the study field drops below a certain threshold. The Modflow software [23] simulates the flow field and advective–dispersive mass transport, while a simple binary elitist GA handles optimization, namely minimization of the pumped volume of water during the variable per solution remediation period. The coupled simulation–optimization (S–O) model, "Modflow-GA", complemented by a sophisticated post-processing results analysis, produces optimal and alternate sub-optimal remediation strategies for the decision makers to select from. These strategies may feature pump-and-treat (PAT), hydraulic pollution control (HC), or a combination of the two remediation techniques. The methodology that is implemented in the current research is graphically presented step-by-step in Figure 1. Figure 1 should not be seen as a typical graphical abstract, as it is not a general summary of the methodology proposed, but a rather over-detailed delineation of the proposed process that the reader could fully comprehend only after reading through all the sections of this paper.

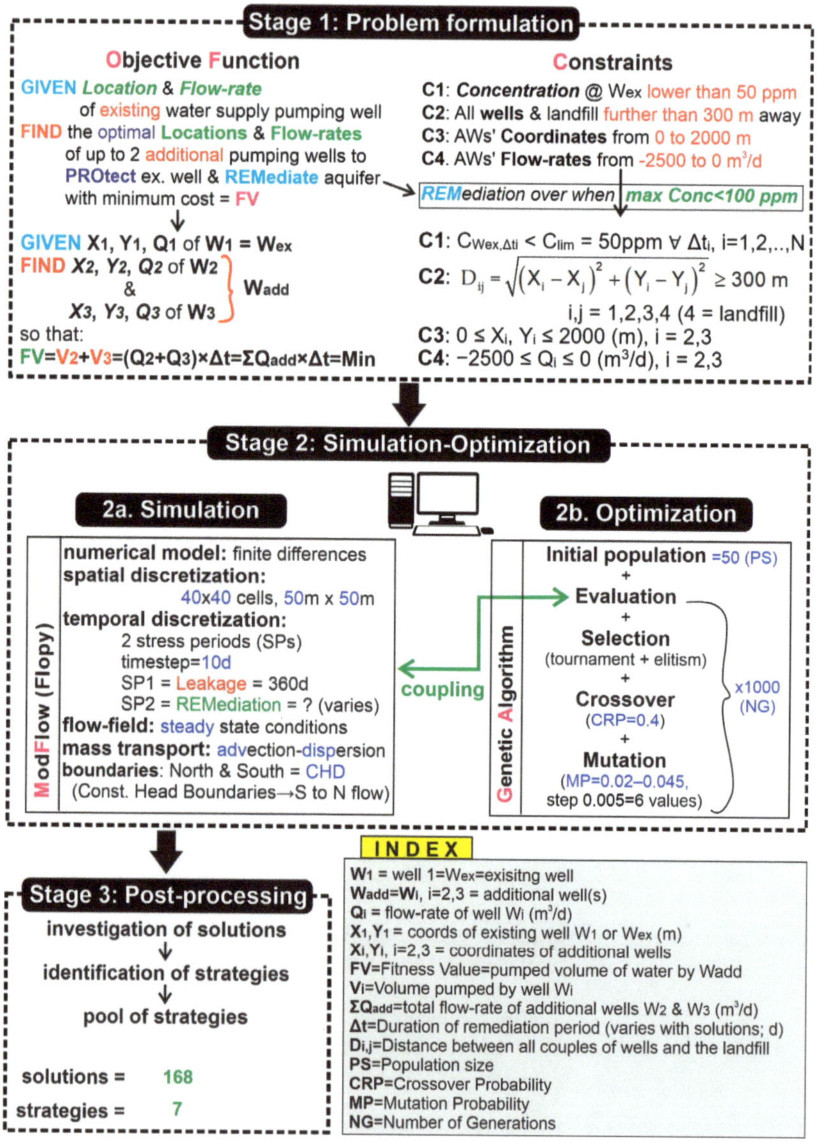

Figure 1. An extremely detailed graphical abstract of the proposed methodology to investigate optimal remediation strategies of a polluted aquifer via a proposed simulation–optimization model, coupling a genetic algorithm and MODFLOW.

2. Materials and Methods

2.1. Theoretical Problem

The 2 km × 2 km top view of the theoretical flow field is presented in Figure 2. It includes an existing fresh water supply well (W_{ex}), a landfill (LF), a pipeline conveying leachate from the LF to the wastewater treatment plant (WTP) that is positioned deep enough to leak directly in the confined aquifer, and two leaking sections of the pipe (Leak 1 and Leak 2). It also includes the southern boundary with a constant hydraulic head of 135 m and the northern boundary with a constant head of 124 m. These two boundaries combined define a south-to-north natural hydraulic gradient and flow. Finally, the field includes the western and eastern no-flow impermeable boundaries. The aquifer is assumed to be directly polluted by constant concentration load leaks (1000 ppm) by two sources, Leak 1 and Leak 2 in the pipeline, respectively. The leaks assumedly persisted unnoticed for approximately 1 year (370 d). W_{ex} should uninterruptedly supply fresh drinking water to an adjacent settlement at a constant flow rate of 950 m^3/d. For this purpose, the solution of the construction and operation of two AWs is promoted by the authorities/decision makers. The two wells should employ PAT or HC or a combination of the two to protect W_{ex} while also remediating the polluted aquifer.

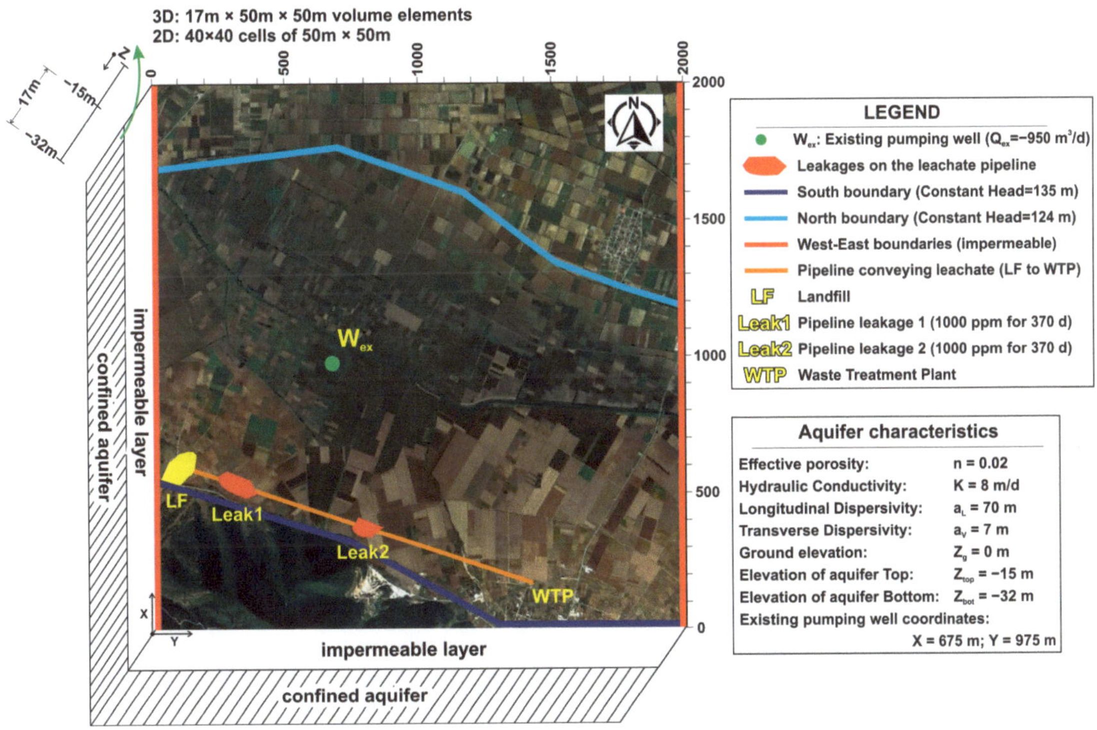

Figure 2. Theoretical study field including the top view of the aquifer and its layering and characteristics, the existing pumping well, the landfill, the pipeline conveying the leachate to the wastewater treatment plant, the leaks, and the boundaries.

The practical goal is to find the locations and flow rates of the AWs (max possible flow rate = 2500 m^3/d) so that the management (protection and remediation) cost is minimal. The cost is simplistically represented here by the volume of water pumped by the AWs. The W_{ex} is assumedly protected provided that the pollutant concentration does not exceed C_{lim} = 50 ppm. Aquifer remediation is assumed to be concluded when the aquifer's maximum pollutant concentration, $C_{aq,max}$, is lower than C_{rem} = 100 ppm.

The computationally greedy nature of GAs dictates a certain balance between computational accuracy and efficiency [24]. Thus, a simplified 2D flow field is studied in a homogenous, isotropic, confined aquifer with a plane, single-phase, steady-state (per stress period) flow. The hydraulic conductivity of the aquifer is assumed to be $K_x = K_y = K_z = 8$ m/d; aquifer thickness is b = 17 m, efficient porosity is $n_e = 0.2$, longitudinal dispersivity is $a_L = 70$ m, and transverse dispersivity is $a_V = 7$ m, throughout its extent. Diffusion is not considered significant and, hence, is not included in the conceptual model. As this is clearly a theoretical problem and this paper focuses on the coupling of Modflow with the optimization method of GAs, the hydrogeological configuration of the assumed studied aquifer is quite simplistic. However, the proposed methodology can easily address any complex hydrogeological setup; Modflow is capable of doing so, as documented in numerous scientific papers. In real case studies, well-log and/or stratigraphic information should be utilized in order to address possible heterogeneity of the aquifer, at least in the proximity of the wells and the pipeline [25,26].

The prediction of the leaked pollution's spread in the assumed current state (a year after the leaks' initiation), simulated in a way presented in Section 2.2, is presented in Figure 3a (see Supplementary Material SM1 for the respective video of the full spread progression; SM0 is a list of all Supplementary Material Files together with a relevant description). If no countermeasures or remediation strategies are implemented, the pollution plumes will spread and pollute, and be at least partially pumped by, W_{ex}. Assuming that the detection threshold for declaring a well polluted is $C_{Wex} > 1$ ppm, in the "do nothing scenario", W_{ex} is polluted approximately 240 d after the leaks' discovery (see SM1). The maximum (max) pollutant concentration in the aquifer would be detected at approximately day 1000 (Figure 3b). The diagram in Figure 3c presents the temporal predicted evolution of pollutant concentration at W_{ex} and the max aquifer concentration in the "do nothing" scenario.

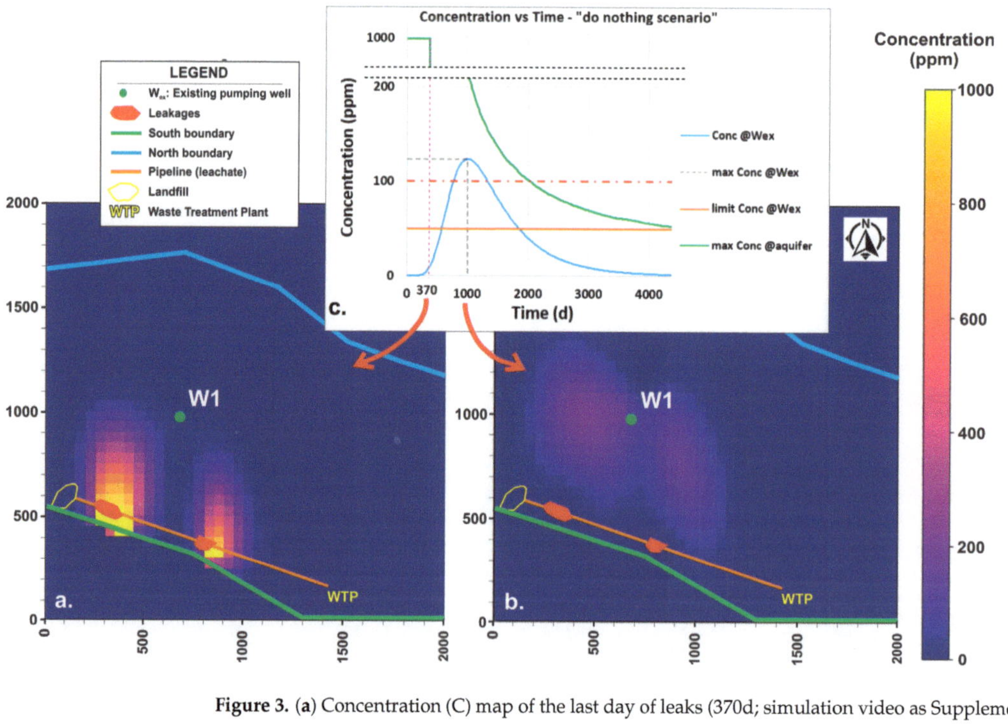

Figure 3. (**a**) Concentration (C) map of the last day of leaks (370d; simulation video as Supplementary Materials SM1). (**b**) Concentration map of the predicted day of max C at W_{ex} (1000th). (**c**) C_{Wex} and max $C_{aquifer}$ vs. Time.

2.2. Simulation Model Based on Modflow 6

The simulation of the flow field and mass transport are handled by the Modflow 6 software of USGS [23]. It is an established finite differences tool. The automated call of recurring simulations due to the coupling of the simulation model with an optimization tool (GAs) of an iterative nature dictates the use of the "Flopy: Python Package for creating, running, and post-processing MODFLOW-Based Models" [27–31]. The 2 km × 2 km flow field is discretized in a square grid of 40 × 40 cells (2D top view), each with a 50 m side. The 17 m thick confined aquifer comprises a single layer; hence, a simplified 2D model is simulated (see Figure 2).

The initial piezometric surface is assumed to be 129.5 m high. The western and eastern boundaries are assumed to be impermeable, while the southern and northern ones (see Figure 2) are assumed to be constant head boundaries (135 m and 124 m, respectively) simulated in Modflow as constant head boundaries (CHD; [31]). This way, a south–north (S–N) natural flow is defined due to the hydraulic gradient of approximately (135 m − 124 m)/2000 m = 5.5 ‰.

2.3. Optimization Tool—Genetic Algorithms

This is a typical, constrained, nonlinear, stochastic, multi-objective optimization problem [8]. It is formed as a minimization of the fitness value (FV) problem. FV stands for the aquifer remediation cost of each solution, here simplistically represented by the total pumped volume of water by the AWs. The constraints include the following:

C1. Concentration at W_{ex} must always be lower than 50 ppm;
C2. All wells and the landfill must retain a distance of more than 300 m between them;
C3. All AWs' coordinates (X_2, Y_2, X_3, Y_3) must retain values from 0 to 2000 m;
C4. The values of the flow rates of the AWs must vary between −2500 m^3/d and 0 (pumping wells' flow rates are negative in Modflow).

2.3.1. Objective Function

Mathematically, the objective function to be minimized can be stated as:

$$\text{Find: } X_i, Y_i, Q_i, i = 2, 3 \text{ so that } FV = V_{tot} + \text{Penalty} = \text{Min} \tag{1}$$

Subject to the constraints:

$$\text{C1: } C_{Wex,\Delta ti} < C_{lim} = 50 \text{ ppm for any } \Delta t_i, i = 1, 2, \ldots, N \tag{2}$$

$$\text{C2}: D_{ij} = \sqrt{(X_i - X_j)^2 + (Y_i - Y_j)^2} \geq 300 \text{ m}, \; i,j = 1,2,3,4 \; (4 = \text{landfill}) \tag{3}$$

$$\text{C3: } 0 \leq X_i, Y_i \leq 2000 \text{ (m)}, i = 2, 3 \tag{4}$$

$$\text{C4: } -2500 \leq Q_i \leq 0 \; (m^3/d), i = 2, 3 \tag{5}$$

given that:

$$V_{tot} = t_{rem} \cdot Q_{AWs} = \left(\sum_{i=1}^{N} \Delta t_i\right) \cdot \sum_{i=2}^{3} Q_i \tag{6}$$

$$t_{rem} = \sum_{i=1}^{N} \Delta t_i, \text{ while } C_{aq,max} \geq C_{rem} = 100 \text{ ppm} \tag{7}$$

where X_i, Y_i are the coordinates of the ith AW (m; i = 2, 3); Q_i is the flow rate of the ith AW (m^3/d; i = 2, 3); FV is the fitness value to be minimized, here, simplistically represented by V_{tot} (m^3; see Equation (6)), the total volume of water pumped by the AWs throughout the varying-per-solution duration of remediation t_{rem} (see Equation (7)); Penalty is the total penalty comprised by the separate constraint-specific penalties imposed on solutions that violate one or more of the four constraints C1–C4 (see Section 2.3.3); $C_{Wex,\Delta ti}$ is the concentration at W_{ex} during timestep Δt_i; N is the last timestep of the remediation period; Q_{AWS} is the sum of the flow rate of the additional wells; $C_{aq,max}$ is the max concentration anywhere in the aquifer at any timestep of the remediation period (which varies with the solution); C_{rem} is the concentration threshold below which the remediation is considered complete.

As far as constraint C3 is concerned, the decision variables are not actually coordinates but rather column (col) and row numbers since the numerical simulation model is based on finite differences (Modflow); the col/row numbers refer to the 2D top view grid. Hence, C3 can be written:

$$C3': 1 \leq col_i, row_i \leq 40, i = 2, 3 \quad (8)$$

2.3.2. Genetic Algorithm Configuration

Simple GAs are utilized with a binary genetic representation. That means that chromosomes, namely proposed solutions, are encoded as binary strings. Each chromosome represents the AWs' coordinates (practically, the respective col and row numbers) and flow rates. Coordinates and flow rates can obtain certain max values in the decimal numeral system; their binary form dictates the max number of digits (string length) each chromosome part requires. The total number of digits (total chromosome/string length) can be calculated. According to C3′ and C4, for AWs it is $1 \leq col_i, row_i \leq 40$ (nominal; i = 2, 3) and $0 \leq |Q_j| \leq 2500$ (m^3/d; j = 3, 4).

Table 1 presents the max values of the decision variables in decimal and binary numeral form, the string length of the chromosome part that represents each decision variable, and the decimal value of the max binary number that the specific string part can obtain. The length of each string part that represents the consecutive number of a column/row of an AW is SLX = 6. Thus, the length of the part of all coordinates of AWs is SL1 = 4 · SLX = 24. The length of each part that represents the flow rate of an AW is SLQ = 12. Thus, the length of the part of all flow rates of AWs is SL2 = 2 · SLQ = 24. The total length of a typical binary chromosome is SL = SL1 + SL2 = 24 + 24 = 48. Its structure is presented in Figure 4.

Table 1. Max values of decision variables in decimal form; number of digits of each decision variable (string part length); decimal value of max binary number of respective length.

Decision Variable	Decimal Form	Binary Form	String Length (SL)	Max Decimal Value for SL
Max col_i/row_i (i = 2, 3; nominal)	40	101000	6	63
Max Q_j (j = 2, 3; m^3/d)	2500	100111000100	12	4095

The genetic operators used are (a) selection, featuring the tournament procedure and elitism for the fittest chromosome, (b) crossover, and (c) mutation. The chromosome population is PS = 50, the number of generations is NG = 1000, the selection constant is SC = 3, and the crossover probability is CRP = 0.4. Mutation probability MP values used are based on our previous research [16]. It suggests that values ranging from 2/SL to 2.5/SL (here, $2/48 \approx 0.042$ to $2.5/48 \approx 0.052$) are more efficient in converging to the optimal solution. Since the goal of the optimization process is the identification of several alternative good (sub-optimal) solutions and strategies that may constitute local minima, a broader range of MP values are tested: MP = 0.020–0.045 with a step of 0.005. Thus, six separate runs of the GA are carried out (Run1–Run6).

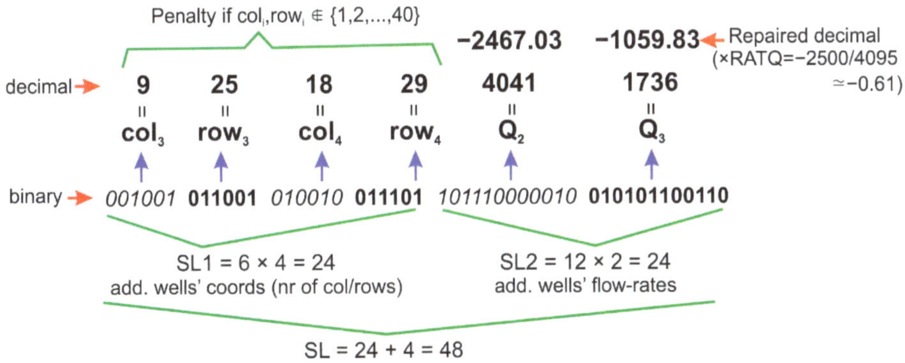

Figure 4. The chromosome of the best-produced solution, representing col/row Nr (coords) and flow rates of additional wells corresponding to certain decimal values. Not acceptable coords entail a penalty while flow rates are repaired; see Section 2.3.3).

2.3.3. Constraint Handling

There are several known techniques to handle constraints in GAs. They can be classified as methods of (a) penalization, (b) repair of infeasible solutions, (c) preservation of feasibility, and (d) hybrid techniques [32]. Here, constraints C1–C4 (Equations (2), (3), (5), and (6)) are handled as follows.

C1: The existing well (W_{ex}) must not pump polluted water of a concentration higher than C_{lim} = 50 ppm. This constraint is handled by a category (a) constraint handling technique: during each timestep Δt_i, the algorithm checks the concentration of the pollutant $C_{Wex,\Delta ti}$ in the respective cell of W_{ex}. If it exceeds C_{lim}, a dynamic penalty (Pen1) is imposed. Pen1 is proportional to the concentration excess over the limit and includes a constant term (C^{con}_{PEN1}) and a variable term. The latter is the sum of all excessive concentration values over C_{lim} during the study period, multiplied by the constant coefficient C^{var}_{PEN1}. Pen1 is:

$$\text{Pen1} = C^{con}_{PEN1} + C^{var}_{PEN1} \cdot \left[\sum_{t=1}^{NS} (C_{Wex,t} - C_{lim}) \cdot \delta_t \right] \tag{9}$$

where $C^{con}_{PEN1} = 20{,}000$ is the constant part of Pen1; $C^{var}_{PEN1} = 200$ is the constant coefficient of the variable part of Pen1; NS is the number of timesteps (it varies from solution to solution as the remediation completion criterion is temporally dynamic: remediation is assumed to conclude when concluded when the aquifer max pollutant concentration $C_{aq,max} < C_{rem}$ = 100 ppm); $C_{Wex,t}$ is the pollutant concentration in the existing water supply well during timestep t; C_{lim} is the concentration threshold that should not be exceeded at W_{ex}; δ_t is a coefficient that is equal to 1 or 0, depending on whether $C_{Wex,t} > C_{lim}$ or not, respectively.

C2: All distances between the existing and additional wells and the landfill, in all combinations, should not exceed 300 m. This constraint is also handled by constraint handling technique (a), penalty imposition. The dynamic penalty depends on the number of C3 violations (number of distances larger than 300 m) and on the magnitude of the violation (remaining of each distance after subtraction by D_{min} = 300 m):

$$\text{Pen2} = \sum_{i=1}^{N_D} \{ [C^{con}_{PEN2} + C^{var}_{PEN2} \cdot (300 - D_i)] \cdot \delta_i \} \tag{10}$$

where N is the number of distances calculated; $C^{con}_{PEN2} = 400{,}000$ is the constant part of Pen2; $C^{var}_{PEN2} = 4000$ is the constant coefficient of the variable part of Pen2; D_i is the ith out of the N total distances calculated; δ_i is a coefficient that is equal to 1 or 0, depending on whether D_i < 300 or not, respectively. N_D = 4 is actually the number of all combinations of distances without repetition between W_{ex}, the two AWs (W2 and W3), and the landfill. In the general case, where there are N wells and landfills:

$$N = N_{ex} + N_{aw} + N_L \quad (11)$$

where N_{ex} is the number of existing wells; N_{aw} is the number of AWs; N_L is the number of landfills, it is:

$$N_D = (N - 1)! \quad (12)$$

C3: Each coordinate (nominal number, consecutive number of col/row in Modflow 40 × 40 grid top view) of AWs should obtain an integer value of 1 to 40. However, after the decoding (binary to decimal) of the binary chromosome, it can obtain integer values from 0 to 63 (Table 1). For this constraint, constraint handling technique (a), penalization, is again utilized: if any coordinate acquires a value equal to zero (0), it instantly converts to a random integer number $N_i > 40$. Subsequently, for each solution that includes several coordinates with a value of $N_i > 40$, a dynamic penalty (Pen3) is imposed. Pen3 is proportional to the number of C3 constraint violations:

$$\text{Pen3} = C_{PEN3} \cdot \left(\sum_{i=1}^{Naw} \text{int}(X_i/\text{maxX}) + \sum_{j=1}^{Naw} \text{int}(Y_j/\text{maxY}) \right) \quad (13)$$

where C_{PEN3} = 500,000 is a coefficient for adjusting Pen1 values to reach the order of magnitude of other penalty items in the total penalty function (see Equation (12)); N_{aw} is the number of AWs (nominal); X_j and Y_j are the abscissa (col number in the 80 × 80 Modflow top view grid) and the ordinate (row number in the Modflow grid) of AW_j (j = 3, 4), respectively (nominal); maxX and maxY are the max physical value of X_i and Y_j, respectively (nominal; here, 40).

C4: Each flow rate of AWs (W2, W3) should obtain a maximum physical value of 2500 m^3/d. However, after the decoding (binary to decimal) of the binary chromosome, it can reach 4095 m^3/d (Table 1). This possible constraint violation is handled via a category (b) handling technique, repairing infeasible chromosomes/solutions: each $Q_j \in [0,1,2,\ldots,4095]$ (i = 2, 3) value decoded by the chromosome is multiplied by the RATQ = − 2500/4095 ratio, repairing its value, so that $Q_{j,repair} \in [0,1,2,\ldots,2500]$ (j = 2, 3). The negative sign is used as Modflow assigns pumping wells with negative flow rates. Figure 4 actually presents the repair of the flow rates of the AWs in the best solution produced in the six runs.

The total penalty of each solution violating at least one constraint of C1, C2, C3, is given:

$$\text{Penalty} = C_{f1} \cdot \text{Pen1} + C_{f2} \cdot \text{Pen2} + C_{f3} \cdot \text{Pen3} \quad (14)$$

where Pen1 is the Penalty item concerning solutions that entail a violation of constraint C3; Pen2 is the Penalty item relating to solutions that involve a violation of constraint C4; C_{f1} and C_{f2} are weighting factors of Pen1 and Pen2 in the total penalty, respectively (here, $C_{f1} = 15$; $C_{f2} = 30$; $C_{f3} = 30$).

The optimal assignment of the values of the penalty function parameters (C^{con}_{PEN1}; C^{var}_{PEN1}; C^{con}_{PEN2}; C^{var}_{PEN2}; C_{PEN3}; C_{f1}; C_{f2}; C_{f3}) entails a large set of tests. The criteria that need to be fulfilled are:

1. Pen1 parameters' values (C^{con}_{PEN1}; C^{var}_{PEN1}) must ensure that the penalty term imposed on solutions violating constraint C1 (no pollution of W_{ex}) is of a dynamic nature and, hence, proportional to the extent of the violation, sorting C1-related unacceptable ("bad") solutions fairly.
2. Pen2 parameters' values (C^{con}_{PEN2}; C^{var}_{PEN2}) value must ensure that the penalty term that deals with solutions violating constraint C2 (no distance between wells or a well and the landfill must be less than D_{min}) is of a dynamic nature and, hence, proportional to the extent of the violation, sorting C2-related bad solutions fairly.

3. Pen3 parameter (C_{PEN3}) value must ensure that the penalty term that deals with solutions violating constraint C3 (no AW's coordinate out of range) is of a similar order of magnitude as the other penalty terms (Pen1 and Pen2).
4. Parameters that are related to the relevant weightings of Pen1, Pen2, and Pen3 in the total Penalty function (C_{f1}; C_{f2}; C_{f3}) as well as the penalty term-specific parameters (C^{con}_{PEN1}; C^{var}_{PEN1}; C^{con}_{PEN2}; C^{var}_{PEN2}) including C_{PEN3} should be set in such a way that individual penalties (Pen1, Pen2, and Pen3), hence constraints' importance, are prioritized. Solutions that violate C4 are infeasible and unrealistic due to out-of-range coordinates. They are less welcome than feasible but bad solutions, namely solutions that allow wells to be closer than 300 m to each other or to the landfill (violating C2) or solutions that allow W_{ex} to be polluted (violating C1). The latter two are unfavorable but could occur in a real physical scenario.
5. Parameters that are related to the relevant weightings of Pen1, Pen2, and Pen3 in the total Penalty function (C_{f1}; C_{f2}; C_{f3}) as well as the penalty term-specific parameters (C^{con}_{PEN1}; C^{var}_{PEN1}; C^{con}_{PEN2}; C^{var}_{PEN2}) including C_{PEN3} must be assigned values that also favor acceptable but not optimal solutions (exhibiting high fitness value, namely expensive groundwater resources' management solutions) over unacceptable (violating C1 or C2) solutions and over infeasible (violating C3) solutions.
6. Finally, the minimum penalty rule must apply: the most suitable is the minimum penalty function that can lead to penalty-free optimal solutions in a quick and consistent fashion (e.g., [33]).

3. Results and Discussion

The software application created and used here, "Modflow-GA", is written in Python, utilizing the "Flopy" package for controlling Modflow. It is part of the "OptiManage" v4 suite for optimal groundwater resources management. "Modflow-GA" is implemented six times (Run 1–6). Each simulation needs 7–8 s. For a population of 50 chromosomes and 1000 generations, the average computational time of each run is approximately 100 h (Intel Core i7 7700 @3.60 GHz; 16 GB RAM @1197 MHz).

The algebraically best solutions of Runs 1–6 (MP values from 0.020 to 0.045, step 0.005) are presented in Table 2. The overall best solution (FV1) is produced by Run 2 (identified solution Nr 19; see Supplementary Material SM4, Sheet 6) in generation 425 of the algorithm. It proposes the operation of two AWs, one positioned at (425 m, 775 m) pumping $Q_w \approx 2467.0\ m^3/d$ and the other one at (875 m, 575 m) pumping $Q_e \approx 1059.8\ m^3/d$ with a ratio of $Q_w/Q_e = 2.3/1$. All AWs are from now on referred to as "western" and "eastern" instead of W_{ex} or W2 in order not to misidentify identical solutions that only differ in the enumeration of AWs as different. The duration of the remediation process is 310 d (370 + 310 = 680 d after the start of the leak, at the start of the 69th 10-day timestep). The fitness value (FV) of the best solution is FV1 = ($2467.033\ m^3/d + 1059.829\ m^3/d) \times 310\ d \approx 1{,}093{,}327\ m^3$. The algebraically optimal solution is not the solution with the lowest total flow rate pumped by additional wells (RUN4; identified solution Nr 24; $\Sigma Q = 2373.6\ m^3/d$) but deviates by +48.66% (+1153.3 m^3/d). It is, however, the solution with the lowest remediation duration.

Figure 5a graphically presents the overall best solution. The graph (Figure 5b) presents the temporal progression of the max pollutant concentration C_{Wex} at the existing water supply well (W_{ex}) against the limit C_{Wex} must not exceed (50 ppm), as well as the max aquifer pollutant concentration $C_{aq,max}$ against the remediation completion threshold C_{rem} = 100 ppm. Supplementary Material SM2 presents the respective video of the full spread progression concerning the algebraically optimal (overall best) solution.

Table 2. Best solutions of Runs 1–6.

Run	Gen	FV (m^3)	X_W (m)	Y_W (m)	X_e (m)	Y_e (m)	Q_W ($-m^3/d$)	Q_e ($-m^3/d$)	ΣQ ($-m^3/d$)	Q_W/Q_e	Dur (d)
1	975	1,326,984	325	875	875	625	2389.5	626.4	3015.9	3.8	440
2	425	1,093,327	425	775	875	575	2467.0	1059.8	3526.9	2.3	310
3	993	1,327,521	325	875	875	625	2391.9	625.2	3017.1	3.8	440
4	880	1,329,231	325	925	875	675	427.4	1946.3	2373.6	0.2	560
5	682	1,115,409	375	875	875	625	−2134.3	−801.0	2935.3	2.7	380
6	936	1,326,716	325	875	875	625	−2389.5	−625.8	3015.3	3.8	440

Abbreviations: gen: generation of solution; FV: fitness value; X_w, Y_w, X_e, Y_e: coordinates of west and east additional wells; Q_w, Q_e: flow rates of west and east additional well; ΣQ: total flow rate of additional wells; Dur: remediation duration.

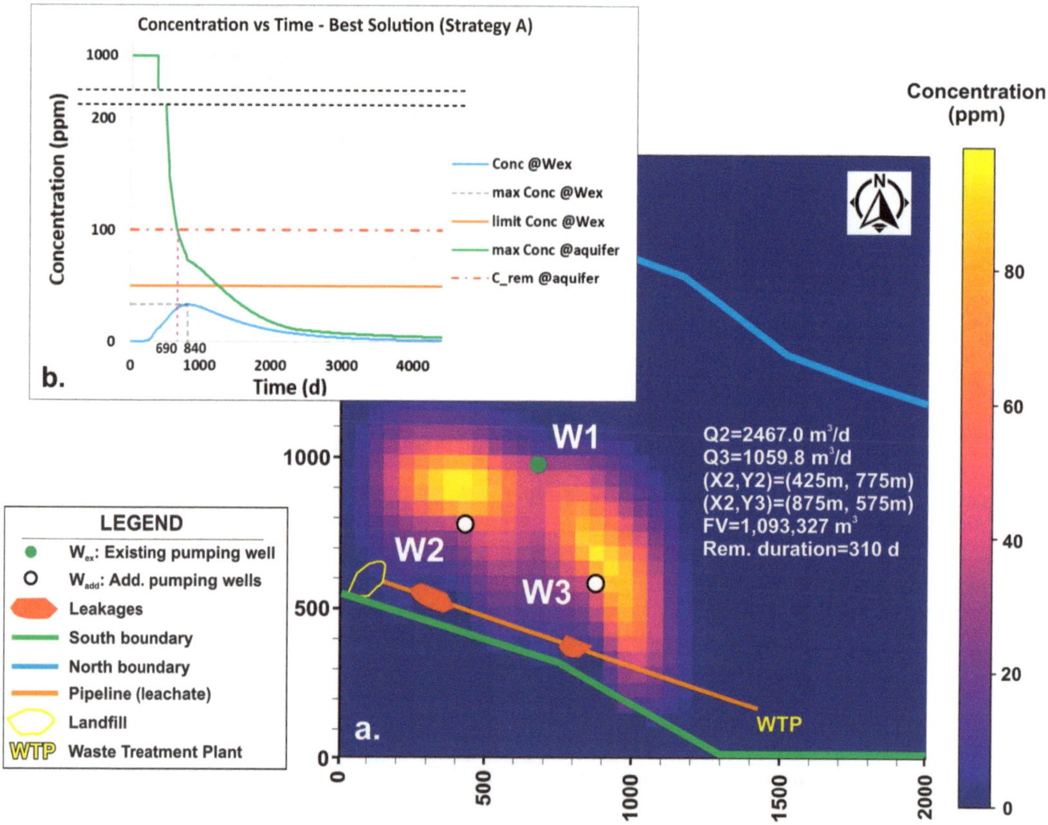

Figure 5. (**a**) Best solution of all runs (Strategy A); simulated concentration (C) map of the predicted day of max C at W_{ex} (840d; simulation video as Supplementary Materials SM2). (**b**) C_{Wex} and max $C_{aquifer}$ vs. Time.

Supplementary Materials SM3 includes all raw result files of all six Modflow_GA applications concerning Runs 1–6 (three ".txt" files per run). Supplementary Materials SM4 presents the same results in an Excel file (sheets 1, 4, 7, 10, 13, and 16 for Runs 1, 2, 3, 4, 5, and 6, respectively), also including relevant FV-generations, ΣQ-generations, and Dur-generations graphs (sheets 2, 5, 8, 11, 14, and 17 for Runs 1, 2, 3, 4, 5, and 6, respectively). The respective graphs for Run 2 that produced the optimal solution are presented in Figure 6. SM4 also includes acceptable only solutions, namely solutions without penalties (sheets 3, 6, 9, 12, 15, and 18 for Runs 1, 2, 3, 4, 5, and 6, respectively).

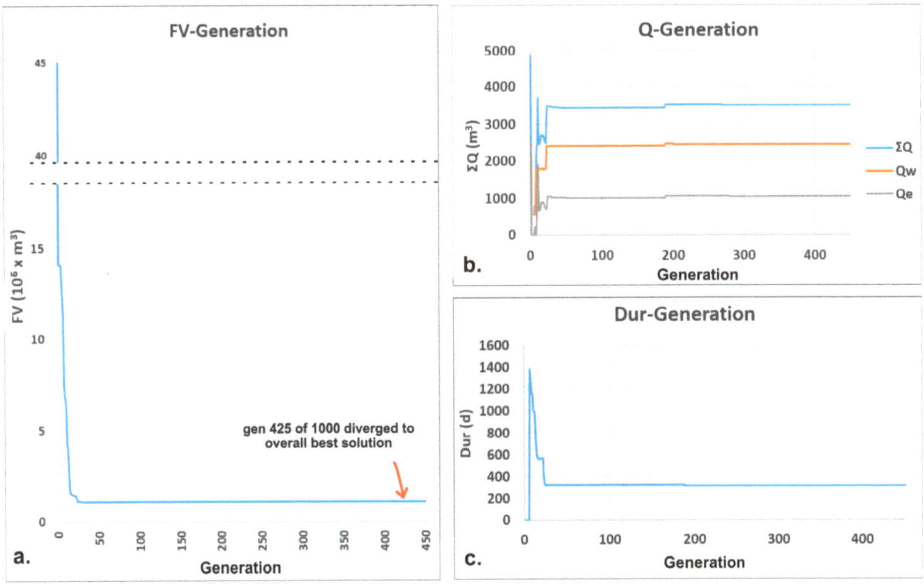

Figure 6. (**a**) FV vs. generations for Run 2 (produced overall best solution). (**b**) Flow rates of western and eastern additional wells and their sum vs. generations. (**c**) Remediation duration vs. generations.

Further study of all results reveals various versions/alterations of the optimal solution, as well as other management concepts. For example, solution Run2-Nr 18 (see SM4, sheet 6 or SM5, sheet 1), which is actually the 2nd best solution of Run 2 and in general (FV2), exhibits FV2 = 1,093,516 m^3 (+189 m^3 or +0.02% compared to FV1), identical AWs' locations, and a trivially higher Q_W by 0.6 m^3/d, leading to the same remediation duration = 310 d. While this is a meaningless variation of FV1, other variations actually constitute alternative interesting versions of FV1, e.g., Run2-Nr 9, 11th best Run 2, and overall solution (FV11). It exhibits FV11 = 1,107,692 m^3 (+14,365 m^3 or +2.32% compared to FV1), eastern AW (W_e) positioned 50 m northern compared to FV1, pumping 1.56% less (−38.5 m^3/d), western AW (W_W) in the same location as FV1, pumping 2.53% less (−26.9 m^3/d). Total pumping is 1.85% reduced (−65.3 m^3/d), while the flow rate distribution between AWs is similar to FV1 Q_W/Q_e = 2.4 instead of 2.3, and the remediation duration is only 10 d higher (320 instead of 310 d). This is just a small variation of FV1. It could be useful if stored, though, as it would be a great alternative solution if, for example, the proposed location for the construction of W_W was not available due to a posteriori changes (e.g., problems in expropriating or using specific private land). Other solutions are even more diverse and can be thought to constitute different concepts/strategies. For example, Run2-Nr 7, which is the 13th best Run 2 and 45th best overall solution, while retaining the AWs' locations of FV1, exhibits a 10% reduction in the total pumping flow rate by AWs, with a highly more uneven distribution of flow rates (Q_W/Q_e = 3.3 instead of 2.3), leading to 100 more days of remediation and a 20% increase in FV (and hence, approximate cost). This solution, which may be classified as a different strategy, also holds value. Unforeseen future constraint variations may occur, e.g., an additional constraint concerning low preferred hydraulic head drawdown in the eastern part of the study area may restrict max pumping flow rates in that area; hence, solutions with lower Q_e will be favored.

The above dictates the need for a systematic investigation of all "acceptable" (no penalties) and "good" (e.g., here, exhibiting FV < median) solutions. This way, different management strategies are given explicitly stated criteria that can be identified in various alternative algebraical versions. A detailed post-processing of solutions is presented in Sections 3.1 and 3.2.

3.1. Systematic Investigation of Solutions

Run 1 produced 41 acceptable solutions (SM4, sheet 3), Run 2 produced 19 (SM4, sheet 6), Run 3 produced 22 (SM4, sheet 9), Run 4 produced 24 (SM4, sheet 12), Run 5 produced 39 (SM4, sheet 15), and Run 6 produced 23 (SM4, sheet 18). In total, Runs 1–6 produced 168 different acceptable (no penalty) solutions. Figure 7 presents the proposed locations of all the additional pumping wells of all 168 acceptable solutions (see SM5, sheet 2), classified only regarding their west–east spatial feature. The radius of each well is proportional to the respective flow rate.

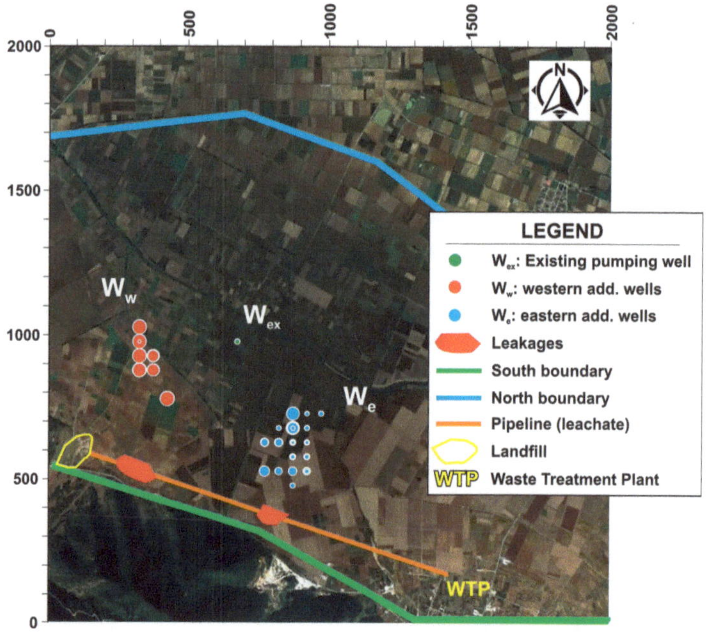

Figure 7. Proposed locations of all additional wells of the 168 acceptable solutions (see Supplementary Materials SM5, sheet 2; radius of each well is proportional to the respective flow rate).

In order to investigate the impact of each decision variable (X_w, Y_w, X_e, Y_e, Q_w, Q_e) and total AWs' flow-rate ΣQ, as well as the remediation duration Dur, the Pearson correlation coefficient (Cc) of each one against FV is calculated together with their standard statistics, median, mean, min, max, standard deviation (Table 3). This is calculated based on all acceptable solutions (Figure 8) and then based on the "best" solutions (Figure 9); these are assumed to be the solutions that exhibit FV < median FV.

Table 3. Pearson correlation coefficients (Cc) for all scenarios between FV and variables.

Statistics	FV (m^3)	X_w (m)	Y_w (m)	X_e (m)	Y_e (m)	Q_w ($-m^3/d$)	Q_e ($-m^3/d$)	ΣQ ($-m^3/d$)	Dur (d)
median	1,355,913	325	875	875	625	2340.0	875.5	3055.6	440
mean	1,390,516	348	875	892	615	2275.0	840.6	3115.9	449
min	1,093,327	325	775	775	475	1798.6	427.4	2373.6	310
max	2,221,325	425	1025	975	725	2498.8	1782.1	3882.8	590
st.dev	211,091	34	45	34	57	207.9	247.1	355.4	67
* r(FV vs. ?) all	N/A	N/A	N/A	N/A	N/A	0.10995	0.48574	0.40204	0.66275
** r(FV vs. ?) best	N/A	N/A	N/A	N/A	N/A	0.27564	−0.47675	−0.10505	0.67096

* for all solutions of RUN 1–6, ** for all solutions with FV < median FV, Abbreviations: X_w, Y_w, X_e, Y_e: coordinates of west and east additional wells; Q_w, Q_e: flow rates of west and east additional wells; ΣQ: total flow-rate of additional wells; Dur: remediation duration.

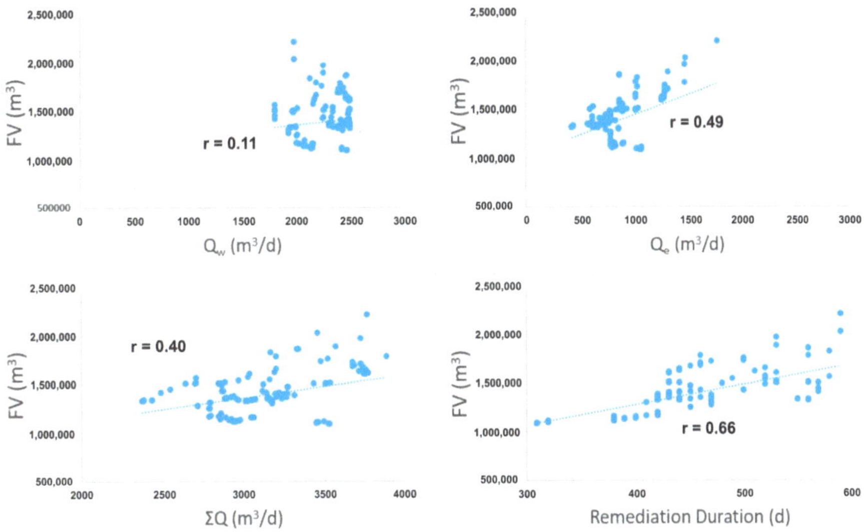

Figure 8. Pearson correlation coefficients of decision variables and FV for all 168 acceptable solutions of all runs (Runs 1–6; see SM5, sheet 5).

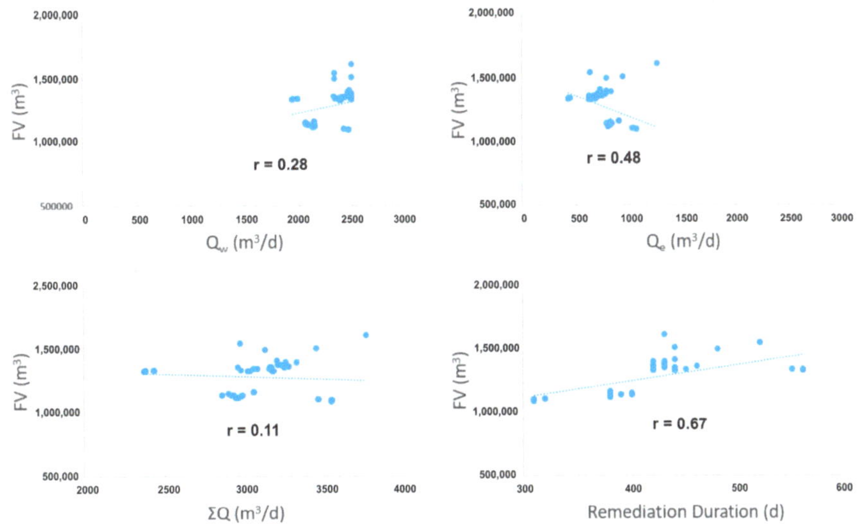

Figure 9. Pearson correlation coefficients of decision variables and FV for "best" (FV < median FV) solutions of all runs (Runs 1–6; see Supplementary Materials SM5, sheet 6).

3.2. From Solutions to Strategies

Depending on the criteria set, various management strategies for the contaminated aquifer can emerge. If, for example, one compares the best overall solution FV1 with the 13th best overall solution (see SM4 and SM5), that is, to compare the best of Run 2 with the best of Run 5, it is recognized that they are likely different strategies. On the other hand, by comparing the best overall solution with the 11th best overall solution, which are essentially the best (1st) and 11th solution of RUN 2, one observes that they belong to the same strategy; it makes sense to store many different solutions even of the same assumed strategy, as they have, e.g., variations in the location of the additional wells.

The process of identifying different strategies in complex management optimization problems is not straightforward and cannot be easily a priori planned or/and generalized. The various different solutions certainly do not constitute different strategies, while local minima do not necessarily represent different strategies. Even standard classification methods that may identify clusters of solutions based on specific rules do not guarantee a categorization process that reflects real logical management diverse strategies. The critical skill of an expert's post-processing of results and solutions is required. The systematic investigation of all acceptable, or at least "best" or "good" solutions, irrelevant to the way they are selected, is a required step-by-step process [22].

1. Generally identify different strategies;
2. Propose the criteria for the initial classification of solutions into strategies;
3. Identify the management strategies;
4. Re-evaluate the criteria and classification ID needed (e.g., if a strategy includes only a single solution or if some solutions are practically very similar);
5. Ultimately, classify solutions into the final strategies.

This process may vary from expert to expert and can be quite objective, but the final result of storing diverse solutions remains unchangeable and is the most important. The studied problem with the current configuration does not offer very diverse optimal solutions and, hence, strategies. The criteria for the strategy identification set here concern the size of the total AWs' flow rate (ΣQ) and its distribution to the AWs (Q_w/Q_e), while the locations of the wells, given that they are not very different (see Figure 7), are considered to simply slightly alter the version of each general strategy.

SM5 presents the results of the effort to identify strategies from the solutions that emerged from the series of runs. The 6 runs produced 168 identified different solutions, which are classified as variants of 7 polluted aquifer remediation strategies. SM5 sheet 1 presents the strategies of all 168 acceptable solutions per run, while sheet 2 presents all acceptable solutions stacked. SM5 sheet 3 presents all 168 acceptable solutions classified per strategy. Table 4 (and SM5, sheet 2) presents the best solution for each one of the 7 identified strategies. The best solution is actually the best of Run 2 and is classified as the best Strategy A version. Each solution in Table 4 corresponds to a video of the pollution mass spread (mass transport) and the remediation process's temporal progression (Supplementary Materials SM2, SM6–SM11). Figure 10 presents the locations of AWs of all solutions that are presented in Figure 7 but classified per strategy. The radius of each well is proportional to its flow rate, while the coupled AWs per solution are connected with a line. Overall statistics per strategy are presented in Table 5, while the frequency of occurrence of each strategy and the run during which it appeared are graphically presented in Figure 11.

Table 4. Best solution of each one of the seven strategies (A–G). See respective videos of the solutions' corresponding remediation process in Supplementary Materials SM2 and SM6–SM11.

RUN	Nr	FV (m^3)	Rank (of 168)	X_w (m)	Y_w (m)	X_e (m)	Y_e (m)	Q_w ($-m^3/d$)	Q_e ($-m^3/d$)	ΣQ ($-m^3/d$)	Q_w/Q_e	Dur (d)	Strategy
2	19	1,093,327	1	425	775	875	575	2467.0	1059.8	3526.9	2.3	310	A
2	7	1,302,589	45	425	775	875	575	2431.0	746.0	3177.0	3.3	410	B
2	6	1,418,388	112	425	775	875	575	1803.4	685.0	2488.4	2.6	570	C
1	22	1,610,269	140	325	875	925	525	2493.3	1251.5	3744.8	2.0	430	D
3	5	1,628,700	147	325	925	925	575	2175.8	1017.7	3193.5	2.1	510	E
4	24	1,329,231	59	325	925	875	675	1946.3	427.4	2373.6	4.6	560	F
5	39	1,115,409	12	375	875	875	625	2134.3	801.0	2935.3	2.7	380	G

Abbreviations: FV: fitness value; X_w, Y_w, X_e, Y_e: coordinates of west and east additional wells; Q_w, Q_e: flow rates of west and east additional wells; ΣQ: total flow rate of additional wells; Dur: remediation duration.

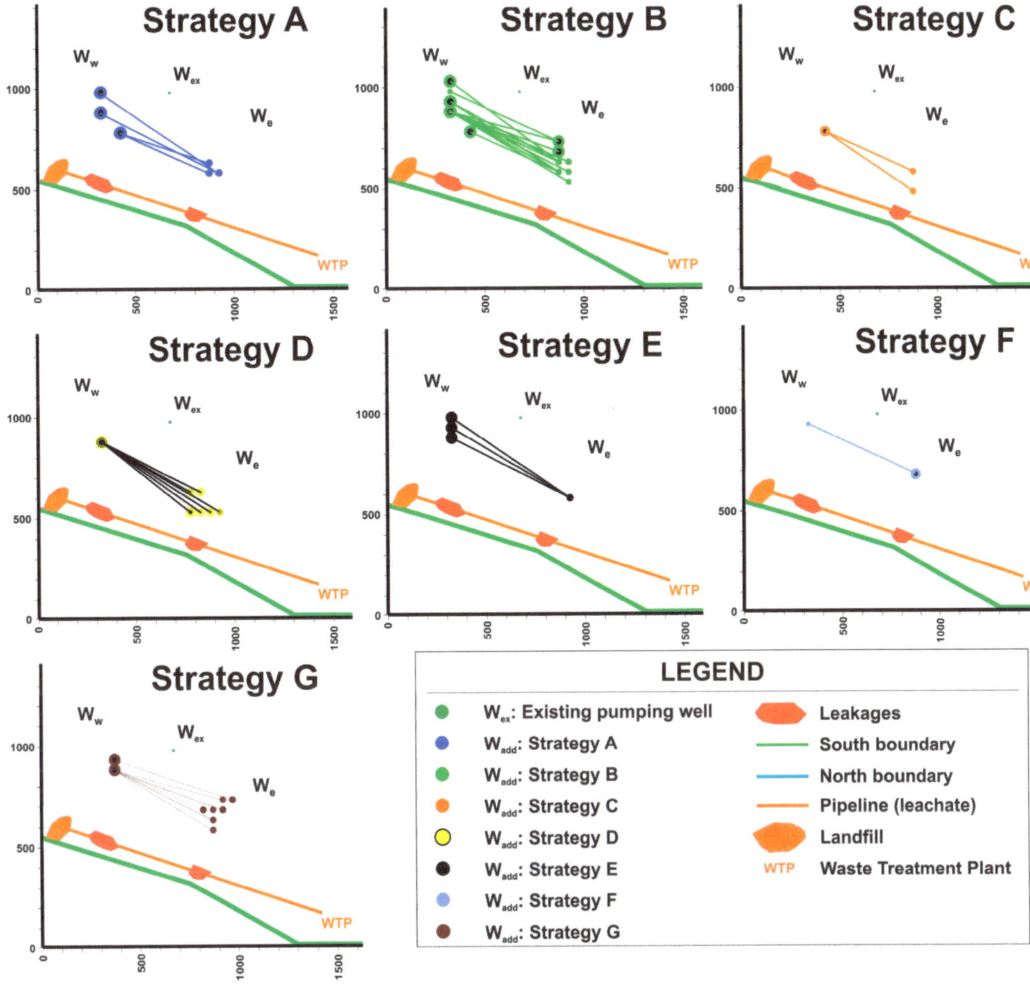

Figure 10. Proposed locations of all additional wells of the 168 acceptable solutions (Figure 7) classified per strategy (see SM5, sheet 4). Wells radii are proportional to their flow rates, while coupled AWs per solution are connected with a line.

Table 5. Overall statistics per strategy.

7 Strategies (168 Versions)	Stats	FV (m^3)	X_W (m)	Y_W (m)	X_e (m)	Y_e (m)	Q_W ($-m^3/d$)	Q_e ($-m^3/d$)	ΣQ ($-m^3/d$)	Q_W/Q_e	Dur (d)
Strategy A (17 versions)	mean	1,234,592	396	810	890	581	2459.6	1044.4	3504.1	2.4	352
	median	1,104,957	425	775	875	575	2467.6	1037.2	3511.0	2.3	320
	min	1,093,327	325	775	875	575	2422.5	1016.5	3448.1	2.3	310
	max	1,739,621	425	975	925	625	2498.2	1068.4	3535.4	2.5	500
	st.dev	214,194	46	59	23	16	25.8	19.0	31.2	0.1	61
Strategy B (69 versions)	mean	1,443,741	325	896	888	600	2411.1	744.0	3155.1	3.5	458
	median	1,396,319	325	875	875	625	2434.1	784.5	3154.8	3.6	440
	min	1,326,716	325	875	875	525	2160.0	625.2	2952.4	2.5	420
	max	1,870,427	325	1025	925	625	2487.2	882.2	3340.0	4.0	560
	st.dev	145,411	0	43	22	29	79.5	88.5	123.6	0.4	40

Table 5. *Cont.*

7 Strategies (168 Versions)	Stats	FV (m^3)	X_w (m)	Y_w (m)	X_e (m)	Y_e (m)	Q_w ($-m^3/d$)	Q_e ($-m^3/d$)	ΣQ ($-m^3/d$)	Q_w/Q_e	Dur (d)
Strategy C (6 versions)	mean	1,495,433	425	775	875	558	1801.8	829.7	2631.5	2.2	568
	median	1,508,938	425	775	875	575	1803.4	872.4	2670.9	2.1	570
	min	1,418,388	425	775	875	475	1798.5	685.0	2488.4	2.0	560
	max	1,568,620	425	775	875	575	1803.4	901.1	2704.5	2.6	580
	st.dev	47,998	0	0	0	37	2.3	85.1	84.3	0.2	7
Strategy D (22 versions)	mean	1,730,032	325	875	884	543	2388.6	1322.6	3711.3	1.8	467
	median	1,668,437	325	875	925	525	2459.7	1281.1	3739.6	1.9	445
	min	1,610,269	325	875	775	525	1982.9	1251.5	3457.9	1.1	430
	max	2,221,325	325	875	925	625	2498.8	1782.1	3882.8	2.0	590
	st.dev	159,148	0	0	58	39	151.0	122.4	89.0	0.2	49
Strategy E (4 versions)	mean	1,731,232	325	938	925	575	2169.4	1022.3	3191.7	2.1	543
	median	1,729,780	325	950	925	575	2180.7	1017.7	3198.4	2.1	540
	min	1,628,700	325	875	925	575	2130.6	1017.7	3166.7	2.1	510
	max	1,836,667	325	975	925	575	2185.6	1036.0	3203.3	2.1	580
	st.dev	86,375	0	41	0	0	22.7	7.9	15.0	0.0	29
Strategy F (11 versions)	mean	1,333,812	325	925	875	675	1960.4	433.3	2393.7	4.5	557
	median	1,332,308	325	925	875	675	1946.9	432.2	2379.1	4.5	560
	min	1,329,231	325	925	875	675	1946.3	427.4	2373.6	4.5	550
	max	1,339,408	325	925	875	675	1992.7	442.6	2435.3	4.6	560
	st.dev	3576	0	0	0	0	20.0	5.2	25.0	0.0	4
Strategy G (39 versions)	mean	1,216,487	375	881	906	658	2075.6	837.8	2913.4	2.5	418
	median	1,159,719	375	875	925	675	2091.0	821.1	2912.1	2.5	390
	min	1,115,409	375	875	825	575	1929.2	784.5	2714.3	2.2	380
	max	1,609,737	375	925	975	725	2164.2	907.8	3062.9	2.7	530
	st.dev	131,585	0	17	39	51	78.9	45.0	99.5	0.1	51

Abbreviations: FV: fitness value; X_w, Y_w, X_e, Y_e: coordinates of west and east additional wells; Q_w, Q_e: flow rates of west and east additional wells; ΣQ: total flow rate of additional wells; Dur: remediation duration.

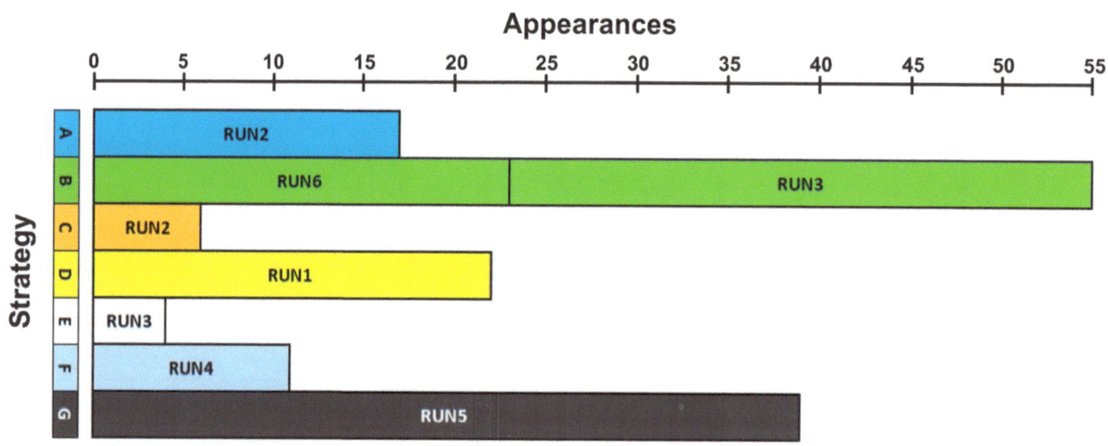

Figure 11. Frequency of occurrence of each strategy and number of appearances per run.

3.3. Discussion of Results

From the results and their post-processing, it is evident that the positioning of the additional wells (AWs) is key for the protection of the existing water supply well W_{ex}. Each AW is carefully positioned along the flow paths connecting each leak to W_{ex}, with a deviation to the external direction (especially the western W_w to the west and the eastern W_e lightly to the east) so that the operation of the AWs does not accelerate the plumes directly towards W_{ex}, also accounting for the given south to north natural groundwater flow. Specifically, W_w is of great importance, as expected, since the western leakage is more extensive. The position of W_w does not vary greatly in the various solutions, as indicated by the value of the standard deviation (see Tables 3 and 5 and Supplementary Materials SM4 and SM5). The position of W_e has its own high but lower than W_w significance and, therefore, shows a greater deviation of its coordinate values in the produced solutions (see Tables 3 and 5 and Supplementary Materials SM4 and SM5). This is also particularly evident in Figure 7.

A similar situation occurs with the flow rates of AWs, but when studying the results of the Pearson correlation coefficient values (Cc or r), one must be careful in their interpretation (Table 3 and Figures 8 and 9). Cc or r for FV-Q_w is 0.10995 for all solutions and 0.27564 for the "best" selected ones (FV < median FV). Therefore, there is an absence of a unilateral correlation of the FV value with Q_w in both cases. On the other hand, Cc or r for FV-Q_e is calculated to be 0.48574 and −0.47675 for all solutions and the "best" ones, respectively. This means that in the full solutions dataset, FV tends to decrease relatively proportionally with Q_e, but in the best solutions dataset, FV tends to be relatively inversely proportional to the Q_e value. To avoid misinterpretation of these results, the correlation values must be studied under the following perspective: the values used to calculate the correlation coefficients are not random and do not represent the entire search space, hence the full domain of the definition of each variable. These are a group of solutions, each one being the best solution of each generation of the six runs. That is, they are acceptable solutions but relatively poor quality-wise at the beginning (with a large FV in a minimization optimization problem), while later, there are solutions around a local minimum until a new local minimum is found. In this context, the odd conflicting values mentioned indicate the complex nature of the optimization problem with the highly entangled variables. The odd correlations can also be explained by the fact that the problem's configuration (western leakage larger and differently positioned against W_{ex} than the eastern leak) leaves no room for many locations' testing for W_w, while the algorithm can more easily test more locations for W_e (see Figure 7).

Regarding ΣQ and its correlation with FV, it is r(FV-ΣQ) = 0.40204 and −0.10505 for all solutions and the "best", respectively. No serious conclusion emerges, and only with indirect assumptions/hypotheses, as above, could one interpret these results. On the other hand, the duration of the remediation process (Dur) seems to exhibit higher importance than ΣQ in the hunt for minimized FV. The respective correlation coefficient for FV-Dur is calculated to be 0.66275 and 0.67096 for all solutions and the "best", respectively. This indicates a consistency for the remediation duration to affect the FV value proportionally in the entire spectrum of directed randomness tests of the GA.

Regarding the values of MP and the degree it affects the convergence to the optimal solution, or at least to the best algebraic solution found in the series of runs of this thesis, no safe conclusions can be drawn, as an extensive series of runs was not performed. However, the best solution emerged for RUN 2 with MP = 0.025, i.e., approximately equal to 1.7/SL, where SL is the chromosome length, supporting our previous research [16] that proposed values of 2–2.5/SL as the best approximations of MP value.

In order for anyone to replicate the results or adjust the methodology in similar or more complex problems, some final remarks are needed. The complexity of the multi-objective optimization problem is proposed to be reduced by the simplification of the representation of the remediation cost by simply linking it to the groundwater volume pumped by additional wells. This converts the problem into a single-objective one. Even simplified, the minimization problem still requires the simplification of the groundwater flow and mass transport model due to the computationally greedy nature of metaheuristic algorithms, like genetic algorithms. This is why a simplified 2D flow and advection–dispersion model is simulated in Modflow (Flopy).

Simple elitist binary genetic algorithms may be used, but they are complemented by a sophisticated post-processing of results/solutions that delivers alternative management strategies. The full proposed methodology includes:

1. Conceptual model setup (i.e., steady-state flow, 2D flow field, boundary conditions, advection–dispersion mass transport mechanisms);
2. Numerical model set up (Flopy controlling Modflow model in Python);
3. Decision variables' definition (coordinates and flow rates of additional wells);
4. The objective function to be minimized delineation (total groundwater volume pumped by additional wells);
5. Identification of physical, numerical, and other constraints and selection of suitable constraint handling techniques (penalty imposition, repair of infeasible solutions etc.);
6. Development of the simulation–optimization model coupling a metaheuristic optimization method (here, simple elitist genetic algorithms with the Flopy-controlled Modflow groundwater flow and mass transport numerical model);
7. Execution of a series of test simulations to find the fittest Penalty parameters;
8. Execution of final simulations storing all results/solutions.
9. Post-processing of results, defining and selecting "acceptable" and "good" or "best" solutions (e.g., solutions exhibiting FV < median FV);
10. General identification of various strategies in the selected solutions' dataset;
11. Proposal of criteria for the initial classification of solutions into strategies and identification of management strategies. Re-evaluation of criteria and classification as much as needed, and classification of solutions into final strategies.

Constrained, nonlinear, stochastic, multi-objective water resources management optimization problems, like the one studied here, are extremely complex. Thus, state-of-the-art metaheuristic methods cannot guarantee that the algebraically optimal solution will be discovered. Moreover, the necessary simplification of the conceptual and numerical models, due to the computation load metaheuristics entail, adds more uncertainties. These are further enhanced by the sensitivity of optimization results to external environmental, socioeconomic, political, and geopolitical factors, e.g., climate change, political unrest, conflicts, poor or non-existent management policies, and lack of central planning. Thus, the search for the algebraically optimal solution is myopic.

In this context, the real practical goal of this research and similar strategy-based environmental management problems is not just to find a single optimal solution shadowed by the intrinsic and acquired uncertainties mentioned above but to create a pool of viable alternative (sub)optimal strategies. This approach provides a robust optimization and decision support tool for management authorities, especially useful in the design phase of pollution control measures but also to be prepared in case of emergency remediation planning due to sudden pollution detection. The methodology developed here generates numerous algebraic variations of pollution control and pump-and-treat strategies, offering decision-makers a broad spectrum of choices, varying in remediation duration, positioning of additional pumping wells, and their operation (flow rates) in a simplified but inclusive and easy-to-comprehend fashion. This variety is particularly advantageous when initial design parameters change, such as modifications in constraints or budget alterations, allowing for adaptive and flexible management without the need for initiating new costly and time-consuming investigations. Examples of such variations in external factors in the studied problem could be the following (for more, investigate SM5, sheet 2):

1. Constructing a pumping well at the location proposed by the optimal solution (FV1; Strategy A; see Supplementary Materials SM5, sheet 2) in the east (W3 in Figure 3) could be prohibited. The reasons could be problems in expropriating or using specific private land or false geological data of specific locations. In this situation, the managing authorities could easily select another solution with the eastern well W_e in a different area, like the solution of Run2-Nr11, which is a Strategy A identical solution with the only exception of We positioned 50 m to the east, with an additional burden of FV +1.31% and a prolonged remediation period of just 10 days while practically

retaining the flow rates of additional wells. If this is not enough, the decision makers can select a more diverse solution, e.g., the solution of Run5-Nr24, which is a Strategy G solution with +5.92 FV value and W_e positioned 50 m to the north and 100 m to the east (see SM5, sheet 2).

2. Solution Run5-Nr24 also features W_w positioned 50 m to the west and 100 m to the north, exhibiting lower flow rates of additional wells (−18% in ΣQ values). Thus, it could serve as an alternative solution/strategy if the construction of a well in the west area that FV1 suggests is prohibited or if additional hydraulic head drawdown constraints or an energy cost increase (like the one experienced in recent years) dictated decreased pumping costs and, hence, flow rates.

4. Conclusions

Current research demonstrates that genetic algorithms can effectively optimize the location and operation of additional pumping wells in a confined aquifer remediation scenario. This optimization significantly reduces the remediation cost while maintaining the operational integrity of existing water resources. The simulation–optimization Modflow_GA modeling approach provides valuable insights into the dynamics of pollutant dispersion in confined aquifers, highlighting the critical factors influencing remediation effectiveness and efficiency in pump-and-treat or/and hydraulic pollution control. The research confirms the practical applicability of simulation–optimization models in real-world scenarios, where decision-makers can utilize these strategies to plan and execute aquifer remediation with optimized resource utilization. Findings indicate that using advanced simulation techniques combined with genetic algorithms can substantially lower the costs associated with extensive groundwater remediation projects by simply minimizing the volume of water that needs to be treated. Research illustrates that the adoption of simulation–optimization frameworks in environmental policy and management strategies can ensure the sustainable and cost-effective remediation of polluted aquifers. It also produces the benefits of using sophisticated post-processing analyses of simulation-optimization results to identify optimal and sub-optimal remediation strategies, offering a robust decision-support tool for complex environmental management problems.

As far as future research is concerned, the methodology of the coupled simulation–optimization model, complemented by detailed expert-based post-processing and meta-analysis of results/solutions, creating a pool of (sub)optimal solutions classified into strategies can be updated and enhanced. The simplified objective function to be minimized can be more realistic, representing the real remediation cost, including additional pumping well construction and operation cost and the pipe network conveying the pumped polluted water to the wastewater treatment facility. An investigation could be carried out concerning the energy cost and how its variation can affect the optimal and sub-optimal solutions and strategies. Also, there could be a provision for the return of a part of the treated/filtered water to the aquifer to prevent a significant drop in water levels, using recharging wells, where the positions and flows of these should be subject to optimization. Additionally, constraints can be added concerning the max limit of hydraulic head drawdown in the aquifer, while solutions could be filtered by testing them in emergency potential scenarios of sudden short interruptions in the operation of one of the additional pumping wells so that solutions are cost-effective but also safe. The simulation of a complex coupled shallow and confined leaky aquifer (divided by a semi-permeable layer) or the addition of fractures would also add to the realism and, of course, the complexity of the problem. Moreover, further research should be carried out into integrating more complex hydrogeological variables into the modeling process to enhance the accuracy and applicability of the results in diverse geological settings. Finally, it would be beneficial to explore the application of multi-objective optimization algorithms, such as NSGA-II [33], as a means to evaluate their performance in comparison to the current methodology within the context of this intricate aquifer management problem.

Supplementary Materials: The supporting information can be downloaded at: https://drive.google.com/drive/folders/13WP_Our3w_YeX3R_PMf0LahHkAMfisMp?usp=sharing (accessed on 7 March 2024), SM0: List of Supplementary Materials with titles and descriptions.pdf; SM1: Video1-do_nothing_scenario.wmv; SM2: Video2-Best_sol_Strategy_A.wmv; SM3: raw output files of Modflow_GA runs.zip; SM4: Runs_1-6_raw_results.xlsx; SM5: Post-processed_results.xlsx; SM6: Video3-Best_sol_Strategy_B.wmv; SM7: Video4-Best_sol_Strategy_C.wmv; SM8: Video5-Best_sol_Strategy_D.wmv; SM9: Video6-Best_sol_Strategy_E.wmv; SM10: Video7-Best_sol_Strategy_F.wmv; SM11: Video8-Best_sol_Strategy_G.wmv.

Funding: This research received no external funding.

Data Availability Statement: All data used in this paper are synthetically simulated as the studied problem and aquifer are theoretical. All parameters and relevant input variable values are cited in the paper.

Conflicts of Interest: The author declare no conflicts of interest.

References

1. GIDEON: Global Infectious Diseases and Epidemiology Online Network. Available online: www.gideononline.com (accessed on 1 April 2024).
2. Kulinkina, A.V.; Shinee, E.; Guzmán Herrador, B.R.; Nygård, K.; Schmoll, O. *The Situation of Water-Related Infectious Diseases in the Pan-European Region*; World Health Organization. Regional Office for Europe: Copenhagen, Denmark, 2016. Available online: https://apps.who.int/iris/handle/10665/329534 (accessed on 2 April 2024).
3. Craun, G.F.; Brunkard, J.M.; Yoder, J.S.; Roberts, V.A.; Carpenter, J.; Wade, T.; Calderon, R.L.; Roberts, J.M.; Beach, M.J.; Roy, S.L. Causes of outbreaks associated with drinking water in the United States from 1971 to 2006. *Clin. Microbiol. Rev.* **2010**, *23*, 507–528. [CrossRef]
4. Williams, P.T. *Waste Treatment and Disposal*; John Wiley & Sons: Hoboken, NJ, USA, 2005.
5. Gupta, S.; Mohan, K.; Prasad, R.; Gupta, S.; Kansal, A. Solid waste management in India: Options and opportunities. *Resour. Conserv. Recycl.* **1998**, *24*, 137–154. [CrossRef]
6. Renou, S.; Givaudan, J.; Poulain, S.; Dirassouyan, F.; Moulin, P. Landfill leachate treatment: Review and opportunity. *J. Hazard. Mater.* **2008**, *150*, 468–493. [CrossRef]
7. Tatsi, A.A.; Zouboulis, A.I. A field investigation of the quantity and quality of leachate from a municipal solid waste landfill in a Mediterranean climate (Thessaloniki, Greece). *Adv. Environ. Res.* **2002**, *6*, 207–219. [CrossRef]
8. Mayer, A.S.; Kelley, C.T.; Miller, C.T. Optimal design for problems involving flow and transport phenomena in saturated subsurface systems. *Adv. Water Resour.* **2002**, *25*, 1233–1256. [CrossRef]
9. Medina, M.A., Jr. Modeling ground water contamination and surface-subsurface interactions. In *Ground Water Pollution: Proceedings of the International Conference on Water and Environment (WE-2003), December 15–18, 2003, Bhopal, India*; Singh, V.P., Yadava, R.N., Eds.; Allied Publishers: Bhopal, India, 2003; p. 401e18.
10. Mulligan, A.E.; Ahlfeld, D.P. Advective control of groundwater contaminant plumes: Model development and comparison to hydraulic control. *Water Resour. Res.* **1999**, *35*, 2285–2294. [CrossRef]
11. Guan, J.; Aral, M.M. Optimal remediation with well locations and pumping rates selected as continuous decision variables. *J. Hydrol.* **1999**, *221*, 20–42. [CrossRef]
12. Gorelick, S.M. Sensitivity analysis of optimal groundwater contaminant capture curves: Spatial variability and robust solutions. In Proceedings of the National Water Well Association Conference: Solving Groundwater Problems with Models, Denver, CO, USA, 10–12 February 1987; National Water Well Association: Westerville, OH, USA, 1987; p. 133e46.
13. Bayer, P.; Finkel, M.; Teutsch, G. Reliability of hydraulic performance and cost estimates of barrier-supported pump-and-treat systems in heterogeneous aquifers. In Proceedings of the Calibration and Reliability in Groundwater Modelling: A Few Steps Closer to Reality, Prague, Czech Republic, 17–20 June 2002; Ka, K., Hrkal, Z., Eds.; IAHS Publisher: Oxon, UK, 2002. no 227. p. 331e8.
14. Nagkoulis, N.; Katsifarakis, K.L. Cost minimization of groundwater supply to a central tank. *Water Supply* **2021**, *22*, 2055–2066. [CrossRef]
15. Antoniou, M.; Theodosiou, N.; Karakatsanis, D. Coupling groundwater simulation and optimization models, using MODFLOW and Harmony Search Algorithm. *Desalination Water Treat.* **2017**, *86*, 297–304. [CrossRef]
16. Kontos, Y.N.; Katsifarakis, K.L. Optimization of Management of Polluted Fractured Aquifers Using Genetic Algorithms. *Eur. Water* **2012**, *40*, 31–42.
17. Kontos, Y.N.; Katsifarakis, K.L. Optimal Management of a Theoretical Coastal Aquifer with Combined Pollution and Salinization Problems, Using Genetic Algorithms. *Energy* **2017**, *136*, 32–44. [CrossRef]
18. Kontos, Y.N.; Katsifarakis, K.L. Genetic Algorithms in Polluted Aquifers' Management. Transactions on Information Science and Applications. *World Sci. Eng. Acad. Soc.* **2017**, *14*, 190–200.

19. Nagkoulis, N.; Kontos, Y.N.; Katsifarakis, K.L. Pumping schedule assignment using chromatic graphs to reduce groundwater pumping energy consumption. *Water Supply* **2022**, *22*, 7618–7634. [CrossRef]
20. Nagkoulis, N.; Katsifarakis, K.L. Using Alternate Pumping and Cooperative Game Theory to Reduce Sea Water Intrusion. *Groundwater* **2023**, *61*, 35–43. [CrossRef] [PubMed]
21. Kontos, Y.N.; Kassandros, T.D.; Perifanos, K.; Karampasis, M.; Katsifarakis, K.L.; Karatzas, K.D. Machine Learning for Groundwater Pollution Source Identification and Monitoring Network Optimization. *Neural Comput. Appl.* **2022**, *34*, 19515–19545. [CrossRef] [PubMed]
22. Kontos, Y.N.; Rompis, I.; Karpouzos, D. Optimal pollution control and pump-and-fertilize strategies in a nitro-polluted aquifer, using Genetic Algorithms and Modflow. *Agronomy* **2023**, *13*, 1534. [CrossRef]
23. Langevin, C.D.; Hughes, J.D.; Banta, E.R.; Provost, A.M.; Niswonger, R.G.; Panday, S. *MODFLOW 6 Modular Hydrologic Model Version 6.2.2*; U.S. Geological Survey Software Release; U.S. Geological Survey: Reston, VA, USA, 2021.
24. Katsifarakis, K.L. Flow Simulation and Optimal Management of Groundwater Resources. The Balance between Accuracy and Computational Efficiency. *Groundw. Model. Manag. Contam.* **2008**, *10*, 291–308.
25. Janetti, E.B.; Guadagnini, L.; Riva, M.; Guadagnini, A. Global sensitivity analyses of multiple conceptual models with uncertain parameters driving groundwater flow in a regional-scale sedimentary aquifer. *J. Hydrol.* **2019**, *574*, 544–556. [CrossRef]
26. Schiavo, M. The role of different sources of uncertainty on the stochastic quantification of subsurface discharges in heterogeneous aquifers. *J. Hydrol.* **2023**, *617*, 128930. [CrossRef]
27. USGS. Python Package for Creating, Running, and Post-Processing MODFLOW-Based Models. USGS Flopy. 2021. Available online: https://www.usgs.gov/software/flopy-python-package-creating-running-and-post-processing-modflow-based-models (accessed on 20 August 2022).
28. USGS. Flopy Documentation 3.3.4. 2021. Available online: https://flopy.readthedocs.io/en/3.3.4/index.html (accessed on 20 August 2022).
29. Bakker, M.; Post, V.; Langevin, C.D.; Hughes, J.D.; White, J.T.; Starn, J.J.; Fienen, M.N. *FloPy: Python Package for Creating, Running, and Post-Processing MODFLOW-Based Models*; U.S. Geological Survey: Reston, VA, USA, 2016.
30. Bakker, M.; Post, V.; Langevin, C.D.; Hughes, J.D.; White, J.T.; Starn, J.J.; Fienen, M.N. Scripting MODFLOW Model Development Using Python and FloPy. *Groundwater* **2016**, *54*, 733–739. [CrossRef]
31. Bakker, M.; Post, V.; Langevin, C.D.; Hughes, J.D.; White, J.T.; Leaf, A.T.; Paulinski, S.R.; Larsen, J.D.; Toews, M.W.; Morway, E.D.; et al. FloPy v3.3.4. USGS 2021. Available online: https://doi.org/10.5066/F7BK19FH (accessed on 10 November 2023).
32. Le Riche, R.; Knopf-Lenoir, C.; Haftka, R.T. A segregated genetic algorithm for constrained structural optimization. In Proceedings of the 6th International Conference on Genetic Algorithms, San Francisco, CA, USA, 15–19 July 1995; Morgan Kaufmann Publishers Inc.: Burlington, MA, USA, 1995; pp. 558–565.
33. Deb, K.; Pratap, A.; Agarwal, S.; Meyarivan, T. A fast and elitist multiobjective genetic algorithm: NSGA-II. *IEEE Trans. Evol. Comput.* **2002**, *6*, 182–197. [CrossRef]

hydrology

MDPI

Article

Determination of Contaminant Transport Parameters for a Local Aquifer by Numerical Modeling of Two Plumes: Trichloroethylene and Hexavalent Chromium

Mahade Ibn Salam [1,*,†], Brian Waldron [1,†], Scott Schoefernacker [2] and Farhad Jazaei [1]

[1] Department of Civil Engineering, University of Memphis, Memphis, TN 38152, USA; bwaldron@memphis.edu (B.W.); fjazaei@memphis.edu (F.J.)
[2] Department of Earth Science, University of Memphis, Memphis, TN 38152, USA; scott.s@memphis.edu
* Correspondence: mahaderony@outlook.com
† These authors contributed equally to this work.

Abstract: The municipal wellfield in Collierville, Tennessee, is contaminated with trichloroethylene (TCE) and hexavalent chromium (Cr (VI)) due to industrial operations dating back to the 1970s and 1980s. This study aims to elucidate the aquifer's contaminant transport mechanisms by determining longitudinal and transverse dispersivities through inverse modeling. Utilizing MT3DMS for contaminant transport simulation, based on a well-calibrated groundwater flow model, and leveraging Python's multiprocessing library for efficiency, the study employs a trial-and-error methodology. Key findings reveal that longitudinal dispersivity values range from 5.5 m near the source to 20.5 m further away, with horizontal and vertical transverse dispersivities between 0.28 m and 3.88 m and between 0.03 m and 0.08 m, respectively. These insights into the aquifer's dispersivity coefficients, which reflect the scale-dependent nature of longitudinal dispersivity, are crucial for optimizing remediation strategies and achieving cleanup goals. This study underscores the importance of accurate parameter estimation in contaminant transport modeling and contributes to a better understanding of contaminant dynamics in the Collierville wellfield.

Keywords: dispersivity; contaminant transport; groundwater pollution; numerical modeling; parameter estimation

Citation: Ibn Salam, M.; Waldron, B.; Schoefernacker, S.; Jazaei, F. Determination of Contaminant Transport Parameters for a Local Aquifer by Numerical Modeling of Two Plumes: Trichloroethylene and Hexavalent Chromium. *Hydrology* **2024**, *11*, 88. https://doi.org/10.3390/hydrology11070088

Academic Editors: Pantelis Sidiropoulos and Augustina Clara Alexander

Received: 5 April 2024
Revised: 6 June 2024
Accepted: 8 June 2024
Published: 21 June 2024

1. Introduction

Predicting the propagation of dissolved contaminants in an aquifer requires adequate characterization of transport parameters. Due to the unknown complexity of the subsurface system, these parameters are often assumed and rarely quantified at the required spatial resolution. Many transport models implement the advection–dispersion equation, which considers advection, mechanical dispersion, and diffusion along with other processes such as reaction and retardation. This dispersion parameter is crucial in describing the spreading of solute due to heterogeneity and is difficult to quantify due to its scale-dependent nature [1].

In many investigations, dispersivity is estimated through adjustments until the modeled concentrations best approximate observed concentrations [2–12]. An alternative approach to estimating dispersivity is through field and laboratory experiments conducted at different spatial scales [12–22]. However, field tracer tests are often time-consuming and very expensive. Some researchers attempted to upscale the laboratory-derived values to apply in field experiments [23], as well as to examine the scale-dependent nature [24] with comparison to field values [25]. However Refs. [26,27] showed that field-scale macrodispersion is intricately linked to subsurface heterogeneity, particularly affected by the integral scale and variance of log-conductivity and variance; thus, it cannot be assessed from the laboratory-scale experiment, which provides only a local dispersion.

Among the works conducted on dispersivities, ref. [28] conducted a critical review of 59 different field sites and found that longitudinal dispersivity ranged from 0.01 to 10^4 m for observational scales ranging from 0.1 to 10^5 m. Improving on that, [11] compiled 307 values of longitudinal dispersivity from 109 authors consisting of lab experiments, aquifer tests, and numerical models representing various types of aquifer media. They found that longitudinal dispersivity increases following a power law with the scale of measurement. Also highlighted in their study was a wide range of reported longitudinal dispersivity, 0.0003 to 30.5 m for an observation distance of 0.2 to 8800 m in unconsolidated sediments, and 0.0016 to 48.7 m at an observation distance of 0.035 and 3066 m in consolidated sediment [11]. These findings underscore the significant variability in longitudinal dispersivity across different scales and aquifer types, highlighting the complexity of accurately predicting solute transport in groundwater.

Numerical models offer a robust way to simulate contaminant transport on a larger scale in all three dimensions [3,4,29–34]. For a numerical model to simulate contaminant transport, a well-calibrated groundwater flow model is required to represent the flow field adequately. Given a properly defined boundary condition and starting concentration, a transport model can then predict the movement of a contaminant using the flow model results. Contaminants usually infiltrate groundwater systems via recharge mechanisms, subsequently dispersing through advection and dispersion processes. Calculating the hydrodynamic dispersion coefficient necessitates having data on the initial contaminant (or source) concentration and observed concentrations over time, derived from either an active contaminant plume or tracer studies. In numerical modeling, it is standard practice to apply a Dirichlet boundary condition (fixing concentration) at certain nodes of the grid, thereby generating a solute flux that results from both advective and dispersive movements [35]. Defining this initial source condition is crucial to avoid non-uniqueness against different dispersivity values, which is often very difficult to obtain in field studies.

In Shelby County, Tennessee, the primary drinking water aquifer is the semi-confined unconsolidated Memphis aquifer [36,37]. The aquifer is threatened by the presence of localized preferential pathways (i.e., breaches) in the overlying aquitard (the upper Claiborne confining unit—UCCU), warranting concern that contaminants could easily bypass the aquitard's natural protection [38–48]. Very little is known about the contaminant transport properties of the Memphis aquifer. In Shelby County, Tennessee, one of the Town of Collierville's five wellfields has been impacted by two contaminants: trichloroethylene (TCE) and hexavalent chromium (Cr(VI)). There exists limited source concentration data and few periodic measures of contaminant concentrations in downgradient monitoring wells. This site was chosen to ascertain estimated dispersivity values for the Memphis aquifer that may be used as initial values elsewhere in Shelby County owing to historical industrialization since the early 1900s and the potential contamination threat to other production wells (there are approximately 200 production wells in Shelby County). Scaled dispersivities are determined through a trial-and-error calibration approach between two plumes in the same area, something that is novel compared to other studies on scaled dispersivity where only a single contaminant plume was modeled. Results will show a relative increase in dispersivity with distance; however, there is some discrepancy in dispersivities over the same distance between plumes which is of interest. MODFLOW-NWT serves as the groundwater model and FloPy enables numerous MT3D simulations to arrive at scaled dispersivity values.

2. Materials and Methods

2.1. Study Area

In southeastern Shelby County, Tennessee, the Town of Collierville was selected because of the impact on one of Collierville's five wellfields by two U.S. Environmental Protection Agency (EPA) Superfund sites, which include TCE and Cr (VI) as contaminants of concern (Figure 1). As shown in Figure 1, Collierville's Wellfield #2 (WF2) is positioned north-northwest and downgradient from the Carrier facility and is 0.8 km to the west and also downgradient from Smalley Piper. WF2 came online in 1967, producing 6166 m^3/d of water for domestic consumption from two wells that are screened in the Memphis aquifer at depths of 87.5 m (COLL-201) and 98.7 m (COLL-202).

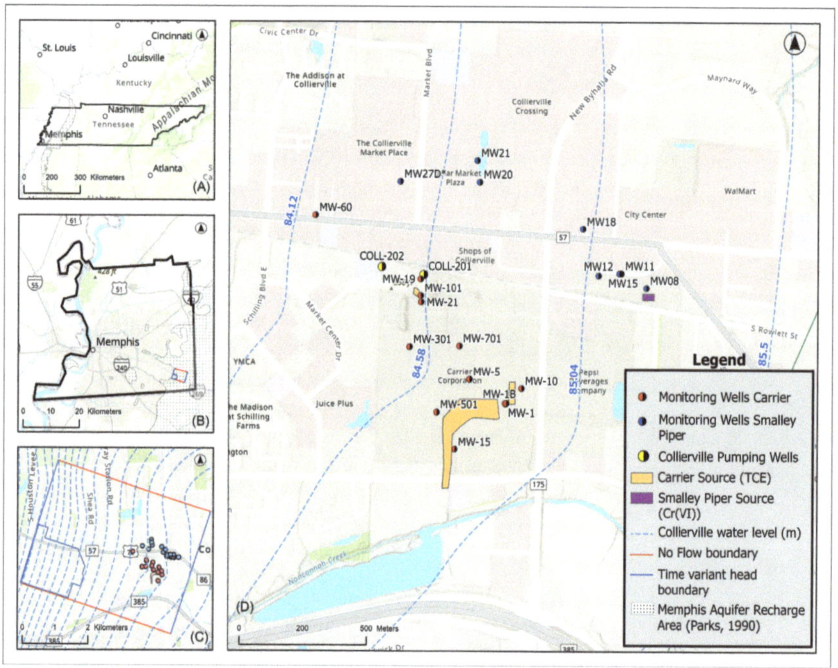

Figure 1. All monitoring well locations for Carrier and Smalley Piper: (**A**) State of Tennessee, (**B**) CAESER II model boundary, (**C**) Submodel Boundary, (**D**) Study Area.

Shelby County sits within nearly square in the middle of the the northern Mississippi embayment, a basin filled with over a thousand meters of unconsolidated sediments from the Cretaceous to Quaternary periods, as detailed by [27,49]. This results in a complex geological structure featuring gently dipping early to mid-Tertiary geologic units of unconsolidated sand, silt, clay, and minor lignite, as described in the Coastal Plain setting, with Pleistocene and Pliocene fluvial-terrace deposits of sand and gravel, 0 to 20 m thick, atop them [50]. Loess deposits, 3 to 20 m thick, excluding the valleys filled with up to 15 m of late Pleistocene to Holocene alluvium, cover the landscape, as shown by [51].

Hydrostratigraphically, the area is divided into several units: the loess and upper alluvium act as a leaky, weak confining layer, with similar textural and hydraulic properties [51], while the fluvial-terrace and lower alluvium deposits form the Shallow aquifer [27]. Below this, the Cockfield and Cook Mountain formations create the upper Claiborne confining unit to the sand-dominated Memphis aquifer, which has a significant transmissivity (average 3.25×10^3 m^2/day) or hydraulic conductivity (average 14.9 m/day) [38,51–53]. The Flour Island Formation then serves as the lower confining unit to the Memphis aquifer and the upper confining unit to the Fort Pillow aquifer, further highlighting the region's intricate hydrogeological structure.

The Upper Claiborne Confining clay (UCCU), upper aquitard to the Memphis aquifer, exists under much of Shelby County except to the east where there are subcrops and the Memphis aquifer transitions from confined to unconfined—this transition occurs beneath Collierville [27,54], thereby making it easier for anthropogenic contaminants to reach the aquifer [27,38].

2.1.1. Carrier Corporation

The Carrier site began operations in 1967, producing air conditioning units. In 1979, a vapor degreaser unit failed, leaking approximately 7500 to 19,000 L of TCE on the southeast side of the property (Figure 1). Carrier also had an unlined 6 m^3 lagoon operational since 1972 (Figure 1) for storing TCE-contaminated paint sludge. This lagoon was filled in 1980. After a heavy rainfall in 1985, Carrier discovered that an unknown quantity of TCE leaked from underground pipes associated with an above-ground TCE storage tank (Figure 1), all of which was removed within a month after the spill. Approximately 2052 L of TCE were recovered. The following year, in 1986, low levels of TCE were detected in the nearby wellfield (WF2), subsequently listed as a U.S. EPA Superfund site in 1990. WF2 continued to operate until December 2003, until hexavalent chromium from the nearby Smalley Piper was detected in the WF2 production wells.

Forty-three monitoring wells were installed between 1986 and 2016 for monitoring and remediation purposes at the site. An expansion of the main building in 2004 closed many wells whose historical locations could not determined. Among the remaining wells, 19 wells were determined to have either concentration data, groundwater level data, or both at various times. The locations of all the wells are shown in (Figure 1). Among those 19 wells, 12 were selected to use in this study. Monitoring wells MW-3, MW-4, and MW-601 are hydrologically upgradient wells (Figure 1). Based on historical concentration measurements onsite, MW-3 had a concentration one order of magnitude higher than MW-4 in the period of 1987–1989; however, these two wells are spatially 2.34 m apart, are separated 9 m vertically, and are screened within the same hydrogeologic unit. Subsequently, no reason can be offered as to their drastic differences in concentration, thereby making them not suitable to use in the numerical model as they fall in the same grid cell where an average concentration is calculated. Six wells among the twelve, namely, MW-1, MW-1B, MW-10, MW-15, MW-19, and MW-21, were used to determine the source concentration of the three sources (1979 spill, 1985 spill, and unlined lagoon). The remaining six wells (MW-5, MW-101, MW-301, MW-501, MW-701, and MW-60) were used to find dispersivity values and are shown in (Figure 1).

2.1.2. Smalley Piper

The second Superfund site, Smalley Piper, manufactured magnesium battery casings from 1970 to 1981. This site had two tarpaulin-lined ponds for treating chromic acid-induced wastewater [55] (Figure 1). Approximately 2 m^3 of chromic acid were discharged once a week via an underground pipe into the treatment ponds, where liquid sulfur dioxide (SO_2) was injected twice per week to precipitate chromic sulfide as less toxic trivalent chromium. However, the reaction efficiency depended on the quantity of SO_2 and mixing, and incomplete reactions resulted in hexavalent chromium present in the wastewater [55].

The remaining wastewater (100 m^3/day) containing hexavalent chromium was discharged into onsite drainage ditches that flowed into Nonconnah Creek, approximately 0.4 km south of the facility. The manufacturing operation ceased in 1981–1982, and the treatment ponds were closed. Since 2007, the former industrial processing buildings on the western portion of the site remained, while the eastern portion of the site had been redeveloped and is currently operating as a public mini-storage facility [56].

A groundwater remediation investigation conducted in 2004 determined that the highest concentrations of Cr(VI) were found 23 to 30.5 m below land surface (bls), with concentrations ranging from 20,000 to more than 250,000 micrograms per liter (µg/L). Groundwater samples from depths greater than 30.5 m bls exhibited total chromium concentrations below the 100 µg/L maximum contaminant level (MCL) specified by the World Health Organization and U.S. EPA [57]. Chromium contamination was not detected in groundwater collected from the eastern part of the property boundary, which is hydrologically upgradient of the site (Figure 1). A total of 21 monitoring wells were installed between 2002 and 2006. Among those twenty-one wells, eight wells (Figure 1) were selected for this study due to the quality of data. Among these eight wells, MW-8 was used to define the source concentration and the remaining seven wells (MW-11, MW-15, MW-12, MW-18, MW-20, MW-21, and MW-27D) were used to determine dispersivity values.

2.1.3. Numerical Flow Model

The Collierville area was included in a prior, calibrated groundwater model (i.e., CAESER II) that simulated groundwater conditions beneath Shelby County between 1960 and 2021 within the shallow Memphis and Fort Pillow aquifers [58]. CAESER-II is an updated model based on [59] (i.e., CAESER-I), which simulates the multi-layer aquifer system in the Shelby County area with a cell size of 250 m by 250 m using MODFLOW-NWT. In CAESER-II, the shallow aquifer was represented as one layer, the UCCU as one layer, the Memphis aquifer as four layers, the Flour Island confining unit as one layer, and the Fort Pillow aquifer as one layer. The authors of [58] extended CAESER-I from 2005 back to 1960 to account for flow conditions during the time when most contaminant spills in the area occurred, recalibrating the model with updated Memphis aquifer properties based on [60], historical pumping data, and hydraulic conductivities measured on a borehole sample obtained from a UCCU breach [58]. All adjusted parameters and calibration metrics for CAESER-II can be found in [58].

A submodel was created from CAESER-II to conduct contaminant transport modeling. The Town of Collierville is isolated from other pumping from the Memphis aquifer and the projection of contaminant migration does not extend far within the model (e.g., westward to the City of Germantown, which has its own production wells); hence, using a submodel was appropriate. Additionally, preliminary testing of contaminant transport using CAESER-II indicated extraordinary long run times due to the size of the model; therefore, using a submodel would reduce runtime without necessitating areas simulated by CAESER-II not relevant to the Collierville area.

The submodel's extents greatly exceeded anticipated distances that contamination may travel to avoid boundary effects. Following the regional groundwater gradient of the Memphis aquifer, the submodel paralleled the orientation of CAESER-II to a southeast to northwest direction II (see Figure 1, insets B and C). Using the same stress periods as CAESER-II, the resulting heads from CAESER-II defined the eastern and western boundary conditions as time-variant heads. Due to the submodel's orientation to align with the Memphis aquifer gradient [61], the northern and southern boundaries were set as no-flow. The submodel has 736 stress periods spanning from 1960 to 2021. Unlike CAESER-II having square cells of 250 m, submodel cell sizes were reduced to 50 m square, resulting in 79 rows by 101 columns. The vertical representation increased from 8 to 24 layers (Figure 2). These modifications to cell dimensions and number of layers were made to accommodate various contaminant source locations (i.e., isolating them to a single cell or the flexibility to be active across multiple cells) and to obtain a more discretized movement of contamination vertically within the system based on monitoring well observations. The shallow aquifer was represented by two layers, the UCCU remained as one layer, the Memphis aquifer went from four layers to fourteen, the Flour Island confining unit was four layers, and the Fort Pillow aquifer was also four layers (Table 1).

Table 1. Average layer thicknesses in submodel.

Layer (s)	Hydrostratigraphic Unit	Average Thickness (m)
Layer 1	Upper Shallow Aquifer	10.10
Layer 2	Lower Shallow Aquifer	9.19
Layers 3–4	Upper Memphis Aquifer	30.16
Layers 5–16	Lower Memphis Aquifer	12.60
Layers 17–20	Flour Island Formation	17.71
Layers 21–24	Fort Pillow Aquifer	25.74

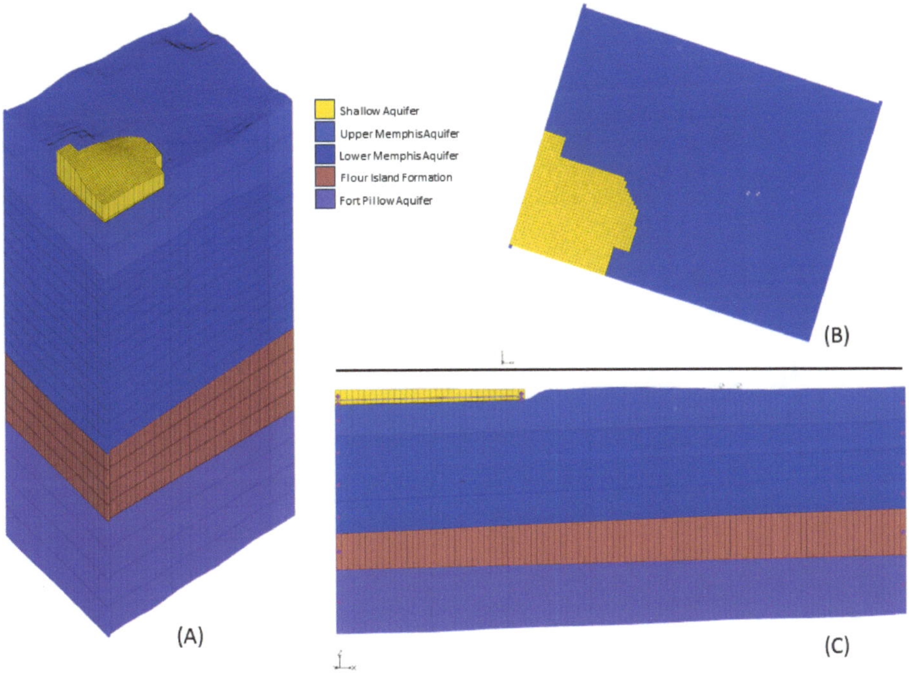

Figure 2. Plan view of the model extent and model grid with different hydrogeological units. (**A**) The 3D Grid, (**B**) Plan View, and (**C**) Cross-section of model (10× vertical exaggeration).

As would be expected, changing cell dimensions (spatially and vertically) from a parent model with larger cell sizes (i.e., 5 time larger) requires redistribution of aquifer characteristics. As CAESER-II represents a calibrated model using PEST, resultant values of hydraulic conductivity (K), specific storage (Ss), and specific yield (Sy) varied from cell to cell. To define these parameters to the submodel cell configuration, values were interpolated using inverse distance weighting (IDW) from CAESER-II onto the finer grid of the submodel (Table 2).

Table 2. Parameters range for submodel simulation.

Layer Group	Parameter	Range/Value	Unit
L1	Hydraulic Conductivity (Hk)	1.35×10^{0}–4.74×10^{0}	m/d
L2	Hydraulic Conductivity (Hk)	5.60×10^{-5}–6.57×10^{-5}	m/d
L3–16	Hydraulic Conductivity (Hk)	1.73×10^{1}–2.44×10^{1}	m/d
L17–20	Hydraulic Conductivity (Hk)	6.43×10^{-5}–2.06×10^{-4}	m/d
L21–24	Hydraulic Conductivity (Hk)	3.36×10^{0}–1.12×10^{1}	m/d
L1	Specific Storage (Ss)	0	1/m
L2	Specific Storage (Ss)	3.24×10^{-3}–4.39×10^{-3}	1/m
L3–16	Specific Storage (Ss)	3.06×10^{-3}–8.13×10^{-3}	1/m
L17–20	Specific Storage (Ss)	1.29×10^{-3}–1.77×10^{-3}	1/m
L21–24	Specific Storage (Ss)	7.96×10^{-4}–1.44×10^{-3}	1/m
L1	Specific Yield (Sy)	3.20×10^{-1}	dimensionless
L2	Specific Yield (Sy)	8.00×10^{-2}	
L3–16	Specific Yield (Sy)	3.00×10^{-1}	
L17–20	Specific Yield (Sy)	8.00×10^{-2}	
L21–24	Specific Yield (Sy)	2.50×10^{-1}	

A time-dependent specified head boundary condition was implemented for the shallow aquifer. For the Memphis aquifer, a time-variant head was implemented in the eastern and western boundaries, while the north and south boundaries were kept no-flow based on historical water level maps [61]. The Flour Island and Fort Pillow boundaries were mimicked as as no-flow at the submodel scale following that of the parent model [58,59]. To determine the accuracy of submodel conditions to that of the parent model (CAESER-II), two analyses were performed: a zonal flow budget comparison of the submodel to the same area represented in CAESER-II and an error analysis of heads on a cell-by-cell basis. As can be seen from Table 3, budget values between the two models shifted between 0 and 0.16 m^3/day.

Table 3. Flow budget comparison between CAESER-II and the submodel.

Flow Budget Component	Volume (m^3/day)	
	CAESER-II	**Submodel**
In:		
Storage	125,670.49	125,670.50
Constant Head	16,974.97	16,974.90
Wells	0.00	0.00
Recharge	119.36	119.36
Total IN	**142,764.93**	**142,764.76**
Out:		
Storage	126,884.20	126,884.00
Constant Head	15,880.72	15,880.75
Wells	0.00	0.00
Recharge	0.00	0.00
Total Out	**142,764.92**	**142,764.76**
Summary:		
In–Out	0.01	−0.00
Percent Discrepancy	0.0000%	0.0000%

Submodel heads were simulated from 1960 to 2021 using MODFLOW-NWT [62]. A cell-by-cell head comparison between the submodel and CAESER-II was performed and, submodel heads were within 0.17 m of CAESER-II; therefore, recalibration was deemed unnecessary.

2.1.4. Contaminant Transport Model

The contaminant transport model was developed using MT3DMS. Assuming conservative conditions, as there was no data to suggest otherwise, the contaminant transport model was calibrated and verified against the observed concentrations in the monitoring wells. Adjusted input parameters were longitudinal and transverse (horizontal and vertical) dispersivities. The source boundary conditions and initial concentration values were defined based on historical contaminant data from the monitoring wells adjacent to the source.

Source concentration plays a vital role in concentration distribution, and a time-variant specified concentration boundary (Dirichlet) was defined for both sites. These boundaries are termed "source function" and mimic the increase, saturation, and decrease in concentration in groundwater through a rising limb, a plateau, and a receding limb, respectively (Figure 3). The length of each limb was determined from field data adjacent to the source. Monitoring well MW-15 was used to define the source function for the 1979 spill, MW-19 and MW-21 were used for the unlined lagoon, and MW-1, MW-1B, and MW-10 were used for the 1985 spill at the Carrier site. For the Smalley Piper site, data from MW-8 were used.

Regarding spatial distribution of the sources, the unlined lagoon, Smalley Piper, and the 1985 Carrier spills each were represented by a single model cell. Conversely, the source for the 1979 TCE spill at the Carrier site spanned across three model cells. All sources were applied in layer 3 of the Memphis aquifer, which represented the first layer of entry either below its upper confining unit (i.e., UCCU) or from the shallow aquifer where the Memphis aquifer was locally unconfined.

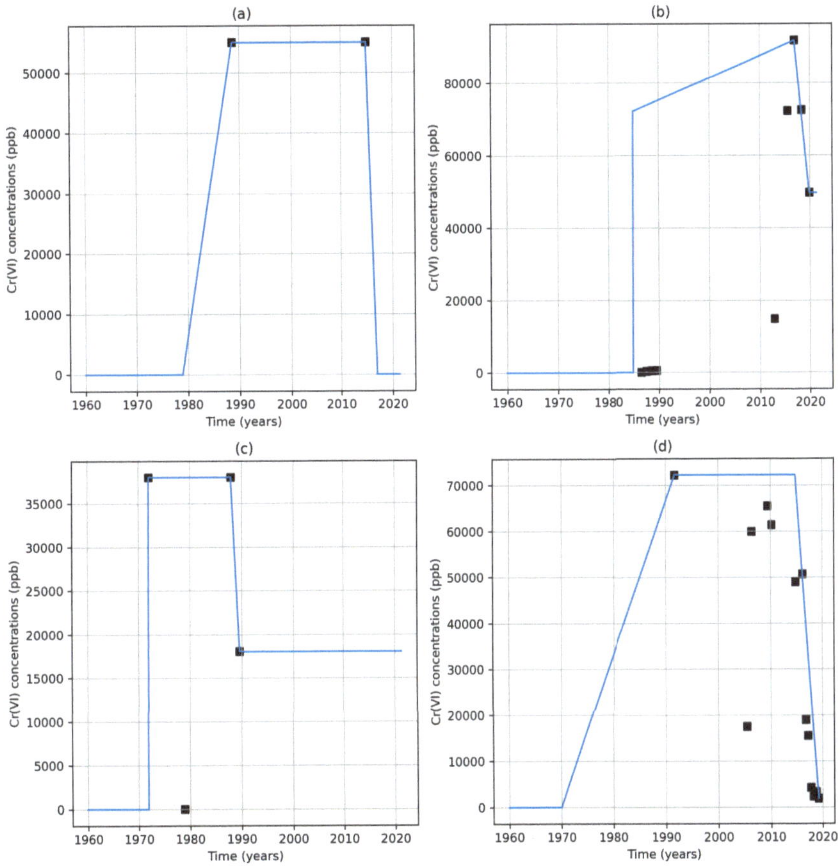

Figure 3. Contaminant source functions for (**a**) 1979 Carrier Spill, (**b**) 1985 Carrier Spill, (**c**) Unlined lagoon, and (**d**) 1970 Smalley Piper Source. Black Squares represent measured concentrations.

Of note, Ref. [63] infers the existence of the UCCU (termed Jackson Clay in report) near the vicinity of the Carrier plant. Ref. [38] shows the Carrier facility and all monitoring wells within the Memphis aquifer unconfined zone, absent of the UCCU; however, ref. [64] indicates the opposite being most confined. With greater detail of the subsurface provided by [63], contaminant sources at Carrier fell either on the fringe of the UCCU subcrop (by 1–2 cell sizes) or outside its influence. Hence, it was assumed that dissolved phase TCE entered the Memphis aquifer and was not impacted by the UCCU.

Advection in MT3DMS was solved using a third-order TVD (total-variation-diminishing) scheme which is mass conservative and minimizes numerical dispersion and artificial oscillation. It is understood that porosity does change slightly, simply owing to the range of hydraulic conductivities shown in Table 2. It was also realized that the option of cell-by-cell alteration of porosity was constrained by the lack of comprehensive field data required for such detailed adjustments. Therefore, following the works by [64,65], who defined porosity locally in the area, a sensitivity analysis on porosity was not deemed necessary and a value of 0.3 was applied. For each iteration in defining dispersivities, a combination of longitudinal, transverse horizontal, and transverse vertical dispersivity values were assigned across the model domain and RMSE (root mean squared error) values at each monitoring well were calculated against observed concentrations.

To gauge initial values of dispersivity for use in the model, a range of possible values were applied until a reduction in RMSE was realized. Once a range was obtained (see Table 4), a straightforward trial-and-error method was implemented for calibration instead of using optimization tools like PEST or UCODE because there are not enough supporting studies that use these tools for this type of calibration. Additionally, the approximated source concentrations were based on nearby monitoring well concentrations. As will be shown, there is much scatter in the observed readings of contaminant concentrations; hence, an analysis based on point value comparisons was not conducted. Instead, an overall concentration trend match was attained. This type of calibration has been successfully demonstrated by [4,34,66,67]. Of note, no dry cells were observed in the MODFLOW-NWT solutions, which eliminated the possibility of contaminant mass accumulation in cells.

Table 4. Dispersivity values for MT3DMS simulation.

Parameters	Lower Bound (m)	Upper Bound (m)	Step
Longitudinal Dispersivity	0.5	60	0.5
Ratio of Transverse Horizontal Dispersivity to Longitudinal Dispersivity (TRPT)	0.01	0.5	0.01
Ratio of Transverse Vertical Dispersivity to Longitudinal Dispersivity (TRVT)	0.001	0.05	0.001

The inverse transport simulations were performed using the FloPy Python package (v3.3.4) [68] and MT3DMS (v5.3) [69]. The FloPy script executed the MT3DMS transport model for both TCE and Cr(VI) plumes for different combinations of dispersivity values following a workflow like [11]. The scripted workflow is illustrated in Figure 4. An iterative process was conducted by implementing enumerable combinations of longitudinal, transverse horizontal, and transverse vertical dispersivities within the ranges shown in Table 4 and incremented using the step jumps of 0.5, 0.01, and 0.001, respectively, for each match to a monitoring well, resulting in tens of thousands of independent model runs. Hence, a suite of dispersivity combinations would be applied globally to the model and a comparison made between the modeled concentration and observed concentration for each well, then the same process would be performed for each monitoring well thereafter to provide a better estimation (and realization) of scale-dependent dispersivity.

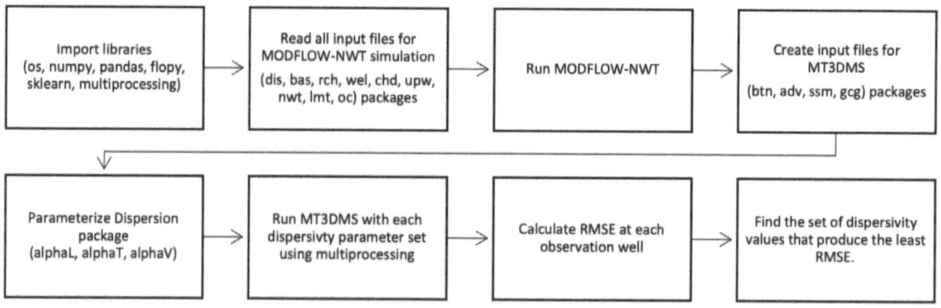

Figure 4. Inverse MT3DMS workflow.

To improve the computational efficiency of conducting numerous MT3DMS model runs, the Python multiprocessing library was used. The runtime was significantly reduced from 30 h to 3.5 h using an 8-core Intel Core i7-9700 CPU.

3. Results

3.1. Simulated vs. Observed Concentration

Simulated and observed concentration plots for the Carrier and Smalley Piper sites are shown in Figures 5 and 6, respectively. The resulting dispersivity values for each well for both sites are given in Table 5 (Carrier) and Table 6 (Smalley Piper).

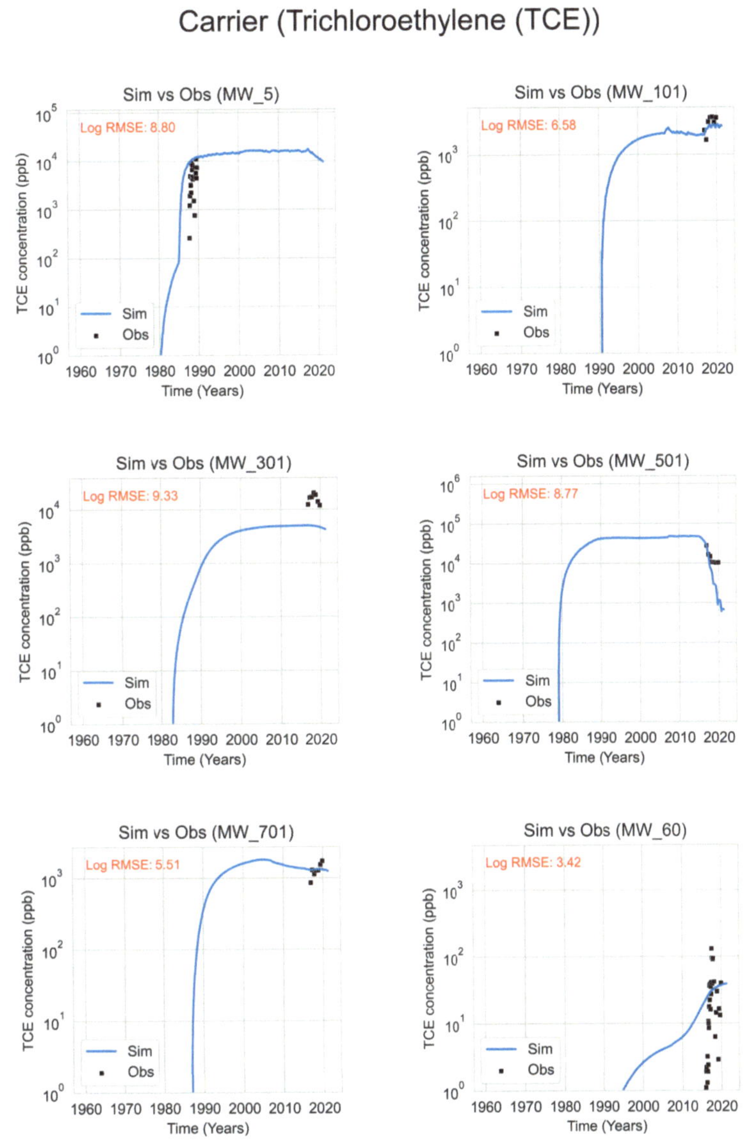

Figure 5. Carrier simulation vs. Observation data (MW-5, MW-101, MW-301, MW-501, MW-701, and MW-60). Concentrations are in log scale with well locations shown in Figure 1.

Table 5. Dispersivity Values (Carrier Site).

Well ID	L(1979) m	L(1985) m	L(Lagoon 1979) m	Longitudinal Dispersivity (m) α_L	Transverse Horizontal Dispersivity (m) α_T	Transverse Vertical Dispersivity (m) α_V
MW-5	232.94	182.75	-	10.5	4.72	0.42
MW-101	540.88	506.32	31.00	55.5	24.97	2.22
MW-301	379.69	432.76	-	55.5	24.97	2.22
MW-501	119.90	303.37	-	25.5	1.27	0.13
MW-701	346.34	267.72	-	50.5	5.05	0.25
MW-60	963.03	993.77	448.28	20.5	8.2	0.82

Table 6. Dispersivity values (Smalley Piper site).

Well ID	L (m)	Longitudinal Dispersivity (m) α_L	Transverse Horizontal Dispersivity (m) α_T	Transverse Vertical Dispersivity (m) α_V
MW-11	80.06	5.5	0.28	0.03
MW-15	86.90	5.5	2.48	0.08
MW-12	156.15	55.5	24.98	1.39
MW-18	219.19	20.5	9.23	0.10
MW-20	695.10	15.5	3.88	0.08
MW-21	750.91	55.5	24.98	2.50
MW-27D	954.42	15.5	3.88	0.08

Figure 5 shows the simulated vs. observed concentration of TCE in six monitoring wells. Among them, MW-5 is one of the earliest wells installed onsite with data available from 1987 to 1988. The other wells have concentration data from 2010 to 2019. From the analysis, longitudinal dispersivity values at the Carrier site range from 10.5 to 55.5 m, transverse horizontal values range from 1.27 to 24.97 m, and vertical dispersivity from 0.42 to 2.22 m (Table 5). The distance between the source and monitoring well is given as L(m) in Table 5. The closest well, MW-5, had the lowest longitudinal dispersivity among all wells; however, its transverse horizontal dispersivity at first seems high yet is impacted by two sources making deciphering values to distance more challenging. MW-101 is only 31 m from the lagoon source in 1979, which would imply a lower longitudinal dispersivity than MW-5, whose shortest distance is nearly 183 m from the 1985 source. However, samples in MW-101 were not taken until 2010, well after any of the three sources started leaking. MW-501 is closer to sources than MW-101 and MW-301, resulting in lower dispersivity values. MW-701 was slightly further away from sources than MW-501 but with less distance than MW-101 and MW-301, resulting in dispersivities higher and lower, respectively. Interesting is MW-60, which is further from the 1979 and 1985 sources than all other wells, yet has the second lowest longitudinal dispersivity among the wells; however, this well is also only 448.28 m from the 1979 lagoon source. MW-60 had the largest concentration spread than the other wells, ranging over two orders of magnitude; however, it had the lowest log RMSE at 3.42.

Figure 6 shows the simulated vs. observed hexavalent chromium concentration at the Smalley Piper site. All seven observation wells have data ranging from 2005 to 2019. From the analysis, longitudinal dispersivity values at this site range from 5.5 m to 55.5 m, transverse horizontal values range from 0.28 m to 24.98 m, and transverse vertical dispersivity from 0.03 m to 2.5 m (Table 6). The two closest wells, MW-11 and MW-15, each just under 90 m from the source, indicated low longitudinal dispersivity, as would be expected for a scale-dependent parameter. Interestingly, MW-12 at 156.15 m from the source had a higher longitudinal dispersivity (more than twice) and transverse dispersivity (just less than three times) than the next furthest well, MW-18, at nearly 220 m from the source; however, it matched these same values for MW-21 at 750.91 m from the source, or almost five times the distance. The furthest well, MW-27D, had dispersivity values lower than the third furthest well, MW-12.

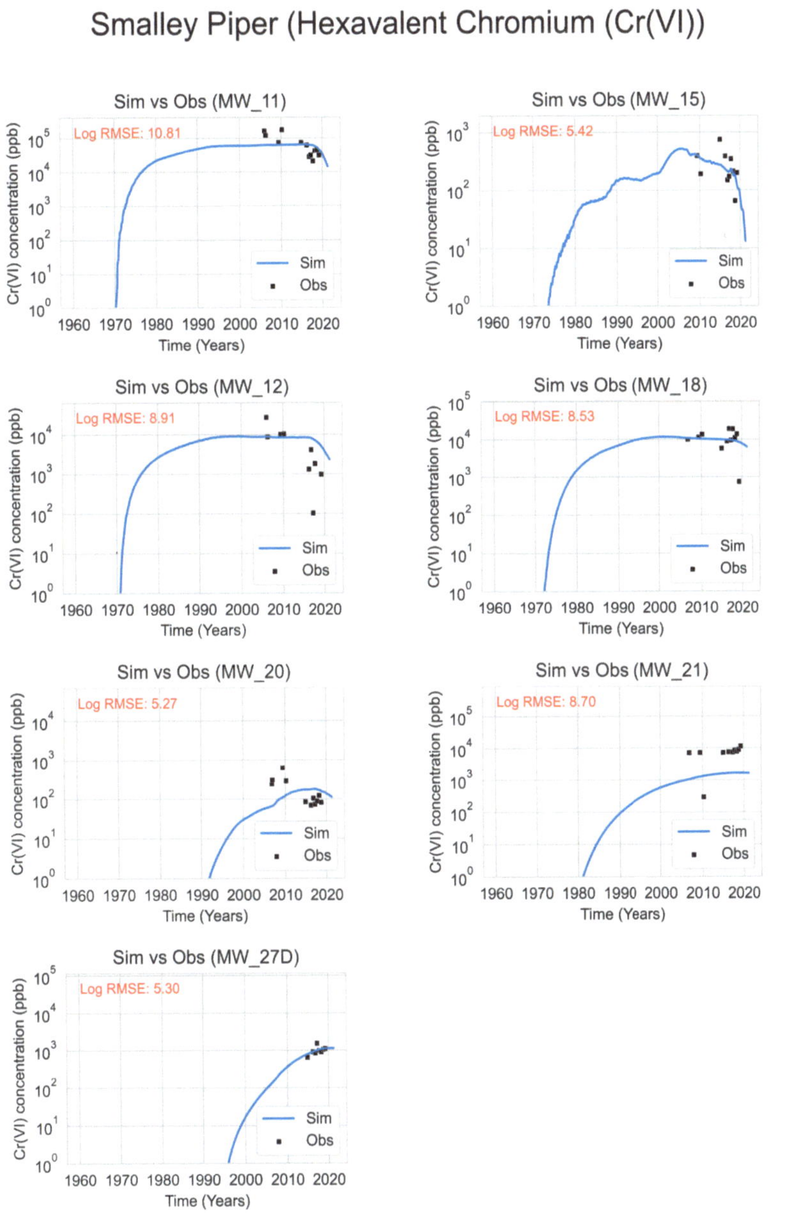

Figure 6. Smalley Piper simulation vs. Observation data (MW-11, MW-15, MW-12, MW-18, MW-20, MW-21, and MW-27D). Concentrations are in log scale with well locations shown in Figure 1.

3.2. Comparison with Empirical Dispersivity Equations

Dispersivity values obtained for both sites were compared against three empirical equations shown in Table 7. Among the equations used, Gelhar and Schulze-Makuch imply a direct proportional relation between observed distance and longitudinal dispersivity where Schulze-Makuch's implies a slower increase of longitudinal dispersivity with distance. Xu and Eckstein's equation also implies slowing down at a larger scale and captures non-linear behavior natural aquifers. Both Schulze-Makuch and Xu–Eckstein's equation suggests a plateau at larger distances where the sensitivity to scale of longitudinal dispersivity diminishes.

Table 7. Empirical equations of longitudinal dispersivity.

Source	
Gelhar (1992) [28]	$\alpha_L = 0.1L$
Xu & Eckstein (1995) [70]	$\alpha_L = 0.83(\log_{10}(L))^{2.414}$
Schulze-Makuch (2005) [11]	$\alpha_L = 0.085L^{0.81}$

Tables 8 and 9 show the calculated and optimized values of longitudinal dispersivity for the Carrier and Smalley Piper sites.

Table 8. Comparison of longitudinal dispersivity values (m) with empirical equations and effective lengths (Carrier site).

Well ID	Effective Length (m)	Gelhar, 1992 [28]	Xu & Eckstein, 1995 [70]	Schulze-Makuch, 2005 [11]	AlphaL
MW-5	160.10	16.01	5.59	5.19	10.5
MW-101	418.82	41.88	8.50	11.30	55.5
MW-301	327.92	32.79	7.70	9.27	55.5
MW-501	186.32	18.63	6.01	5.87	25.5
MW-701	236.05	23.61	6.68	7.10	50.5
MW60	872.39	87.24	11.22	20.48	20.5

Table 9. Comparison of longitudinal dispersivity values (m) with empirical equations (Smalley Piper site).

Well ID	L (m)	Gelhar, 1992 [28]	Xu & Eckstein, 1995 [70]	Schulze-Makuch, 2005 [11]	AlphaL
MW-11	80.06	8.006	3.93	2.96	5.5
MW-15	86.90	8.69	4.10	3.16	5.5
MW-12	156.15	15.615	5.53	5.08	55.5
MW-18	219.19	21.919	6.47	6.69	20.5
MW-20	695.10	69.51	10.33	17.04	15.5
MW-21	750.91	75.091	10.63	18.14	55.5
MW-27D	954.42	95.442	11.58	22.03	15.5

The effective length, $L_{\text{effective}}$, was calculated as a weighted average of peak concentrations of the three sources, using the following formula:

$$L_{\text{effective}} = L_{1979} \cdot w_{1979} + L_{1985} \cdot w_{1985} + L_{\text{Lagoon}} \cdot w_{\text{Lagoon}}$$

where the weights (w) are proportional to the concentrations of contaminants measured in those years, such that:

$$w_{1979} = \frac{C_{1979}}{C_{\text{total}}}, \quad w_{1985} = \frac{C_{1985}}{C_{\text{total}}}, \quad w_{\text{Lagoon}} = \frac{C_{\text{Lagoon}}}{C_{\text{total}}}$$

and $C_{\text{total}} = C_{1979} + C_{1985} + C_{\text{Lagoon}}$.

From the analyses presented in Tables 8 and 9, it is evident that there is not a consensus that one empirical model better matches modeled longitudinal dispersivities to the other equations. At the shorter distances, such as for MW-5, MW-11, and MW-15, values are reasonable comparable across the three empirical equations. At greater distances, Gelhar (1992) [28] overestimates hydraulic conductivity where Xu and Eckstein (1995) [70] and

Schulze-Makuch (2005) [11] are within reason (see MW-60 in Table 8 and MW-27D in Table 9).

4. Discussion

The primary goal of this study was to derive the range of longitudinal, transverse horizontal, and transverse vertical dispersivities by matching observed contaminant data against modeled concentrations and to assess the presence of longitudinal dispersivity scale dependency. Additionally, FloPy was employed in a novel way to concurrently run parameterized MT3DMS simulations, reducing the overall simulation runtime by a factor of eight. What is interesting about this investigation compared to others in the literature is the presence of two distinct plumes from two separate industries, Carrier and Smalley Piper, which for similar distances between source and monitoring well indicated notably different dispersivities.

As with many investigations into contaminant transport, obtaining the original source concentration proves difficult. This investigation is no exception, whereby no original source concentrations were available and concentrations at the closest monitoring wells were used to derive source concentrations. For the Carrier 1979 spill, only two readings existed near the source, spanning nearly 30 years between the readings. The Carrier 1979 unlined lagoon source function was complicated as concentrations from MW-19 and MW-21 offered a wide spread in concentrations, varying from 0 to over 35,000 mg/L bracketing the lagoon discharge of 1979 by plus-minus a decade. The 1985 Carrier spill had three nearby monitoring wells with multiple concentration readings enabled, except for a single, low reading in 2012. The Smalley Piper was a newer incident and therefore had more available data, though scattered to a degree.

The Carrier site was also complicated by having three sources separated by six years and of varying size, with the 1979 spill covering the largest area on the southern portion of the property. Carrier wells, namely, MW-5, MW-101, MW-501, and MW-701 demonstrated a relatively good agreement between observed and model simulation concentrations whereby the simulated concentration trend fell within the cluster of concentration readings, with the slight exception of MW-5 (log RMSE of 8.80) due to a predicted early arrival. The log RMSE values corroborate this assessment with MW-101, MW-501, and MW-701 having lower log RMSE values than the under-matched trends on MW-301 and MW-601.

Surprisingly, MW-60 had the lowest log RMSE (3.42) though the largest scatter with observed concentrations bracketing the modeled trend above and below. The undershot of the model concentration trend to the observed concentrations in wells MW-301 and MW-601 (Figure 5) could be improved by a higher advective component, which would bring into question the accuracy of the simulated heads for the Memphis aquifer. In CAESER-I, Ref. [59] used PEST to calibrate the model. Unfortunately, no control points existed in the Collierville area. Likewise, Ref. [58] for CAESER-II used the same control points as [59] but extended some control further back in time to 1960 where data existed. Hence, the heads in CAESER-II and subsequently the submodel received questionable calibration. There does exist some historical water level data in [71] that contrast the resulting submodel generated heads (Table 10), with the prior being consistently higher.

Among the historical water level data, MW-1, MW-1A, and MW-1B are within 3.5 m of each other and are of similar depth, but they exhibit a head difference of only 3 m, thereby raising questions into these readings. MW-5 indicated a water level of 94.28 m above mean sea level (amsl) in 1988, where under predevelopment conditions (1886), the groundwater level in downtown Collierville (about 2.3 km upgradient) was estimated to be 90 m amsl [72]. Furthermore, ref. [61] estimated the water level in the Collierville area to be 85.34 m which is a mere 4.66 m change in 101 years.

Submodel data show a water level difference of 2 to 5 m compared to the historical data in most wells, except for MW-5. More recently, a groundwater assessment performed for the Town of Collierville by the Center for Applied Earth Science and Engineering Research (CAESER) in 2018/2019 showed groundwater levels to be between 84 and 85 m, which is consistent with the submodel generated head values [73]. Ignoring the historical data obtained from the reports (i.e., [71]) due to questionable measurement disparities and relying more on these other measures [61,72], there is no reason to assume a possible shift in submodel gradients and, hence, change in advection.

Table 10. Water level data.

Well ID	Elevation of the Center of Well Screen (m amsl)	Date	Historical Water Level Data (m amsl)	Submodel Head (m amsl)	Difference (m)
MW-15	87.57	18 April 1988	88.16	84.76	3.40
MW-1	71.97	18 April 1988	89.7	84.88	4.82
MW-1A	89.29	18 April 1988	86.7	84.88	1.82
MW-1B	72.58	18 April 1988	89.69	84.82	4.87
MW-5	87.57	18 April 1988	94.28	84.62	9.66
MW-6	79.21	18 April 1988	86.4	84.62	1.78
MW-11	88.82	18 April 1988	89.1	84.78	4.32
MW-12	72.22	18 April 1988	86.5	84.78	1.72

The various values of dispersivities by well (Tables 5 and 6) illustrate scale-dependency of this parameter, where consistently between Carrier and Smalley Piper the shortest distance wells (MW-5, MW-ll, and MW-15) have much lower longitudinal dispersivities than for those well further away. However, the premise of scale-dependency of an increasing value with distance is a challenge to observe from our results. Taking the furthest two wells (MW-60 and MW-27D), the calculated longitudinal dispersivities were less than the second furthest wells in each dataset, MW-501 at 119.90 m (ignoring MW-101) and MW-12 at 156.15 m, respectively. With the highest longitudinal dispersivity at each site at 55.5 m, it would stand to reason that MW-60 and MW-27D, with longitudinal dispersivities of 20.5 and 15.5, respectively, would be higher than 55.5 m.

MW-60 at Carrier is over 900 m from the two spills on the south end of the plant, but is 448.28 m from the lagoon source (1979). MW-301, being approximately 380 m and 430 m from the two southern sources, hence, near the same distance as MW-60 from the lagoon, had a longitudinal dispersivity of more than twice MW-60 at 55.5 m. Being that the 1979 lagoon source had a concentration half that of the other spills and spanned fewer decades than the larger spills, adds to the complexity of MW-60, as can it be observed very large scatter of concentration readings between 2016 and 2019 (see Figure 5 for MW-60). Hence, obtaining a good longitudinal dispersivity proved challenging.

Though not incorporated at the time of modeling due to missing screen depth information, two observation wells, MW1101 and MW1201, located 360 m and 440 m, respectively, southwest of MW19 (near COLL201 and COLL202, the Collierville Wellfield #2, see Figure 1), indicated near non-detect (ND) TCE concentrations between March 2018 and November 2019. This would suggest that pumping from Wellfield #2 was capturing the TCE plume. However, ignoring dispersion and relying solely on groundwater gradients and wellfield pumping as simulated in the calibrated numerical model, TCE concentrations are not captured by the wellfield and instead there are elevated concentrations of TCE at MW1101 and MW1201. We do not dispute the observed concentrations yet have observed wide variations in measured TCE concentrations, for example, see Figure 3c. Such fluctuations in observed readings are not uncommon, as [74] observed in TCE concentrations where they dropped four orders of magnitude in a 3-month period to near ND only to rise back to prior concentrations a few months later. In such cases, TCE concentration trends were used to define plumes; hence, we cannot always match acute TCE concentrations such as those observed in MW1101 and MW1201.

The Smalley Piper well, MW-27D, at 954.42 m from the source had the second lowest log RMSE with the lowest being MW-27, although MW-27 had much more scatter in its readings over a period of nearly 12 years than MW-27D with readings over a smaller time period (5 years). However, Xu and Eckstein's (1995) and Schulze-Makuch's (2005) empirical equations suggest near similar longitudinal dispersivities to MW-27D (15.5 m) at 11.58 m and 22.03 m, respectively. This does not necessarily discount Gelhar's empirical equation, which produced at longitudinal dispersivity of approximately 95 m where the maximum [28] found from their review was 104 m. Certainly, all three empirical equations adhere to the scale-dependency concept of increasing dispersivity with distance, making a direct comparison difficult with the results of this study.

At Carrier, aside from MW-60, the other monitoring wells do adhere more closely to an increased dispersivity with increasing distance, although having three spatially and temporally different sources add a lot of complexity to this site. Interestingly, MW-101 and MW-301 share the same dispersivity values. This is because these wells are reaching the threshold constraints as stated in Table 4. MW-101 seems to match the observed data well (see Figure 5) and has the third lowest log RMSE at 6.58. The readings are clustered tightly like MW-301; however, the concentration trend fails to pass through its cluster of readings.

If the thresholds in Table 4 were to be expanded, which fall between [11,28], then the transverse horizontal and transverse vertical could be reduced to move more mass longitudinally and possible have the concentration trend pass through the MW-3-1 cluster. However, doing so would result in a counter-intuitive longitudinal dispersivity as it would be larger than MW-101 even though MW-101 is further away from MW-301 by 70–160 m from the southern sources. Then again, the fact that MW-101 is a mere 31 m from the lagoon source adds reason for a possible lower longitudinal dispersivity than MW-301, yet why is it not much lower like MW-5, which is further from its sources (between 182.75 and 232.94 m) and has the lowest longitudinal dispersivity (10.5 m)?

Lastly, MW-501 had a longitudinal dispersivity of 25.5 m even though it is only 119.9 m from the 1979 source, which, based on scale dependency, would imply a lower longitudinal dispersivity. This may be attributed to high variability of concentrations at MW-15, which was used to define the 1979 source. In the period of record (1987–1989), MW-15 had concentrations varying two orders of magnitude with a peak value of 400,000 ppb in June 1989, yet two months prior (April) had a measured concentration of 140,000 ppb and 5900 ppb in January of that same year. This dramatic fluctuation in concentrations leads to misinterpretation of source history (likely overestimation), therefore manifesting a high dispersivity to match declining concentrations in observation well MW-501.

Smalley Piper's monitoring wells produced similar dispersivity oddities to those seen with Carrier. MW-12 and MW-21 both pressed against the imposed thresholds like with MW-101 and MW-301 at Carrier. Both MW-12 and MW-21 had scattered readings across nearly three orders of magnitude to about two orders of magnitude, respectively, and both had the highest log RMSE errors at 8.91 and 8.70, respectively. Strangely, MW-12 is extremely close to the source at 156.15 m and MW-21 is almost five times further away. Hence, it is difficult to say that one or both of these wells' dispersivities can be trusted. To the contrary, MW-20 and MW-27D are both distant from the source at approximately 695 m and 954 m, respectively, and with the same dispersivities. The concentration trend for MW-27D passes through its cluster of readings (see Figure 6); however, for MW-20 there is a bit more scatter and the concentration trend it is forced to pass between are two clusters around 2007–2010 and 2014–2018. Yet again, we see a mid-distance well, MW-18, at 219.19 m from the source with a higher longitudinal dispersivity than the farther wells, MW-20 and MW-27D, yet lower than MW-12 (longitudinal dispersivity of 55.5 m), which is closer to the source at 156.15 m.

Part of this discrepancy is complicated by the observed concentration readings as some seem either difficult to believe or it is not possible to decipher the reality of the actual values. For example, MW-21 is the second farthest well at 750.91 m and is near enough along the same groundwater gradient line as MW-20 at 695.10 m from the source, or only 33 m from MW-21. Measured during the same time period, MW-21 had an average concentration of 7028 ppb and MW-20 had an average concentration of only 185.8 ppb, despite their depth difference of only 3 m. This 38-times difference in concentration resulted in the two different estimates of longitudinal dispersivity for that location with large disproportionality.

Though the longitudinal dispersivities for the shortest and furthest wells found similarity between the Carrier and Smalley Piper sites, the wells in between these wells had dispersivities that were counter-intuitive to the concept of increasing dispersivity with increasing distance. Because the concentration trends fit relatively close to observed data and depending on the choice of empirical equation, alternative dispersivities would seem to not deviate much from what resulted in this study. Additionally, transverse horizontal and vertical dispersivity values obtained from this study vary by one order of magnitude for both sites and do not exhibit a scale dependency, thereby deviating from the common practice of the heuristic relationship where transverse horizontal is typically 0.1 and transverse vertical is 0.01 times the longitudinal dispersivity. Recent studies by [75] compiled and compared existing reliable estimates of transverse dispersivity to be site-specific and showed a variation of three orders of magnitude with no apparent scale dependency.

The analysis revealed that the concentration time series at observation wells generated by the model were more sensitive to source characterization (location and initial concentration). The Smalley Piper site showed a relatively better match between the two sites in concentrations and subsequent determination of longitudinal dispersivity compared to the Carrier site. One reason was that Smalley Piper had a single source, whereas Carrier had three, two of which, the 1979 spill and unlined lagoon, had questionable data quality. Additionally, a factor not considered were reactions that would only reduce simulated concentrations, which, all other factors remaining the same, would require lower dispersivities to keep the peaks, shown in Figures 5 and 6, from declining below the observed concentrations. There was no additional data to suggest reactions to TCE and Cr (VI) were occurring.

A generalized delineation of the TCE and Cr(VI) plumes was constructed by aggregating the separate MT3DMS model runs which used the dispersivity values shown in Tables 5 and 6, respectively. The median concentration level per grid cell was used as the concentration and TCE and Cr(VI) plumes are depicted for layer 3 (upper Memphis aquifer). This is an illustrative example to show how each contaminant moved over time (i.e., Figure 7 represents last model transport time step). As shown, the plumes move in accordance with the groundwater gradient with elevated concentration signatures west of their respective sources.

A final consideration for the possible reasons mid-distance wells favored higher longitudinal dispersivities was the distribution of hydraulic conductivity that may offer a preferential flowpath between the source and certain downstream wells. Investigation of cell-to-cell hydraulic conductivity for the Carrier site did not suggest such a flowpath where values ranged between 21.09 and 22.54 m/day. Following an eight-cell connectivity, cell-to-cell changes were no greater than 0.7 m/day with the majority less than 0.5 m/day. Similarly, hydraulic conductivities along the Smalley Piper flowpath were also nominal at 22.74 to 23.18 m/day with the largest cell-to-cell change at 0.33 m/day. Therefore, no preferential flow pathways existed. Additionally, if hydraulic conductivity can serve as a proxy for changes in porosity, the ranges of conductivity do not warrant a variable field of porosities, and any model-wide increase or decrease would only shift the entire concentration field rather than impact individual arrivals of contamination to a particular well.

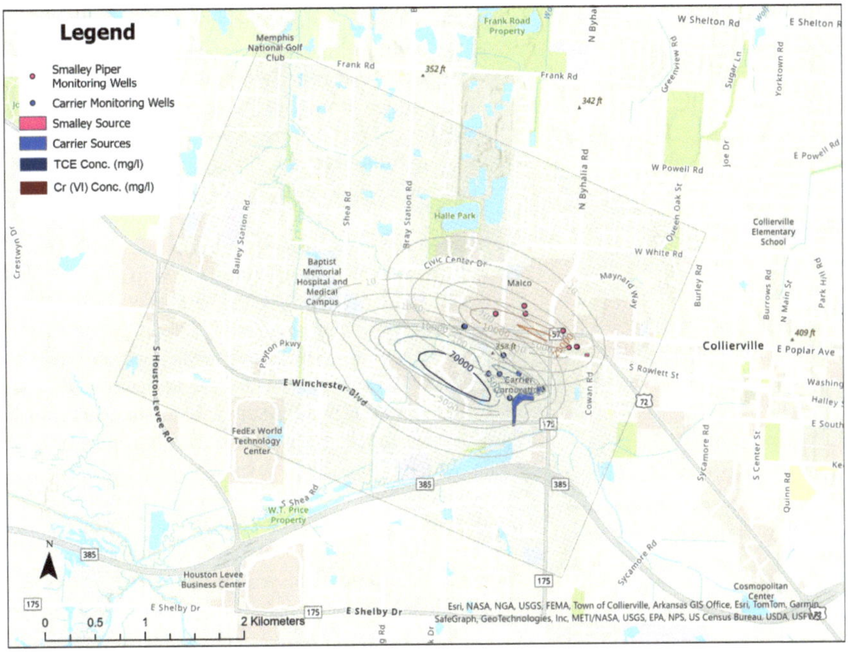

Figure 7. TCE and Cr(VI) plume extent for the Upper Memphis Aquifer (layer-3).

5. Conclusions

This study demonstrates the complexity of quantifying dispersivity in a heterogenous aquifer system with multiple sources and the importance of proper identification of source concentrations as well as contaminant data from monitoring wells. The range of dispersivity values obtained from this analysis provides several benefits over analytical or empirical values as numerical models can estimate all three dispersivity values using a minimal amount of observation data at different depths, provided having a well-calibrated flow model and known source concentration. The Memphis aquifer, although mostly confined, is threatened by anthropogenic contamination coming through the unconfined area or via preferential pathways (i.e., breaches) in the upper Claiborne confining unit [38,44,47,48], several of which are in the vicinity of municipal wellfields. Dispersivity values obtained in this study are crucial to simulating any existing or future contamination arriving at the Memphis aquifer through those breaches.

In this study, contaminant data from two U.S. EPA Superfund sites in the Town of Collierville were used to test the scale-dependent nature of longitudinal dispersivity and evaluate a range of all three dispersivity values using a three-dimensional numerical model. Source boundary conditions were defined using historical observation data near contaminant sources and dispersivity values were optimized by minimizing the root mean squared error (RMSE) between observed and simulated concentration. The following are the main conclusions from this study:

- Values of longitudinal dispersivity from the Smalley Piper and Carrier sites have the lowest values at well near the source and higher at further distances from the source.
- Dispersivities observed at wells midway and downgradient showed an increase; however, this trend was complicated by findings from the monitoring wells furthest from the sources at both sites, where dispersivities decreased below the levels observed at mid-distance wells.
- Concentration trends across all monitoring wells were generally consistent, although a few wells never reached the peak concentrations predicted by the model. The

possibility of preferential flowpaths or errors in interpreting groundwater levels was considered but ultimately discounted.

- High log RMSE errors between modeled and observed concentrations at several wells were likely due to inherent inaccuracies in some of the source concentration data and a broad dispersion of observed concentrations, which made achieving a good match challenging under certain conditions.
- The Carrier site had added complexity over Smalley Piper due to it having three sources.
- Somewhat novel to this study into scale-dependent dispersivity is the presence of two plumes in the same vicinity, wherein, with a constant porosity and low-variable hydraulic conductivity field, the pattern of dispersivity changing with distance should have more closely aligned.
- Out of the thirteen monitoring wells, four had dispersivities that reached the upper limit of the likely range set by this investigation. Although an empirical equation by Gelhar (1992) [28] might suggest raising this threshold, such an adjustment could complicate matters, as it would involve increasing dispersivities for mid-distance wells, which is already problematic.
- With the threat of localized contamination elsewhere at the Memphis aquifer and not having prior knowledge of plausible values of dispersivity, this study has produced a starting range whereby longitudinal dispersivities range from 5.5 m to 20.5 m, transverse horizontal dispersivities range from 0.28 m to 3.88 m, and transverse vertical dispersivities range from 0.03 m to 0.08 m.

The results show the complexity of estimating field scale dispersivity parameters in the presence of multiple sources, multiple plumes, questionable source concentration data, and broad scatter in observed readings. Such challenges prove difficult to remedy; however, the results still offer value insights into scale-dependent dispersivity and the behavior of contaminant migration within the Memphis aquifer.

Author Contributions: Conceptualization, M.I.S. and B.W.; methodology, M.I.S.; validation, M.I.S., B.W. and F.J., formal analysis; M.I.S.; investigation, M.I.S.; resources, B.W. and S.S.; data curation, M.I.S.; writing—original draft preparation, M.I.S.; writing—review and editing, M.I.S., B.W., S.S. and F.J.; visualization, M.I.S.; supervision, B.W. and F.J.; project administration, B.W.; funding acquisition, B.W. All authors have read and agreed to the published version of the manuscript.

Funding: This study funded by Memphis Light Gas and Water (MLGW) Contract No. 12064, and supported by the Center for Applied Earth Science and Engineering Research (CAESER) at the University of Memphis.

Data Availability Statement: Access project data may be released with permission provided by the funder, Memphis Light, Gas, and Water (MLGW).

Conflicts of Interest: The authors declare no conflicts of interest regarding the publication of this paper. This research was supported by Memphis Light Gas and Water (MLGW), but the funders had no role in the design, execution, interpretation, or writing of the study.

References

1. Lee, J.; Rolle, M.; Kitanidis, P.K. Longitudinal dispersion coefficients for numerical modeling of groundwater solute transport in heterogeneous formations. *J. Contam. Hydrol.* **2018**, *212*, 41–54. [CrossRef]
2. Coats, K.H.; Smith, B.D. Dead-end pore volume and dispersion in porous media. *Soc. Pet. Eng. J.* **1964**, *4*, 73–84. [CrossRef]
3. Chapelle, F.H. A solute-transport simulation of brackish-water intrusion near Baltimore, Maryland. *Groundwater* **1986**, *24*, 304–311. [CrossRef]
4. Avon, L.; Bredehoeft, J.D. An analysis of trichloroethylene movement in groundwater at Castle Air Force Base, California. *J. Hydrol.* **1989**, *110*, 23–50. [CrossRef]
5. Chiang, C.Y.; Salanitro, J.P.; Chai, E.Y.; Colthart, J.D.; Klein, C.L. Aerobic biodegradation of benzene, toluene, and xylene in a sandy aquifer—data analysis and computer modeling. *Groundwater* **1989**, *27*, 823–834. [CrossRef]
6. Jensen, K.; Bitsch, K.; Bjerg, P.L. Large-scale dispersion experiments in a sandy aquifer in Denmark: Observed tracer movements and numerical analyses. *Water Resour. Res.* **1993**, *29*, 673–696. [CrossRef]

7. Zou, S.; Parr, A. Two-Dimensional Dispersivity Estimation Using Tracer Experiment Data. *Groundwater* **1994**, *32*, 367–373. [CrossRef]
8. Engesgaard, P.; Jensen, K.H.; Molson, J.; Frind, E.O.; Olsen, H. Large-scale dispersion in a sandy aquifer: Simulation of subsurface transport of environmental tritium. *Water Resour. Res.* **1996**, *32*, 3253–3266. [CrossRef]
9. Hyndman, D.W.; Gorelick, S.M. Estimating lithologic and transport properties in three dimensions using seismic and tracer data: The Kesterson aquifer. *Water Resour. Res.* **1996**, *32*, 2659–2670. [CrossRef]
10. Mallants, D.; Espino, A.; Hoorick, M.V.; Feyen, J.; Vandenberghe, N.; Loy, W. Dispersivity Estimates from a Tracer Experiment in a Sandy Aquifer. *Groundwater* **2000**, *38*, 304–310. [CrossRef]
11. Schulze-Makuch, D. Longitudinal dispersivity data and implications for scaling behavior. *Groundwater* **2005**, *43*, 443–456. [CrossRef]
12. Cupola, F.; Tanda, M.; Zanini, A. Laboratory Estimation of Dispersivity Coefficients. *Procedia Environ. Sci.* **2015**, *25*, 74–81. [CrossRef]
13. Bai, B.; Xu, T.; Guo, Z. An experimental and theoretical study of the seepage migration of suspended particles with different sizes. *Hydrogeol. J.* **2016**, *24*, 2063–2078. [CrossRef]
14. Citarella, D.; Cupola, F.; Tanda, M.G.; Zanini, A. Evaluation of dispersivity coefficients by means of a laboratory image analysis. *J. Contam. Hydrol.* **2015**, *172*, 10–23. [CrossRef] [PubMed]
15. Frippiat, C.C.; Pérez, P.C.; Holeyman, A.E. Estimation of laboratory-scale dispersivities using an annulus-and-core device. *J. Hydrol.* **2008**, *362*, 57–68. [CrossRef]
16. Kret, E.; Kiecak, A.; Malina, G.; Nijenhuis, I.; Postawa, A. Identification of TCE and PCE sorption and biodegradation parameters in a sandy aquifer for fate and transport modelling: Batch and column studies. *Environ. Sci. Pollut. Res.* **2015**, *22*, 9877–9888. [CrossRef] [PubMed]
17. Shamir, U.Y.; Harleman, D.R.F. Numerical solutions for dispersion in porous mediums. *Water Resour. Res.* **1967**, *3*, 557–581. [CrossRef]
18. Sternberg, S.P.; Cushman, J.H.; Greenkorn, R.A. Laboratory observation of nonlocal dispersion. *Transp. Porous Media* **1996**, *23*, 135–151. [CrossRef]
19. Sternberg, S.P.K. Dispersion Measurements in Highly Heterogeneous Laboratory Scale Porous Media. *Transp. Porous Media* **2004**, *54*, 107–124. [CrossRef]
20. Tan, C.; Tong, J.; Liu, Y.; Hu, B.X.; Yang, J.; Zhou, H. Experimental and modeling study on Cr (VI) transfer from soil into surface runoff. *Stoch. Environ. Res. Risk Assess.* **2016**, *30*, 1347–1361. [CrossRef]
21. Zhang, X.; Tong, J.; Hu, B.X.; Wei, W. Adsorption and desorption for dynamics transport of hexavalent chromium (Cr (VI)) in soil column. *Environ. Sci. Pollut. Res.* **2018**, *25*, 459–468. [CrossRef] [PubMed]
22. Viotti, P.; Sappa, G.; Tatti, F.; Andrei, F. nZVI Mobility and Transport: Laboratory Test and Numerical Model. *Hydrology* **2022**, *9*, 196. [CrossRef]
23. Klotz, D.; Seiler, K.P.; Moser, H.; Neumaier, F. Dispersivity and velocity relationship from laboratory and field experiments. *J. Hydrol.* **1980**, *45*, 169–184. [CrossRef]
24. Pickens, J.F.; Grisak, G.E. Scale-dependent dispersion in a stratified granular aquifer. *Water Resour. Res.* **1981**, *17*, 1191–1211. [CrossRef]
25. Taylor, S.R.; Moltyaner, G.L.; Howard, K.W.F.; Killey, R.W.D. A comparison of field and laboratory methods for determining contaminant flow parameters. *Groundwater* **1987**, *25*, 321–330. [CrossRef]
26. Dagan, G. Statistical theory of groundwater flow and transport: Pore to laboratory, laboratory to formation, and formation to regional scale. *Water Resour. Res.* **1986**, *22*, 120S–134S. [CrossRef]
27. Graham, D.D. *Potential for Leakage among Principal Aquifers in the Memphis Area, Tennessee*; US Department of the Interior, Geological Survey: 1986; Volume 85.
28. Gelhar, L.W.; Welty, C.; Rehfeldt, K.R. A critical review of data on field-scale dispersion in aquifers. *Water Resour. Res.* **1992**, *28*, 1955–1974. [CrossRef]
29. Anderson, M.P.; Cherry, J.A. Using models to simulate the movement of contaminants through groundwater flow systems. *Crit. Rev. Environ. Control* **1979**, *9*, 97–156. [CrossRef]
30. Guo, Z.; Fogg, G.E.; Brusseau, M.L.; LaBolle, E.M.; Lopez, J. Modeling groundwater contaminant transport in the presence of large heterogeneity: A case study comparing MT3D and RWhet. *Hydrogeol. J.* **2019**, *27*, 1363–1371. [CrossRef]
31. Hamzaoui-Azaza, F.; Zammouri, M.; Ameur, M.; Baba Sy, M.; Gueddari, M.; Bouhlila, R. Hydrogeochemical modeling for groundwater management in arid and semiarid regions using MODFLOW and MT3DMS: A case study of the Jeffara of Medenine coastal aquifer, South-Eastern Tunisia. *Nat. Resour. Model.* **2020**, *33*, e12282. [CrossRef]
32. Banaei, S.M.A.; Javid, A.H.; Hassani, A.H. Numerical simulation of groundwater contaminant transport in porous media. *Int. J. Environ. Sci. Technol.* **2021**, *18*, 151–162. [CrossRef]
33. Priyanka, B.; Kumar, M.M.; Amai, M. Estimating anisotropic heterogeneous hydraulic conductivity and dispersivity in a layered coastal aquifer of Dakshina Kannada District, Karnataka. *J. Hydrol.* **2018**, *565*, 302–317. [CrossRef]
34. Ansarifar, M.M.; Salarijazi, M.; Ghorbani, K.; Kaboli, A.R. Aquifer-wide estimation of longitudinal dispersivity by the combination of empirical equations, inverse solution, and aquifer zoning methods. *Appl. Water Sci.* **2023**, *13*, 14. [CrossRef]

35. Konikow, L.F.; Sanford, W.E.; Campbell, P.J. Constant-concentration boundary condition: Lessons from the HYDROCOIN variable-density groundwater benchmark problem. *Water Resour. Res.* **1997**, *33*, 2253–2261. [CrossRef]
36. Criner, J.H.; Parks, W.S. *Historic Water-Level Changes and Pumpage from the Principal Aquifers of the Memphis Area, Tennesse 1886–1975*; Geological Survey (US): Reston, VA, USA, 1976.
37. Kingsbury, J.A. *Altitude of the Potentiometric Surfaces, September 1995, and Historical Water-Level Changes in the Memphis and For Pillow Aquifers in the Memphis Area, Tennessee*; Geological Survey (US): Reston, VA, USA, 1996.
38. Parks, W.S.; Carmichael, J.K. *Geology and Ground-Water Resources of the Memphis Sand in Western Tennessee*; Department of the Interior, US Geological Survey: Reston, VA, USA, 1990; Volume 88.
39. Bradley, M.W. *Ground-Water Hydrology and the Effects of Vertical Leakage and Leachate Migration on Ground-Water Quality near the Shelby County landfill, Memphis, Tennessee*; US Geological Survey: Reston, VA, USA, 1991.
40. Parks, W.S.; Mirecki, J.E.; Kingsbury, J.A. *Hydrogeology, Ground-Water Quality, and Source of Ground Water Causing Water-Quality Changes in the Davis Well Field at Memphis, Tennessee*; US Department of the Interior, US Geological Survey: Reston, VA, USA, 1995; Volume 94.
41. Carmichael, J.K. *Hydrogeology and Ground-Water Quality at Naval Support Activity Memphis, Millington, Tennessee*; US Department of the Interior, US Geological Survey: Reston, VA, USA, 1997; Volume 97.
42. Larsen, D.; Gentry, R.W.; Ivey, S.; Solomon, D.K.; Harris, J. Groundwater leakage through a confining unit beneath a municipal well field, Memphis, Tennessee, USA. In Proceedings of the Geochemical Processes in Soil and Groundwater: Measurement, Modelling, Upscaling. GeoProc2002 Conference, Bremen, Germany, 4–7 March 2002; Wiley-VCH Verlag GmbH: Weinheim, Germany, 2003; pp. 51–64.
43. Clark, B.R.; Hart, R.M. *The Mississippi Embayment Regional Aquifer Study (MERAS): Documentation of a Groundwater-Flow Model Constructed to Assess Water Availability in the Mississippi Embayment*; Technical Report; U. S. Geological Survey: Reston, VA, USA, 2009.
44. Waldron, B.A.; Harris, J.B.; Larsen, D.; Pell, A. Mapping an aquitard breach using shear-wave seismic reflection. *Hydrogeol. J.* **2009**, *17*, 505. [CrossRef]
45. Ge, J.; Magnani, M.B.; Waldron, B. Imaging a shallow aquitard with seismic reflection data in Memphis, Tennessee, USA. Part I: Source comparison, walk-away tests and the plus-minus method. *Near Surf. Geophys.* **2010**, *8*, 331–340. [CrossRef]
46. Carmichael, J.K.; Kingsbury, J.A.; Larsen, D.; Schoefernacker, S. *Preliminary Evaluation of the Hydrogeology and Groundwater Quality of the Mississippi River Valley Alluvial Aquifer and Memphis Aquifer at the Tennessee Valley Authority Allen Power Plants, Memphis, Shelby County, Tennessee*; Technical Report; US Geological Survey: Reston, VA, USA, 2018.
47. Jazaei, F.; Waldron, B.A.; Schoefernacker, O.; Larsen, D. Numerical tools for identifying confining unit breaches impacting semi-confined water-supply aquifers. In *AGU Fall Meeting Abstracts*; American Geophysical Union: Washington, DC, USA, 2018; Volume 2018, p. H41J–2212.
48. Torres-Uribe, H.E.; Waldron, B.; Larsen, D.; Schoefernacker, S. Application of Numerical Groundwater Model to Determine Spatial Configuration of Confining Unit Breaches near a Municipal Well Field in Memphis, Tennessee. *J. Hydrol. Eng.* **2021**, *26*, 05021021. [CrossRef]
49. Moore, G.K.; Brown, D.L. *Stratigraphy of the Fort Pillow Test Well, Lauderdale County, Tennessee*; Tennessee Department of Conservation, Division of Geology: Nashville, TN, USA, 1969.
50. Van Arsdale, R.; Bresnahan, R.; McCallister, N.; Waldron, B. Upland Complex of the central Mississippi River valley: Its origin, denudation, and possible role in reactivation of the New Madrid seismic zone. In *Continental Intraplate Earthquakes: Science, Hazard, and Policy Issues*; US Geological Survey: Reston, VA, USA, 2007.
51. Robinson, J.L. *Hydrogeologic Framework and Simulation of Ground-Water Flow and Travel Time in the Shallow Aquifer System in the Area of Naval Support Activity Memphis, Millington, Tennessee*; US Department of the Interior, US Geological Survey: Reston, VA, USA, 1997; Volume 97.
52. Gentry, R. *Novel Techniques for Investigating Recharge to the Memphis Aquifer*; American Water Works Association: Denver, CO, USA, 2006.
53. Waldron, B.; Larsen, D.; Hannigan, R.; Csontos, R.; Anderson, J.; Dowling, C.; Bouldin, J. *Mississippi Embayment Regional Ground Water Study*; United States Environmental Protection Agency: Washington, DC, USA, 2011; Volume 600.
54. Parks, W.S. *Hydrogeology and Preliminary Assessment of the Potential for Contamination of the Memphis Aquifer in the Memphis Area, Tennessee*; Report 90-4092; Department of the Interior, US Geological Survey: Reston, VA, USA, 1990. [CrossRef]
55. Black & Veatch Special Projects Corp. *Groundwater Implementation Status Report 2*; For the Groundwater Sampling Event, November 2016; Technical Report 2; Smalley-Piper Superfund Site Collierville: Shelby County, TN, USA, 2017.
56. Black & Veatch Special Projects Corp. *Remedial Action Report, Smalley-Piper Site*; Technical Report; US Geological Survey: Reston, VA, USA, 2014.
57. Agency for Toxic Substances and Disease Registry (ATSDR). Hazardous Substances Emergency Events Surveillance (HSEES) Annual Report 2006. Available online: https://www.atsdr.cdc.gov/HS/HSEES/annual2006.pdf (accessed on 6 December 2023).
58. Hasan, K. Investigation of Modern Leakage Based on Numerical and Geochemical Modeling Near a Municipal Well Field in Memphis, Tennessee. Ph.D. Thesis, The University of Memphis, Memphis, TN, USA, 2023.
59. Villalpando-Vizcaino, R.; Waldron, B.; Larsen, D.; Schoefernacker, S. Development of a numerical multi-layered groundwater model to simulate inter-aquifer water exchange in Shelby County, Tennessee. *Water* **2021**, *13*, 2583. [CrossRef]

60. Sahagún-Covarrubias, S.; Waldron, B.; Larsen, D.; Schoefernacker, S. Characterization of hydraulic properties of the Memphis Aquifer by conducting pumping tests in active well fields in Shelby County, Tennessee. *JAWRA J. Am. Water Resour. Assoc.* **2022**, *58*, 185–202. [CrossRef]
61. Schrader, T. *Potentiometric Surface in the Sparta-Memphis Aquifer of the Mississippi Embayment*; U.S. Geological Survey Scientific Investigations Map 3014; Technical Report; U.S. Geological Survey: Reston, VA, USA, 2007.
62. Niswonger, R.G.; Panday, S.; Ibaraki, M. *MODFLOW-NWT, A Newton formulation for MODFLOW-2005*; Report 6-A37; U.S. Geological Survey: Reston, VA, USA, 2011. [CrossRef]
63. Brantley, B. *United Technologies Corporation, Carrier Air Conditioning Collierville, Tennessee*; Supplemental Investigation Report; U.S. Geological Survey: Reston, VA, USA, 2017.
64. Larsen, D.; Brock, C.F. Sedimentology and petrology of the Eocene Memphis Sand and younger terrace deposits in surface exposures of western Tennessee. *Southeast. Geol.* **2014**, *50*, 193–214.
65. Lumsden, D.N.; Hundt, K.R.; Larsen, D. Petrology of the Memphis Sand in the northern Mississippi Embayment. *Southeast. Geol.* **2009**, *46*, 121–133.
66. Antonacci, T.; Lee, E.S.; Kim, Y. Characterizing and predicting contaminant transport in the Newport Wellfield aquifer, Ohio. *Geosci. J.* **2013**, *17*, 465–477. [CrossRef]
67. Rai, D.; Eary, L.E.; Zachara, J.M. Environmental chemistry of chromium. *Sci. Total Environ.* **1989**, *86*, 15–23. [CrossRef] [PubMed]
68. Leaf, A.T.; Fienen, M.N. Flopy: The Python Interface for MODFLOW. *Groundwater* **2022**, *60*, 710–712. [CrossRef] [PubMed]
69. Zheng, C.; Wang, P.P. *MT3DMS: A Modular Three-Dimensional Multispecies Transport Model for Simulation of Advection, Dispersion, and Chemical Reactions of Contaminants in Groundwater Systems; Documentation and User's Guide*; U.S. Army Engineer Research and Development Center, Environmental Laboratory: Vicksburg, MI, USA, 1999.
70. Xu, M.; Eckstein, Y. Use of Weighted Least-Squares Method in Evaluation of the Relationship Between Dispersivity and Field Scale. *Groundwater* **1995**, *33*, 905–908. [CrossRef]
71. Environmental and Safety Designs, Inc. *Carrier Corporation Site Investigation Report*; Site Investigation: Memphis, TN, USA, 1988; Volume 1.
72. Waldron, B.; Larsen, D. Pre-Development Groundwater Conditions Surrounding Memphis, Tennessee: Controversy and Unexpected Outcomes. *JAWRA J. Am. Water Resour. Assoc.* **2015**, *51*, 133–153. [CrossRef]
73. Lozano-Medina, D.; Waldron, B.; Schoefernacker, S.; Antipova, A.; Villalpando-Vizcaino, R. Stories of a water-table: Anomalous depressions, aquitard breaches and seasonal implications, Shelby County, Tennessee, USA. *Environ. Monit. Assess.* **2023**, *195*, 953. [CrossRef]
74. Sloto, R. *Changes in Groundwater Flow and Volatile Organic Compound Concentrations at the Fischer and Porter Superfund Site, Warminster Township, Bucks County, Pennsylvania, 1993–2009*; US Geological Survey: Reston, VA, USA, 2010.
75. Zech, A.; Attinger, S.; Bellin, A.; Cvetkovic, V.; Dagan, G.; Dietrich, P.; Fiori, A.; Teutsch, G. Evidence based estimation of macrodispersivity for groundwater transport applications. *Groundwater* **2023**, *61*, 346–362. [CrossRef]

MDPI AG
Grosspeteranlage 5
4052 Basel
Switzerland
Tel.: +41 61 683 77 34

Hydrology Editorial Office
E-mail: hydrology@mdpi.com
www.mdpi.com/journal/hydrology

www.ingramcontent.com/pod-product-compliance
Lightning Source LLC
Chambersburg PA
CBHW041410210926
51850CB00013B/3233
9783725833054